Autonomic Computing

Concepts, Infrastructure, and Applications

Books are to be returned on or before
the last date below.

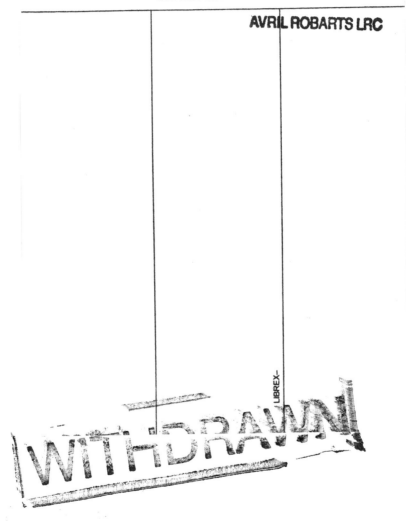

Autonomic Computing

Concepts, Infrastructure, and Applications

Edited by

Manish Parashar

Salim Hariri

CRC Press
Taylor & Francis Group
Boca Raton London New York

CRC Press is an imprint of the
Taylor & Francis Group, an informa business

CRC Press
Taylor & Francis Group
6000 Broken Sound Parkway NW, Suite 300
Boca Raton, FL 33487-2742

© 2007 by Taylor & Francis Group, LLC
CRC Press is an imprint of Taylor & Francis Group, an Informa business

No claim to original U.S. Government works
Printed in the United States of America on acid-free paper
10 9 8 7 6 5 4 3 2 1

International Standard Book Number-10: 0-8493-9367-1 (Hardcover)
International Standard Book Number-13: 978-0-8493-9367-9 (Hardcover)

Library of Congress Cataloging-in-Publication Data

Autonomic computing : concepts, infrastructure, and applications / editor(s):
    Manish Parashar and Salim Hariri.
        p. cm.
    Includes bibliographical references and index.
    ISBN-13: 978-0-8493-9367-9 (alk. paper)
    ISBN-10: 1-4200-0935-4 (alk. paper)
    1. Autonomic computing. I. Parashar, Manish, 1967- II. Hariri, Salim. III. Title.

QA76.9.A97A96 2007
004--dc22   2006024028

Visit the Taylor & Francis Web site at
http://www.taylorandfrancis.com

and the CRC Press Web site at
http://www.crcpress.com

Dedication

To Gowrie and Anushka - Manish Parashar
To Sonia, Lana and George - Salim Hariri

Preface

Introduction: Advances in networking and computing technologies, and software tools have resulted in an explosive growth in applications and information services that influence all aspects of our life. These sophisticated applications and services are complex, heterogeneous, and dynamic. Further, the underlying information infrastructure (e.g., the Internet) globally aggregates large numbers of independent computing and communication resources, data stores, and sensor networks, and is itself similarly large, heterogeneous, dynamic, and complex. The combined scale, complexity, heterogeneity, and dynamism of networks, systems, and applications have made our computational and information infrastructures brittle, unmanageable, and insecure. This has necessitated the investigation of an alternate paradigm for system and application design, which is based on strategies used by biological systems to deal with similar challenges — a vision that has been referred to as autonomic computing.

The *Autonomic Computing Paradigm* has been inspired by the human autonomic nervous system. Its overarching goal is to realize computer and software systems and applications that can manage themselves in accordance with high-level guidance from humans. Autonomic systems are characterized by their self-* properties including self-configuration, self-healing, self-optimization, and self-protection. Meeting the grand challenges of autonomic computing requires scientific and technological advances in a wide variety of fields, as well as new programming paradigms and software and system architectures that support the effective integration of the constituent technologies. The goal of this handbook, titled *"Autonomic Computing: Concepts, Infrastructure and Applications,"* is to give a comprehensive overview of the state-of-the-art of this emerging research area, which many believe to be the next paradigm for designing and implementing future computing systems and services.

Overview of the Handbook: The handbook is organized into four parts. Part I of the handbook focuses on the "The Autonomic Computing Paradigm" and includes chapters that present the vision, underlying concepts, challenges and requirements, and proposed architectures. In Chapter 1, Alan Ganek provides an overview of autonomic computing and its origins, evolution, and direction. This chapter defines autonomic computing, i.e., creating systems that are self-aware and self-managing to help reduce management complexity, increase availability, and enhance flexibility. Starting from the premise that all systems could, or indeed, should be constructed as autonomic systems, in Chapter 2, David W. Bustard and Roy Sterritt present a requirements engineering perspective of autonomic computing and examine the implications of the requirements of autonomic systems from a software engineering perspective. In Chapter 3, Robbert van Renesse and Kenneth P. Birman argue that autonomic systems cannot be built simply by composing autonomic components, but that an autonomic system-wide monitoring and control infrastructure is required as well. In Chapter 4, Manish Parashar investigates the challenges of emerging wide-area Grid environments and applications, and motivates self-management as a means for addressing these challenges. This chapter presents autonomic Grid computing solutions and sample autonomic Grid applications. Finally, in Chapter 5, John Sweitzer and Christine Draper present an architecture for autonomic computing and detail its four key aspects, i.e., process definition, resource definition, technical reference architecture, and application patterns. The chapter highlights the fundamental functions of the "autonomic manager" and "touchpoint" building blocks of the architecture, and summarizes an initial set of application patterns commonly found in autonomic computing systems.

Parts II and III of this handbook focus on achieving self-* properties in autonomic systems and applications. Part II presents different approaches and infrastructures for enabling autonomic behaviors. In Chapter 6, Tom De Wolf and Tom Holvoet present a taxonomy of self-* properties for decentralized autonomic computing systems, and use the taxonomy to guide the design and verification of self-* properties. In Chapter 7, Richard Anthony et al. investigate the use of emergent properties for constructing autonomic systems and realizing self-* properties. In Chapter 8, Sherif Abdelwahed and Nagarajan Kandasamy describes a more formal control theoretic approach using model-based control and optimization strategies to design autonomic computing systems that continually optimize their performance in response to changing workload demands and operating conditions. Since some autonomic applications cannot be built from scratch, it is necessary to transparently introduce autonomic behaviors into existing composite systems without modifying the existing code. In Chapter 9, S. Masoud Sadjadi and Philip K. McKinley address this requirement and describe transparent shaping to enable dynamic adaptations of existing applications. Chapter 10 addresses the design of self-* systems in which adaptive behavior can be specified as a set of externalized adaptation strategies. In this chapter, Peter Steenkiste and An-Cheng Huang present an architecture that separates service-specific knowledge, represented as a service recipe, from generic functionality, and supports automatic service-specific optimizations of a broad class of services.

Part III of the handbook presents core enabling systems, technologies, and services that support the realization of self-* properties in autonomic systems and applications. In Chapter 11, Hua Liu and Manish Parashar et al. describe the Accord programming system that extends existing programming systems to enable the development of self-managing Grid applications by allowing application and system behaviors to be dynamically specified at runtime. In Chapter 12, Thomas Heinis et al. describe a self-configuring composition engine for Grid and Web services that achieves self-configuring, self-tuning, and self-healing behaviors in the presence of varying workloads. Dynamic collaboration among self-managing resources is a key requirement for system-level self-management. In Chapter 13, David Chess et al. establish a set of behaviors, interfaces, and patterns of interaction within the Unity system to support such dynamic collaborations. In Chapter 14, Karsten Schwan et al. describe the AutoFlow project designed to meet the critical performance requirements of distributed information flow applications. In Chapter 15, Robert Adams et al. propose an approach for the management of large scale distributed services based on scalable publish-subscribe event systems, scalable WS-based deployment, and model-based management. The autonomic systems and prototypes described in this handbook have been typically implemented through middleware or through OS modifications. As an alternate, in Chapter 16, Lenitra Durham et al. investigate autonomic computing support at the hardware/physical layer and describe platform support for autonomics.

Part IV, the final part of this handbook, focuses on specific realizations of self-* properties in autonomic systems and applications. Chapter 17, the first chapter in this part, studies how autonomic computing techniques can be used to dynamically allocate servers to application environments in a way that maximizes a global utility function. In this chapter, Daniel A. Menasce et al. present a system that exhibits self-* properties and successfully reallocates servers when workload changes and/or when servers fail. Chapter 18 examines how a managed execution environment can be leveraged to support runtime system adaptations. In this chapter, Rean Griffith et al. describe the Kheiron adaptation framework that dynamically attaches/detaches an engine capable of performing reconfigurations and repairs on a target system while it executes. Kheiron remains transparent to the application and does not require recompilation of the application or specially compiled versions of the runtime. In Chapter 19, Arjav Chakravarti et al. describe the Organic Grid, which is a biologically inspired and fully decentralized approach to the organization of computation. Organic Grid is based on the autonomous scheduling of strongly mobile agents on a peer-to-peer network. Efficient and robust data streaming services are a critical requirement of emerging Grid applications, which are based on seamless interactions and coupling between geographically distributed application components. In Chapter 20, Viraj Bhat et al. present the design and implementation of a self-managing data-streaming service based on online control and optimization strategies. In Chapter 21, Bithika Khargharia et al. discuss the construction of an autonomic data center using autonomic clusters, servers and device components, and demonstrate autonomic power and performance management for a three tier data center. In Chapter 22, Guofei Jiang et al. propose a novel fault detection method based on trace analysis in application servers. The approach uses varied-length n-grams and automata to characterize normal traces and to detect abnormal behaviors and faults. In Chapter 23, Guangzhi Qu and Salim Hariri address the self-management and self-protection of networks, and presents methodologies for effectively detecting network attacks in real time. These methodologies configure network and system resources and services to proactively recover from network attacks and prevent the attacks from propagating in the network.

Once again, the goal of this handbook is to provide readers with an overview of the emerging discipline of autonomic computing. We do hope that it will lead to insights into the underlying concepts and issues, current approaches and research efforts, and outstanding challenges of the field, and will inspire further research in this promising area.

Acknowledgements: This book has been made possible due to the efforts and contributions of many individuals. First and foremost, we would like to acknowledge all the contributors for their tremendous efforts in putting together excellent chapters that are comprehensive, informative, and timely. We would like to thank the reviewers for their excellent comments and suggestions. We would also like to thank Nora Konopka, Jessica Vakili, and the team at Taylor & Francis Group LLC–CRC Press for patiently helping us put this book together. Finally, we would like to acknowledge the support of our families and would like to dedicate this book to them.

Credits: Based on "Emergence: A Paradigm for Robust and Scalable Distributed Applications", by Richard Anthony which appeared in First International Conference on Autonomic Computing (ICAC), New York, USA, May 2004 IEEE.

Based on "Self-Optimization in Computer Systems via Online Control: Application to Power Management," by authors N. Kandasamy, S. Abdelwahed, and J. P. Hayes in the Proceedings of the IEEE Conf. Autonomic Computing (ICAC), pp. 54–61, 2004, C[2004] IEEE.

Based on "Online Control for Self-Management in Computing Systems," by authors S. Abdelwahed, N. Kandasamy, and S. Neema in the Proceeding of the 10th IEEE Real-Time and Embedded Tech. & Application Symp. (RTAS), pp. 368–375, 2004, c[2004] IEEE.

Based on "Building Self-adapting Services Using Service-specific Knowledge", by An-Cheng Huang and Peter Steenkiste, which appeared in Fourteenth IEEE International Symposium on High-Performance Distributed Computing (HPDC-14), Research Triangle Park, NC, July 24–27, 2005, C 2005 IEEE; and based on "Building Self-configuring Services Using Service-specific Knowledge," by An-Cheng Huang and Peter Steenkiste, which appeared in 13th IEEE Symposium on High-Performance Distributed Computing (HPDC'04), IEEE, June 2004, Honolulu, Hawaii, pages 45–54, C 2004 IEEE.

Based on "An autonomic service architecture for self-managing grid application", by H. Liu, V. Bhat, M. Parashar and S. Klasky, which appeared in Proceedings of the 6th IEEE/ACM International Workshop on Grid

Computing (Grid 2005), Seattle, WA, USA, IEEE Computer Society Press, pp. 132–139, November 2005.

Based on Enabling self-management of component-based high-performance scientific applications", by H. Liu and M. Parashar, which appeared in Proceedings of the 14th IEEE International Symposium on High Performance Distributed Computing (HPDC 2005), Research Triangle Park, NC, USA, IEEE Computer Society Press, pp. 59–68, July 2005.

Based on "Accord: A programming framework for autonomic applications", by H. Liu and M. Parashar, which appeared in IEEE Transactions on Systems, Man and Cybernetics, Special Issue on Engineering Autonomic Systems, Editors: R. Sterritt and T. Bapty, IEEE Press, 2005.

Based on "Design and Evaluation of an Autonomic Workflow Engine", by Thomas Heinis, Cesare Pautasso and Gustavo Alonso which appeared in the proceedings of the Second International Conference on Autonomic Computing, 2005 (pp 27–38).

Based on "Implementing Diverse Messaging Models with Self-Managing Properties using IFLOW", by Vibhore Kumar, Zhongtang Cai, Brain F. Cooper, Greg Eisenhauer, Karsten Schwan, Mohamed Mansour, Balasubramaniam Seshasayee, Patrick Widener which appeared in IEEE International Conference on Autonomic Computing, 2006, ICAC'06.

Based on Lenitra M. Durham, Milan Milenkovic, and Phil Cayton, "Platform Support for Autonomic Computing: a Research Vehicle," in *Proceedings of the Third International Conference on Autonomic Computing*, pp. 293–294, June 13–16, 2006, C 2006 IEEE.

Based on Paper title "Resource Allocation for autonomic data centers using analytic performance models" which appeared in the Proceedings of the 2005 IEEE International Conference on Autonomic Computing (ICAC'05), Seattle, Washington, June 13–16, 2005.

Based on their chapter in "Self-organizing Scheduling on the Organic Grid".

Based on "Autonomic Power and Performance Management for Computing Systems", by Bithika Khargharia, Salim Hariri and Mazin S. Yousil which appeared in "IEEE International Conference on Autonomic Computing, 2006, ICAC '06.

Dr. Manish Parashar

Professor, Department of Electrical and Computer Engineering, Rutgers The State University of New Jersey, Piscataway, NJ, USA

Dr. Salim Hariri

Professor, Department of Electrical and Computer Engineering University of Arizona, Tucson, AZ, USA

About the Editors

Dr. Manish Parashar is Professor of Electrical and Computer Engineering at Rutgers University, where he also is director of the Applied Software Systems Laboratory. He has received the Rutgers Board of Trustees Award for Excellence in Research (2004-2005), NSF CAREER Award (1999), and the Enrico Fermi Scholarship from Argonne National Laboratory (1996). His research interests include autonomic computing, parallel and distributed computing (including peer-to-peer and Grid computing), scientific computing, and software engineering.

Manish is a senior member of IEEE and of the executive committee of the IEEE Computer Society Technical Committee on Parallel Processing (TCPP), part of the IEEE Computer Society Distinguished Visitor Program (2004-2006), and a member of ACM. He is the co-founder of the IEEE International Conference on Autonomic Computing (ICAC) and has served as General Co-Chair of ICAC 2004, 2005, and 2006. He is actively involved in the organization of conferences and workshops, and has served as general and program chairs for several conferences/workshops. He also serves on various steering committees and journal editorial boards.

Manish has co-authored over 200 technical papers in international journals and conferences, has co-authored/edited over 15 books and proceedings, and has contributed to several others, all in the broad area of computational science and applied parallel and distributed computing.

Manish received a BE degree in Electronics and Telecommunications from Bombay University, India and MS and Ph.D. degrees in Computer Engineering from Syracuse University. For more information, please visit http://www.caip.rutgers.edu/~parashar/.

Dr. Salim Hariri is Professor of Electrical and Computer Engineering at The University of Arizona. He is the director of the Center for Advanced TeleSysMatics (CAT): Next Generation Network Centric Systems. CAT is a technology transfer center to facilitate the interactions and collaborations among researchers in academia, industry as well as government research labs and is a vehicle for academic research to be transitioned to industry.

Salim is the Editor-In-Chief for the Cluster Computing Journal (Springer, http://www.springer.com/journal/10586) that presents research techniques and results in the area of high speed networks, parallel and distributed computing, software tools, and network-centric applications. He is the founder of the IEEE International Symposium on High Performance Distributed

Computing (HPDC) and the co-founder of the IEEE International Conference on Autonomic Computing. His current research focuses on autonomic computing, self protection and self-healing of networked systems and services, and high performance distributed computing. He has co-authored over 200 journal and conference research papers, and is the co-author/editor of three books, Tools and Environments for Parallel and Distributed Computing (Wiley, 2004), Virtual Computing: Concept, Design and Evaluation (Kluwer, 2001), and Active Middleware Services (Kluwer, 2000).

Salim received his B.S. from Damascus University, M.S. degree in Computer Engineering from The Ohio State University in 1982 and Ph.D. in Computer Engineering from University of Southern California in 1986. For further information, please visit http://www.ece.arizona.edu/~hpdc.

Contributors

Hasan Abbasi
College of Computing
Georgia Institute of Technology
Atlanta, GA, U.S.A.

Sherif Abdelwahed
Institute for Software Integrated
  Systems
Vanderbilt University
Nashville, TN, U.S.A.

Robert Adams
Intel
Hillsboro, OR, U.S.A.

Sandip Agarwala
College of Computing
Georgia Institute of Technology
Atlanta, GA, U.S.A.

Gustavo Alonso
Department of Computer Science
ETH Zurich
Zurich, Switzerland

Richard Anthony
Department of Computer Science
University of Greenwich
Greenwich, London, UK

Gerald Baumgartner
Department of Computer Science
Louisiana State University
Baton Rouge, LA, U.S.A.

Mohamed N. Bennani
Oracle, Inc.
Portland, OR, U.S.A.

Viraj Bhat
Department of Electrical and
  Computer Engineering
Rutgers University
Piscataway, NJ, U.S.A.

Kenneth P. Birman
Department of Computer Science
Cornell University
Ithaca, NY, U.S.A.

Paul Brett
Intel
Hillsboro, OR, U.S.A.

Alun Butler
Department of Computer Science
University of Greenwich
Greenwich, London, UK

David W. Bustard
School of Computing and
  Information Engineering
Faculty of Engineering
University of Ulster at Coleraine
Coleraine, Co. Londonderry
Northern Ireland

Zhongtang Cai
College of Computing
Georgia Institute of Technology
Atlanta, GA, U.S.A.

Phil Cayton
Corporate Technology Group
Intel Corporation
Hillsboro, OR, U.S.A.

Arjav Chakravarti
The MathWorks, Inc.
Natick, MA, U.S.A.

Haifeng Chen
Robust and Secure System Group
NEC Laboratories America
Princeton, NJ, U.S.A.

David M. Chess
Distributed Computing
    Department
IBM Thomas J. Watson
    Research Center
Hawthorne, NY, U.S.A.

Brian F. Cooper
College of Computing
Georgia Institute of Technology
Atlanta, GA, U.S.A.

Tom De Wolf
Department of Computer Science
K.U. Leuven
Leuven, Belgium

Christine Draper
IBM
U.S.A.

Lenitra Durham
Corporate Technology Group
Intel Corporation
Hillsboro, OR, U.S.A.

Greg Eisenhauer
College of Computing
Georgia Institute of Technology
Atlanta, GA, U.S.A.

Alan Ganek
IBM
Autonomic Computing
Somers, NY, U.S.A.

Ada Gavrilovska
College of Computing
Georgia Institute of Technology
Atlanta, GA, U.S.A.

Rean Griffith
Department of Computer Science
Columbia University
New York, NY, U.S.A.

James E. Hanson
Distributed Computing
    Department
IBM Thomas J. Watson
    Research Center
Hawthorne, NY, U.S.A.

Salim Hariri
Department of Electrical and
    Computer Engineering
University of Arizona
Tucson, AZ, U.S.A.

Thomas Heinis
Department of Computer Science
ETH Zurich
Zurich, Switzerland

Tom Holvoet
Department of Computer Science
K.U. Leuven
Leuven, Belgium

An-Cheng Huang
Department of Computer Science
Carnegie Mellon University
Pittsburgh, PA, U.S.A.

Mohammed Ibrahim
Department of Computer Science
University of Greenwich
Greenwich, London, UK

Subu Iyer
HP Labs
Palo Alto, CA, U.S.A.

Guofei Jiang
Robust and Secure System Group
NEC Laboratories America
Princeton, NJ, U.S.A.

Gail Kaiser
Department of Computer Science
Columbia University
New York, NY, U.S.A.

Nagarajan Kandasamy
Electrical and Computer
    Engineering Department
Drexel University
Philadelphia, PA, U.S.A.

Jeffrey O. Kephart
Distributed Computing
    Department
IBM Thomas J. Watson
    Research Center
Hawthorne, NY, U.S.A.

Bithika Khargharia
Department of Electrical and
    Computer Engineering
University of Arizona
Tucson, AZ, U.S.A.

Vibhore Kumar
College of Computing
Georgia Institute of Technology
Atlanta, GA, U.S.A.

Mario Lauria
Dept. of Computer Science
    and Engineering
Dept. of Biomedical Informatics
The Ohio State University
Columbus, OH, U.S.A.

Hua Liu
Xerox
Webster, NY, U.S.A.

Jay Lofstead
College of Computing
Georgia Institute of Technology
Atlanta, GA, U.S.A.

Mohamed Mansour
College of Computing
Georgia Institute of Technology
Atlanta, GA, U.S.A.

Philip K. McKinley
Professor of Computer Science
    and Engineering
Michigan State University
East Lansing, MI, U.S.A.

Daniel A. Menascé
Department of Computer Science
George Mason University
Fairfax, VA, U.S.A.

Milan Milenkovic
Corporate Technology Group
Intel Corporation
Hillsboro, OR, U.S.A.

Dejan Milojicic
HP Labs
Palo Alto, CA, U.S.A.

Manish Parashar
Department of Electrical and
    Computer Engineering
Rutgers University
Piscataway, NJ, U.S.A.

Cesare Pautasso
Department of Computer Science
ETH Zurich
Zurich, Switzerland

Guangzhi Qu
Department of Electrical and
    Computer Engineering
University of Arizona
Tucson, AZ, U.S.A.

Sandro Rafaeli
HP Brazil
Porto Alegre, Brazil

S. Masoud Sadjadi
School of Computing and
    Information Sciences
Florida International
    University
Miami, FL, U.S.A.

Karsten Schwan
College of Computing
Georgia Institute of Technology
Atlanta, GA, U.S.A.

Balasubramanian Seshasayee
College of Computing
Georgia Institute of Technology
Atlanta, GA, U.S.A.

Peter Steenkiste
Department of Computer
    Science and Electrical
    and Computer Engineering
Carnegie Mellon University
Pittsburgh, PA, U.S.A.

Roy Sterritt
School of Computing and
    Mathematics
University of Ulster
Jordanstown campus
Newtownabbey, Co. Antrim
Northern Ireland

John Sweitzer
IBM
U.S.A.

Vanish Talwar
HP Labs
Palo Alto, CA, U.S.A.

Cristian Ungureanu
Robust and Secure System Group
NEC Laboratories America
Princeton, NJ, U.S.A.

Giuseppe Valetto
Department of Computer Science
Columbia University
New York, NY, U.S.A.

Robbert van Renesse
Department of Computer Science
Cornell University
Ithaca, NY, U.S.A.

Ian Whalley
Distributed Computing
    Department
IBM Thomas J. Watson
    Research Center
Hawthorne, NY, U.S.A.

Steve R. White
Distributed Computing
    Department
IBM Thomas J. Watson
    Research Center
Hawthorne, NY, U.S.A.

Patrick Widener
Department of Computer Science
University of New Mexico
Albuquerque, NM, U.S.A.

Matt Wolf
College of Computing
Georgia Institute of Technology
Atlanta, GA, U.S.A.

Kenji Yoshihira
Robust and Secure System Group
NEC Laboratories America
Princeton, NJ, U.S.A.

Mazin Yousif
Corporate Technology Group
Intel Corporation
Hillsboro, OR, U.S.A.

Contents

Part I

The Autonomic Computing Paradigm

1

Overview of Autonomic Computing: Origins, Evolution, Direction

Alan Ganek

CONTENTS

"Civilization advances by extending the number of important operations which we can perform without thinking about them." Said by mathematician and philosopher Alfred North Whitehead almost a century ago, this statement today embodies both the influence and importance of computer technology.

In the past two decades alone, the information technology (IT) industry has been a driving force of progress. Through the use of distributed networks, Web-based services, handheld devices, and cellular phones, companies of all sizes and across all industries are delivering sophisticated services that fundamentally change the tenor of daily life, from how we shop to how we bank to how we communicate. And in the process it's dramatically improving business productivity. Consider that call centers can now respond to customer inquiries in seconds. Banks can approve mortgage loans in minutes. Phone companies can activate new phone service in just one hour.

Yet, while business productivity is soaring, these advances are creating significant management challenges for IT staffs. The sophistication of services has inspired a new breed of composite applications that span multiple resources—Web servers, application servers, integration middleware, legacy systems—and thus become increasingly difficult to manage. At the same time, escalating demand, growing volumes of data, and multi-national services are driving the proliferation of technologies and platforms, and creating IT environments that require tens of thousands of servers, millions of lines of code, petabytes of storage, multitudes of database subsystems, and intricate global networks composed of millions of components.

With physically more resources to manage and increasingly elaborate interdependencies among the various IT components, it is becoming more difficult for IT staff to deploy, manage, and maintain IT infrastructures. The implications are far-reaching, affecting operational costs, organizational flexibility, staff productivity, service availability, and business security. In fact, up to 80 percent of an average company's IT budget is spent on maintaining existing applications.[1] And an increasing number of companies today report that their IT staffs spend much of their time locating, isolating, and repairing problems.

The rapid pace of change—unpredictable demand, growing corporate governance prompted by both regulatory requirements and an increasingly litigious world, and an escalating number of access points (cellular phones, PDAs, PCs)—makes these problems even more acute. In an IBM study of 456 organizations worldwide, only 13 percent of CEOs surveyed believed that their organizations could be rated as "very responsive" to change.[2]

With so much time required to manage core business processes, IT staffs have little time left to identify and address areas of potential growth. To enable companies to focus on the application of technology to new business opportunities and innovation, the IT industry must address this complexity.

That's where autonomic computing comes in.

1.1   Improving Manageability

The term autonomic computing was coined in 2001 by Paul Horn, senior vice president of research for IBM. According to Horn, the industry's focus on creating smaller, less expensive, and more powerful systems was fueling

---

[1] Klein, A., Optimizing enterprise IT. *Intelligent Enterprise*. February 2005. http://www.intelligententerprise.com/showArticle.jhtml?articleID = 60403261 (accessed Dec. 6, 2005).

[2] Your Turn—Global CEO Study 2004. IBM Corporation. Survey of 456 CEOs worldwide. http://www.ibm.com/news/us/2004/02/241.html (accessed Dec. 6, 2005).

the problem of complexity. Left unchecked, he said, this complexity would ultimately prevent companies from "moving to the next era of computing" and, therefore, the next era of business.[3] In response, he issued a "Grand Challenge" to the IT industry to focus on the development of autonomic systems that could manage themselves.

In much the same way as the autonomic nervous system regulates and protects our bodies without our conscious involvement, Horn envisioned autonomic IT infrastructures that could sense, analyze, and respond to situations automatically, alleviating the need for IT professionals to perform tedious systems management tasks and enabling them to focus on applying IT to solve business problems. For example, rather than having to worry about what database parameters were needed to optimize data delivery for a customer service application, IT administrators could instead spend their time extending that application to provide customers even greater conveniences.

For years, science fiction writers have imagined a world in which androids and sentient systems could make decisions independent of human input. This represents neither the spirit nor the goal of autonomic computing. Although artificial intelligence plays an important role in the field of autonomic computing—as some of the research highlighted later in this chapter shows— autonomic computing isn't focused on eliminating the human from the equation. Its goal is to help eliminate mundane, repetitive IT tasks so that IT professionals can apply technology to drive business objectives and set policies that guide decision making. At its core, the field of autonomic computing works to effectively balance people, processes, and technology.

When first introduced, the concept of autonomic computing was sometimes confused with the notion of automation. However, autonomic computing goes beyond traditional automation, which leverages complex "if/then scripts" written by programmers to dictate specific behavior, such as restarting a server or sending alerts to IT administrators. Rather, autonomic computing focuses on enabling systems to adjust and adapt automatically based on business policies. It addresses the process of how IT infrastructures are designed, managed, and maintained. And it calls for standardizing, integrating, and managing the communication among heterogeneous IT components to help simplify processes. In a self-managing autonomic environment, a system can sense a rapidly growing volume of customer requests, analyze the infrastructure to determine how best to process these requests, and then make the necessary changes—from adjusting database parameters to provisioning servers to rerouting network traffic—so that all requests are handled in a timely manner in accordance with established business policies.

Although the target of autonomic computing is improving the manageability of IT processes, it has great implications on a company's ability to transition to on demand business, where business processes can be rapidly

---

[3] IBM. *Autonomic Computing: IBM's Perspective on the State of Information Technology.* http://www.research.ibm.com/autonomic/manifesto/autonomic_computing.pdf (accessed Dec. 6, 2005).

adapted and adjusted as customer demand or market requirements dictate. Without autonomic computing, it would be nearly impossible for the IT infrastructure to provide the level of flexibility, resiliency, and responsiveness necessary to allow a company to shift strategies to realize on demand goals.

1.2   The Road to Autonomic Computing

Autonomic computing represents a fundamental shift in managing IT. Beginning in the 1970s, reliability, availability, and serviceability (RAS) technology worked to help improve performance and availability at the resource level. This technology—primarily developed for mainframe computers—created redundancies within the hardware systems themselves and used embedded scripts that could bypass failing components, enabling IT staffs to easily install spare parts while the system was running, or even detect and make use of new processors automatically. By the late 1980s, as distributed systems began taking hold, IT vendors created management solutions that centralized and automated monitoring and management at the systems level. Examples of such management systems include the use of a single console for managing a particular set of resources or business services, and the automation of problem recovery using scripts for well-known and well-defined problems.

Although these approaches helped streamline many discrete IT activities, they were not based on interoperable standards and, hence, they lacked the ability to integrate business requirements in a holistic manner across all resources in the infrastructure and across IT processes themselves. Moving to this next era of IT management requires more than the work of a single IT vendor. It requires a collaborative effort by a critical mass of vendors to create an industry-standard management architecture that enables heterogeneous resources working in concert to take action based on business policy.

1.3   Defining an Autonomic Environment

Autonomic computing proposes the development of intelligent, open, and self-managing systems; but what does this really mean? At its core, a self-managing system must:[4]

---

[4] IBM. Autonomic Computing: IBM's Perspective on the State of Information Technology. http://www.research.ibm.com/autonomic/manifesto/autonomic_computing.pdf   (accessed Dec. 6, 2005).

- Have knowledge—not only of its components, status, capacity, etc., but also of the context of its activity and those of other resources within the infrastructure.
- Be able to sense and analyze environmental conditions. This includes both the ability to proactively take the pulse of individual components and services, looking for ways to improve its functions, and the ability to notice change and understand the implications of that change.
- Be able to plan for and affect change by altering its own state, effecting changes in other components of the environment, or choreographing IT processes based on business policy.

These characteristics are being applied today in four fundamental areas of self-management to drive significant operational improvements where traditional manual-based processes are neither efficient nor effective. These four areas are:[5]

- *Self-configuring capabilities* that enable the system to adapt to unpredictable conditions by automatically changing its configuration, such as adding or removing new components or resources, or installing software changes without disrupting service.
- *Self-healing capabilities* that can prevent and recover from failure by automatically discovering, diagnosing, circumventing, and recovering from issues that might cause service disruptions.
- *Self-optimizing capabilities* that enable the system to continuously tune itself—proactively to improve on existing processes and reactively in response to environmental conditions.
- *Self-protecting capabilities* that can detect, identify, and defend against viruses, unauthorized access, and denial-of-service attacks. Self-protection also could include the ability for the system to protect itself from physical harm, such as the motion detection capabilities of today's laptops that can temporarily park their disk drive heads if they sense that they are being dropped.

These characteristics are defined in terms of the overall system behavior. However, the system is composed of thousands—or even tens of thousands—of components from numerous vendors that interact in complex ways on various levels. It is not feasible to change all the constituent technologies in concert to achieve autonomic systems. Rather, an approach that delivers significant value while evolving to greater and greater autonomic characteristics is far more pragmatic and promising. This evolution is guided by the *Autonomic*

---

[5] Kephart, J.O. and Chess, D.M., The vision of autonomic computing, *Computer,* Jan. 2003. http://www.research.ibm.com/autonomic/research/papers/AC_Vision_Computer_Jan_2003.pdf (accessed Dec. 6, 2005).

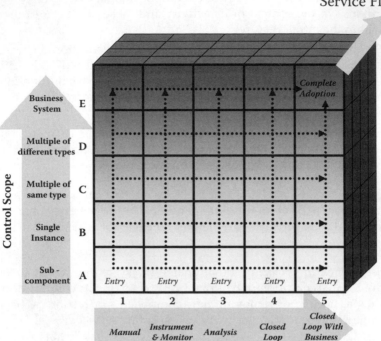

FIGURE 1.1
Autonomic computing adoption model. (Adapted from IBM, *An architectural blueprint for autonomic computing.* http://www-03.ibm.com/autonomic/pdfs/AC%20Blueprint%20White %20Paper%20V7.pdf. With permission.)

Computing Adoption Model, which analyzes the computing environment in three dimensions as depicted in Figure 1.1:

1. Increasing the scope of autonomic control, from a sub-component level to the entire business system, in which self-managing functions encompass all components supporting a business solution (depicted in Figure 1.1 along the y-axis).

2. Increasing autonomic functionality within one or more components, moving from a position where tasks are completed manually to an autonomic level where components of the infrastructure work in concert to dynamically manage processes based on business policy. At each step within this progression, shown along the x-axis in Figure 1.1, increasing levels of automation, correlation, analysis, and adaptation are implemented.

3. Coordinating and integrating functions across IT processes such as incident, change, performance, and security management to enable

the system to autonomically address a service requirement that traverses multiple IT processes. For example, providing an employee access to new services or applications requires the completion of tasks not only within configuration management (to install the new software), but also within security management (to verify user rights), release management (to manage version control), and change management (to authorize deployment of services) as well. In Figure 1.1, this dimension is illustrated as the z-axis of the adoption model.

This model allows both IT vendors and customers to approach the introduction of autonomic computing capabilities in a flexible way. For example, one could evaluate the value of introducing a high level of autonomic behavior to a small scope of IT resources in the context of a specific IT process. In another case, one could consider a less aggressive level of self-managing behavior with respect to a broader set of components or processes. The IT process dimension enables a focus on the tasks and best practices for managing IT resources that are discussed next.

1.4 Applying Autonomic Computing to IT Service Management

As the role of IT has evolved from cost center to profit center, organizations have begun to apply traditional business process management techniques to improve the control, order, and governance of IT infrastructure operations. This has led to the emergence of the *IT Infrastructure Library* (ITIL) best practices as a commonly used reference for designing the approach and governance for managing IT. Forrester Research Group estimates that ITIL adoption will increase from 13 percent in 2004 to 40 percent in 2006 to 80 percent by 2008.[6]

Originally created in the 1980s by the U.K. Central Computer and Telecommunications Agency—now the Office of Government Commerce—ITIL methodologies help govern the processes for effective and efficient IT operations in both service support (including incident management, problem management, change management, configuration management, and release management) and service delivery (including availability management, capacity management, continuity management, and financial management for IT services).[7]

However, it's important to note that ITIL recommendations are simply written guidelines—outlined in more than 2,000 pages of text. IT service

---

[6] Boshoff, T., The pitfalls of ITIL. *ITWorld*. April 18, 2005. http://www.itworld.com/Man/2672/050418itilpitfalls/ (accessed Dec. 6, 2005).

[7] More information about ITIL can be found at http://www.itil.co.uk/ (accessed Dec. 6, 2005).

management (ITSM) is a new focus for technology management that takes these processes and converts them into workflows with definable, repeatable tasks, linked directly to available technology components and products, to help companies implement ITIL best practices.

Although a number of IT process tasks, such as software distribution, inventory, and resource monitoring, can be automated today using traditional systems management tools, ITSM extends the process model to the end-to-end process definitions outlined in ITIL.

Self-managing autonomic technologies are critical tools utilized in ITSM to achieve the levels of efficiency and effectiveness envisioned by ITIL. These technologies provide the core capabilities to:

- Automate and integrate portions of a process flow.
- Provide coordination across different IT processes.
- Align processes with business policy and create an enterprise view that resolves conflicts.
- Provide a knowledge source to enable process managers to understand the state of the system and how to respond to situations.

For example, as described earlier, granting an employee access to a new application can require the coordination of at least four IT processes (change management, configuration management, security management, and release management) as well as the myriad discrete tasks within each process. Through autonomic computing, an autonomic manager can analyze the request, plan how to choreograph the various tasks in and across each process, and then execute each task in the correct order, automating those tasks appropriately.

1.4.1   Relating Autonomic Computing to Grid, Service-Oriented Architecture, and Virtualization Technologies

Autonomic computing represents an entirely new paradigm that both contributes to and benefits from a number of other technologies, such as grid computing, service-oriented architecture (SOA), and virtualization.

1.4.1.1   Grid Computing

Grid computing, which enables organizations to access and integrate heterogeneous distributed resources, forms an important conduit that enables intelligent provisioning of resources regardless of their supplier or location. Via grid technologies, independent system owners from different institutions can establish connectivity with one another and introduce protocols that allow resource sharing. This enables an expansion of systems to groups of systems, greatly extending the capabilities available to any user in the group. With this expanded environment, however, comes the threat of additional complexity as systems participate in grids with many other systems.

Autonomic computing provides the architecture that enables systems, including grids, to be flexible, dynamic, and adaptable. It provides technologies to offset the inherently increased complexity as grids expand the domain of computing. For example, self-managing autonomic systems can help optimize where processes are performed and manage workloads across the grid to ensure that resources are most efficiently utilized. Likewise, they can enable the provisioning of software and configuration specifications so that servers can be allocated or deallocated on the fly. Conversely, grid technologies provide enhancements to facilitate distributed computing that can be leveraged to implement autonomic behavior across distributed, heterogeneous resources.

1.4.1.2 Service-Oriented Architecture (SOA)

Through the use of standards-based interfaces and connections, SOA enables IT organizations to create application building blocks that can be "wired" together as needed. For example, a financial services organization may create a credit-checking capability based on an SOA that can be easily coupled with its mortgage loan application (to approve mortgages), its credit card application (to determine a customer's credit limit), and its customer service application (to determine the appropriate interest rate for each customer).

Autonomic computing leverages this same Web-based services model to facilitate communication among heterogeneous components. At the same time, autonomic computing helps simplify the modeling, assembly, deployment, and management of the discrete processes that are composed together using an SOA to help IT staffs more effectively and efficiently create the required building blocks for service delivery.

1.4.1.3 Virtualization

Similarly, virtualization technology and autonomic computing share an equally symbiotic relationship. Virtualization technology allows an institution to pool its computing resources—servers, networks, storage devices—into a single environment with a common interface so that they can be allocated as needed and managed more efficiently. However, within a virtualized environment, how do IT staffs stay attuned to available capacity? How do they rapidly move workloads as performance wanes? How do they ensure that data from business-critical applications is moved to storage devices with the highest quality of service?

Through the virtualization of resources, autonomic computing can manage IT resources on a more granular level, dynamically provisioning software and managing workloads at multiple levels. At the same time, autonomic computing provides critical "sense and respond" capabilities to reduce the complexity of managing virtualized resources.

1.5   Current State of Autonomic Computing

Since Paul Horn challenged the industry in 2001, autonomic computing has emerged as a strategic initiative for computer science and the IT industry, and progress has been made on a number of fronts to help make the vision of autonomic computing a reality in the data center today.

The most visible sign to IT staffs has been the incorporation of self-managing autonomic capabilities into individual products. Today, at every level of the infrastructure, vendors are embedding autonomic capabilities within their products. Chips can now sense change and alter the configuration of circuitry to enhance processor performance or avoid potential problems. Databases can automatically tune themselves as workload fluctuates and optimize performance as data organization changes. Networking components can intelligently route traffic. Blade servers can automatically populate new blades with the required software as they're plugged in. The list goes on.

However, as described earlier, this is only one aspect of a self-managing autonomic infrastructure. Regardless of the autonomic capabilities built into individual products, unless these components can communicate and effect change in an orchestrated fashion across the infrastructure and across IT processes, the promise of autonomic computing cannot be realized.

Considerable progress has been made in this area as well, including creating an open architecture that provides a common language and design for autonomic computing; developing industry standards that enable communication among heterogeneous components; and producing reference implementations (software development kits, toolkits, proofs-of-concept, etc.) for applying these standards.

1.5.1   Autonomic Computing Architecture

Traditionally, IT operations have been organized based on individual silos—separated by both component type and platform type. For example, a particular administrator might be concerned with managing only databases, or only application servers, or only Linux® servers. The autonomic computing architecture formalizes a reference framework that identifies common functions across these silos and sets forth the building blocks required to achieve autonomic computing. These building blocks include:[8]

- A *task manager* that enables IT personnel to perform management functions through a consistent user interface.
- An *autonomic manager* that can automate common functions and management activities using an autonomic control loop (see

---

[8] An architectural blueprint for autonomic computing. http://www-03.ibm.com/autonomic/pdfs/AC%20Blueprint%20White%20Paper%20V7.pdf (accessed Dec 6, 2005).

FIGURE 1.2
Autonomic control loop. (Adapted from IBM, *An architectural blueprint for autonomic computing.*
http://www-03.ibm.com/autonomic/pdfs/AC%20Blueprint%20White%20Paper%20V7.pdf.
With permission.)

Figure 1.2). Through this control loop, autonomic managers monitor resource details, analyze those details, plan adjustments, and execute the planned adjustments—using both information from humans (administrators) as well as rules and policies both defined (by humans) and learned by the system.

- A *knowledge source* that provides information about the managed resources and data required to manage them, such as business and IT policies.
- An *enterprise service bus* that leverages Web standards to drive communications among components throughout the environment.
- A *touchpoint* that provides a standardized interface for managed resources—servers, databases, storage devices, etc.—enabling autonomic managers to sense and effect behavior within these resources.

The notion of a consistent view of the configuration of the computing environment—from the vantage point of the many technology components as well as from the perspective of the IT processes—is central to the orchestration of IT processes so that they work in concert. ITIL describes the idea of a *configuration management database* (CMDB) as a technology that tracks the configurable elements of a system as well as the relationships among them. The CMDB concept, in conjunction with configuration and change management processes, is emerging as a fundamental capability in the evolution of autonomic systems.[9] Several IT vendors are working on implementations.

---

[9] Making ITIL actionable in an IT service management environment. http://www-306.ibm.com/software/tivoli/resource-center/overall/eb-itil-it-serv-mgmt.jsp (accessed Dec. 6, 2005).

1.5.2   Standards

Open communication among heterogeneous systems is vital to enabling autonomic computing. How can systems sense and effect behavior across disparate resources without such capabilities? The autonomic computing architecture laid the foundation for industry participants to identify the mechanisms by which this communication would take place. This resulted in the identification of existing standards and the development of new standards that are needed to enable the open exchange of management data that can be processed programmatically.

The outcome of this work has been significant. Already, several important standards have been ratified and many more are being developed. The need to enable individual components to communicate with each other and allow autonomic managers to execute management functions (monitoring, provisioning, configuring, etc.) across disparate resources has resulted in the development of the *Web Services Distributed Management* (WSDM) standard ratified by the Organization for the Advancement of Structured Information Standards (OASIS), which now provides a standard for a common interface for managed resources.[10]

The need to correlate and aggregate log file information for use by autonomic managers in problem determination and self-healing functions has resulted in the *WSDM event format* (WEF) standard, which is part of the OASIS WSDM standard just mentioned.[11] Traditionally, IT administrators must pore through hundreds of log files, all generated in different formats, and correlate that information manually. Using the WEF specification, resources can create events and log records in a standard format (either natively or through conversion) that can be correlated programmatically by autonomic managers. Already, this format is helping to relieve some of the manual work associated with problem determination. One company that now converts selected log data into WEF has realized a 40 percent reduction in time spent on problem determination, eliminated critical runtime errors, improved overall quality and availability of its IT services, and minimized delays in customer production schedules.[12] And that's simply through the use of WEF by IT professionals who are still manually handling problem determination and resolution functions. Ultimately, the development and implementation of additional autonomic computing standards will enable the software to manage these processes as well.

Another important standard is the *Solution Deployment Descriptor* (SDD) specification that is currently under review by the OASIS SDD Technical Committee. This specification will provide the framework for standard

---

[10] For more information about WSDM standard, see http://www.oasis-open.org/committees/tc_home.php?wg_abbrev=wsdm (accessed Dec. 6, 2005).

[11] For more information about WSDM event format, see http://www.oasis-open.org/committees/tc_home.php?wg_abbrev=wsdm (accessed Dec. 6 2005).

[12] Results based on data provided to IBM by its customers in 2005.

descriptors that enable individual software entities to identify their dependencies and their relationships so that a system can understand how the pieces fit together and how change will affect them.[13] This information can be used to validate installations (such as conducting the necessary dependency checks and downloading required patches or software prior to installation), support dynamic changes, and enable high availability.

These are clearly the first steps in creating a uniform approach that enables the implementation of self-managing autonomic behavior across heterogeneous systems. Other standards currently in development include:

- The specification of *symptoms* (patterns of events) in a common format for use by autonomic managers in problem determination and self-healing activities.

- A consistent format for capturing IT *policies* so that autonomic systems can ensure that appropriate actions are performed in accordance with corporate policies.

- A consistent method for managing change requests.

- A standard approach for obtaining configuration information from heterogeneous systems so that information can be consolidated in a single configuration management database.

1.5.3   Reference Implementations

Through the use of software development kits, technology toolkits, proofs-of-concept, and other reference implementations, the application of these standards is well underway. Individual vendors have work in progress to enable their products to function as part of an autonomic ecosystem and IT departments are working to facilitate problem determination and configuration management. For example, one software vendor—working to decrease the deployment time for new solutions and free up staff assigned to packaging solution componentry—separated its solution into *installable units* that could be aggregated at the customer site. Using SDD technology, the company's sales and services organizations could build solutions at the customer sites based on each customer's specific business need. By doing so, the organization realized a 30 percent reduction in solution deployment time and experienced a 50 percent gain in productivity of the packaging staff.[14] Key to this success was the availability of open source technology that enables industry-wide collaboration on the development of new software. For example, IBM's delivery of open source self-managing autonomic technologies based on industry standards—technologies for solution deployment,

---

[13] For more information about the OASIS SDD Technical Committee, visit http://xml.coverpages.org/SDD-Announce2005.html (accessed Dec. 6, 2005).

[14] Results based on data provided to IBM by its customers in 2005.

problem determination, integrated consoles, and other areas—is being leveraged by more than sixty software vendors to increase the autonomic capabilities in their products. This type of work has been critical in driving and accelerating the adoption of autonomic computing.

1.5.4 Research Opportunities

Although much work has been accomplished to date to advance autonomic computing, continued research is required to enable organizations to evolve their infrastructures to become truly self-managing autonomic environments. Volumes of work are underway at leading academic institutions worldwide on a wide range of topics, leveraging expertise in a number of fields, from computer science to mathematics to behavioral sciences.[15] This includes research in:

- Continuous operations and predictive problem determination and recovery
- Predictive and continuous optimization
- Self-protecting, self-learning security implementations
- Universal representation for policies and service-level agreements
- Policy negotiation and conflict resolution
- Context awareness and automated derivation of actions
- Designing manageability into applications
- Unified, automated discovery of IT resources and their interdependencies; real-time configuration updates
- Human-computer interface
- Culture change and trust

And there are many exciting areas that still must be explored. For example, within the autonomic environment, it is the knowledge source that provides autonomic managers with detailed information about individual resources and management data such as business policies. Yet, in most cases, current efforts have focused on leveraging "captured" knowledge—that is, information placed into the knowledge database, such as symptom definitions that outline known problems—such that the system understands what actions to take. However, additional research is vital to evolve from this state of captured knowledge to a state of full awareness, in which human knowledge and actions can be easily translated into electronic forms of knowledge, enabling systems to learn from their environment. Several examples of important research to enable this transition include:

---

[15] Kephart, J.O., Research Challenges of Autonomic Computing. IBM. 2005. http://domino. research.ibm.com/library/cyberdig.nsf/1e4115aea78b6e7c85256b360066f0d4/ 5e932dbbecf5ebcf85257067004ee94c?OpenDocument (accessed Dec. 6 2005).

- Applying artificial intelligence to control theory to create "learning systems" that can observe a pattern, prompt the IT administrator for action, and then automatically learn from that action and add it to its own knowledge base for future reference and use.

- Applying artificial intelligence to the handling of IT policy for improved decision making based on business policy. Basic policies today can be hardcoded into a knowledge base to provide a system with cause and effect data that forms the basis for taking action. Future research is needed to help the system itself analyze operations and determine the best course of action based on current data (e.g., if response time is slow, should the system increase the database cache size, add networking bandwidth, or both).

- Applying pattern analysis and pattern recognition science to enable autonomic systems to define normal and abnormal states based on operational patterns rather than captured information.

- Applying human-computer interface science to policy development to identify methods for capturing human knowledge and paper-based information and converting that knowledge into electronic policies that are actionable.

1.6 Conclusion

This chapter opened with a reference to Whitehead's assertion that progress can be made only through automation. Although this statement clearly emphasizes the impact on society of computer technology in general, it also reflects the vital need for autonomic computing in particular. The complexity of IT is dramatically increasing. As of the writing of this handbook, probably more than half a billion people worldwide have access to the Internet. Industry analyst IDC estimates that the volume of Internet traffic worldwide will nearly *double annually* through 2008.[16] With increasing demand comes the need for more servers, databases, storage devices, networking devices, and applications—all of which must be configured, deployed, managed, and maintained—as well as larger workloads that must be managed and greater fluctuations in demand.

Autonomic computing is about creating systems that are self-aware and self-managing to help reduce management complexity, increase availability, and enhance flexibility. Great strides have already been made in this emerging field. Further chapters in this handbook provide in-depth discussions

---

[16] Legard, D., IDC: Internet traffic to keep doubling each year. Network World. March 6, 2003. http://www.networkworld.com/news/2003/0306idcinter.html (accessed Dec. 6, 2005).

on autonomic computing architecture, implementation models, design, and application in both academia and industry to demonstrate what exists today and what will be possible in the future. Through extensive research and collaboration among academia, industry, and government, autonomic computing is helping companies better manage their IT resources and build an infrastructure that is primed for on demand business. As additional standards and new self-managing autonomic technologies are created, the opportunities will only increase.

2

A Requirements Engineering Perspective on Autonomic Systems Development

David W. Bustard and Roy Sterritt

CONTENTS

Starting from the premise that all software could or, indeed, should be constructed as an autonomic system, this chapter examines the resulting implications for the identification of requirements for autonomic systems. The first section covers the systems analysis needed to understand and model the environment in which an autonomic system will be hosted. This is then followed by the consideration of a basic set of generic requirements for autonomic systems. Some research issues are identified and the discussion illustrated using aspects of the development of an autonomic cashpoint.

2.1 Introduction

Autonomic computing is an "emerging discipline." Partly, this is because there are still many research challenges in making it a practical reality [20]. More fundamentally, however, the interpretation of the concept itself is still under discussion [23]. From the dictionary, "autonomic" essentially means "self-governing" (or "self-regulating" or "self-managing"), derived from the noun "autonomy." In launching the autonomic systems concept, IBM proposed eight defining characteristics [18], the main four of which are self-configuring, self-healing, self-optimizing, and self-protecting. The meaning of these characteristics is clear, and no one, so far, has suggested reducing or increasing this core set. Where difficulties arise, however, is in deciding how many of these properties need to be present for a system to be considered autonomic, and to what degree each should be implemented. For example, is a system autonomic if it includes code to protect itself from faulty user data (self-protecting) or is structured to allow for unexpected failure using exception handling (self-healing)? Moreover, is the conclusion any different if such fault handling is coded directly or implemented using machine learning techniques?

IBM has attempted to address this issue by proposing a maturity model for autonomic systems [19]. This suggests that the degree of autonomicity can be measured in three dimensions:

- *Functionality:* the extent to which autonomic behavior is automated, ranging from 'manual' up to automatic decision making, based on knowledge of business processes, policies, and objectives.

- *Control scope:* the breadth of coverage of the autonomic system, from subcomponents up to business-level activities.

- *Service flow:* the level of service provided to information technology (IT) management processes, from internally focused systems to those that help support their own construction and evolution.

It can be concluded from the model that almost all computing systems currently have some autonomic characteristics but that it is desirable to design and develop such systems in ways that move progressively toward higher levels of autonomicity in all three dimensions of maturity.

The purpose of this chapter is to explore the implications for requirements engineering in attempting to support autonomic system development to whatever level of sophistication is required. The first section focuses on systems analysis, examining four basic styles that are available. Soft Systems Methodology (SSM) [9] [31] is then considered in support of a top-down approach that first builds a model of the environment in which the autonomic system will reside. This is followed by a section that considers a possible set of generic requirements for autonomic systems. This discussion reveals a number of areas for further research. The outline development of an autonomic cashpoint is used to help focus the discussion and to illustrate the arguments presented.

2.2   Systems Analysis

Systems analysis can be approached in different ways that are essentially variations of two choices:

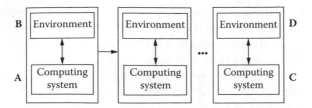

FIGURE 2.1
System Co-evolution model.

- Analysis can either be *top-down*, starting with an environment (or context) analysis, or *bottom-up*, focusing on the computing system to be produced.
- Analysis can either be *goal oriented*, developing a vision of what the computing system (and its aligned environment) could or should become, or *problem oriented*, focusing on current shortcomings (or opportunities) and determining how they could be addressed.

In most cases, the analysis reveals the need for a sequence of systems to be constructed. Figure 2.1 shows a representation of an evolving computing system and its corresponding environment. The arrow between each system and its environment implies that they are 'aligned.' That is, ideally, the computing system provides optimal support for the environment in which it is used, and this alignment is maintained as the situation evolves.

Figure 2.1 identifies four different starting points for systems analysis, indicated by the letters A, B, C, and D in the diagram [7]. These are in an order, representing the extent of their perceived current use:

- *A-Type Analysis:* Software Focus, Immediate Needs: A-type analysis is the classic product-focused approach to software development that has been common practice since software was first developed. Here, the problem situation is mainly understood in terms of its relevance to that software, taking account of any constraints imposed and identifying the services that need to be provided (software functions). An iterative sequence of implementation phases may be planned if the work cannot be completed conveniently in one project. A-type analysis is appropriate when the target environment is well understood and future development is either unimportant or appreciated to an adequate extent. This is true, for example, (i) if software is being maintained and no significant changes have occurred since the original analysis was performed; (ii) the software is not expected to evolve and the developers have a good understanding of the development context; or (iii) the software is already well defined.

- *B-Type Analysis:* Environment Focus, Immediate Needs: B-type analysis is the traditional information systems approach to software development, involving the analysis and modeling of the environment to provide a context for identifying software requirements. It is appropriate when a problem situation needs to be examined but future development is either obvious or of low importance.

- *C-Type Analysis:* Software Focus, Long-Term Goal: C-type analysis helps develop a long-term vision of what the software could or should become. Implementation steps are then defined to advance the software in that direction. One well-known example of this strategy is Gilb's Evolutionary Delivery method [12]. Gilb's approach has been adopted as one of the core practices of eXtreme Programming [1] and is an aspect of agile development, in general [11]. C-type analysis is appropriate when it is important to have a good vision of how software might evolve in a stable environment that is well understood.

- *D-Type Analysis:* Environment Focus, Long-Term Goal: D-type analysis helps develop a long-term vision of the environment in which the software will be used. This can serve as the basis for determining environment changes as well as the identification of the desirable supporting technology. D-type analysis is appropriate when the context for development has to be understood and there is some uncertainty about future needs. This is particularly relevant when the nature of the problem or its computing solution is unclear.

As an illustration of these variations, consider the problem of developing an autonomic bank cashpoint. With an A-type analysis, the focus would be on identifying the required functionality of the cashpoint and constraints on its operation. An appreciation of the environment would then be developed, progressively, through an understanding of the needs of its users and of those who own and service the cashpoint. For a B-type analysis, the first step would be to understand the context in which the cashpoint is used. This would typically involve modeling its environment, with the cashpoint then defined to integrate with that description. In this case, the main function is the withdrawal of deposited cash, but other functionality, such as producing an account statement, might also be supported. A C-type analysis, like the A-type, would focus on the cashpoint but this time consider a much wider set of possible functions, leading to the definition of a sequence of releases implementing those that seem desirable. For example, in the longer term, a camera might be added to the cashpoint to help reduce vandalism and fraudulent transactions; also, if there are problems with a transaction, such as multiple attempts to enter a password, additional constraints might be imposed before cash is released. A D-type analysis, like the B-type, would again focus on the environment of the cashpoint but this time consider longer-term developments, such as using an Internet banking approach rather than the traditional teller model.

From these options, the B-type and D-type forms of analysis seem best for autonomic systems because of their emphasis on understanding the environment. This need is highlighted by IBM as the sixth defining characteristic of autonomic systems [18]:

"An autonomic computing system must know its environment and the context surrounding its activity, and act accordingly. It will find and generate rules for how best to interact with neighboring systems. It will tap available resources; even negotiate the use of other systems of its underutilized elements, changing both itself and its environment in the process—in a word, adapting."

Although this describes the highest level of autonomic maturity expected, use of external knowledge is clearly an important factor for successful autonomic system operation. The choice between B-type and D-type analysis will depend on circumstances. If the desire is simply to support some existing well-established situation, then B-type is adequate. Alternatively, if any improvement to current practice is required, then the more visionary D-type analysis should be used.

Another factor in the decision is having a suitable technique available to support the analysis. One particularly appropriate way to perform a D-type analysis is through SSM [10] [9] [24] [25] [30] [31]. SSM has a long-standing, well-respected pedigree in the systems community. It emerged from work that began at the University of Lancaster (in the UK) in the late 1960s and evolved through action research up to 1990, when it reached its current stable form. It is essentially a general systems improvement technique that helps identify opportunities for beneficial change by promoting a better understanding of a "problem situation" among system stakeholders. This is achieved through the construction of relevant system models. The models and the process through which they are constructed promote debate about possible improvements and lead to recommendations for change. The approach is applicable to any problem situation, but its use in information systems development has received particular attention [29]. In this work, SSM is the first stage of analysis, providing context information for subsequent development through the models it creates. SSM seems particularly relevant to D-type analysis, as its models are abstractions of a desired system unconnected to what currently exists.

Classically, SSM has been described as a seven-stage process [9], as illustrated in Figure 2.2. There are five stages associated with so-called real world thinking: two of them for understanding and finding out about a problem situation (1, 2), and the other three for deriving change recommendations and taking action to improve the problem situation (5-7). There are also two stages (below the dotted line) concerned with "systems thinking" (3, 4), in which root definitions and conceptual models are developed. Each root definition provides a particular perspective of the system under investigation.

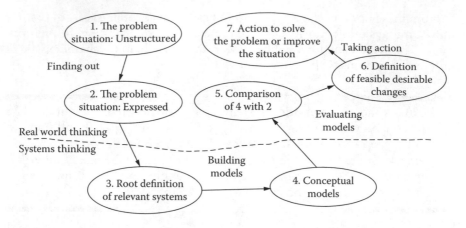

FIGURE 2.2
Seven-stage soft systems methodology model.

A conceptual model then defines activities necessary to implement the perspective given in a root definition.

As an illustration of the modeling involved, the following is a root definition of the main perspective on cashpoints, namely that they provide a cash dispensing service for account holders. Note that other views are also possible, including, for example, a perspective that cashpoints can also support the business, such as a supermarket, near which or in which they are located.

> A Big Bank owned system operated by bank staff and automated cashpoints to provide a convenient cash dispensing service to supply previously deposited cash, on request, to registered account holders, taking account of the demand for cash, the threat of fraudulent transactions, and equipment misuse.

In general, each root definition identifies or implies six particular pieces of information, as listed in Table 2.1.

TABLE 2.1

General Components of a Root Definition

Components	Meaning
Customers	The beneficiaries or victims of a system
Actors	The agents who carry out, or cause to be carried out, the main activities of the system
Transformation	The process by which defined inputs are transformed into defined outputs
Weltanschauung	A viewpoint, framework, image, or purpose which makes a particular root definition meaningful
Owner	Those who own a system (have the power to close it down)
Environment	Influences external to a system that affect its operation.

TABLE 2.2

'Cash Dispensing Service' Root Definition

Components	Meaning
Customers	Registered account holders
Actors	Bank staff and automated cashpoints
Transformation	Supply previously deposited cash
Weltanschauung	A convenient cash dispensing service
Owner	Big Bank
Environment	Demand for cash; possible fraudulent transactions; possible equipment misuse

The Weltanschauung, or worldview, identifies why a system exists, and the "transformation" indicates what the system does to achieve its purpose. These are the two most important elements of the root definition. Table 2.2 identifies the six elements in the cashpoint root definition given above.

Each root definition is developed into a conceptual model, defining the activities necessary for the system to meet the purpose specified and indicating relationships among the activities involved. For example, Figure 2.3 shows a conceptual model based on the cash dispensing service root definition described in Table 2.2. The activities have been labeled for convenience. The model includes the transformation taken directly from the root definition (A1). This is essentially the central activity of the model. Another important activity is A2, which monitors that the defined Weltanschauung (viewpoint) is achieved, taking control action (TCA) if necessary, which can affect any other activity in the model. This is supported by an activity that defines the

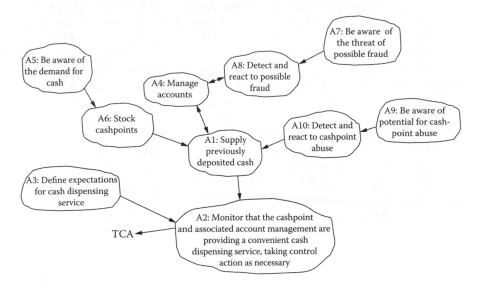

FIGURE 2.3

Conceptual model for cash dispensing service viewpoint.

standards against which the system is assessed (A3). Activities are also included to handle the environmental constraints listed in the root definition (A5, A7-A10) and to cover consequential or implied activities (A4, A6). Note that conceptual models have a basic process control structure which makes them especially suitable for the description of autonomic systems. In particular, each is expected to include a monitoring activity to ensure that the system purpose is being achieved (A2) and that adjustments are made as necessary.

Each activity in a conceptual model can be expanded into a similarly structured lower-level model, thereby producing a hierarchy of descriptions. For example, in the case of the cashpoint, activities A4: manage accounts and A1: supply previously deposited cash are substantial and so need to be elaborated. The direct operations on the cashpoint would be considered in activity A1.

Conceptual models are largely informal. The meaning of each activity identified is described solely by the text displayed in the diagram, and the linking arrows simply imply relationships between activities with no accompanying labels or explanations. Nevertheless, they are sufficient to support discussion about the necessity and sufficiency of the activities, how well they are currently implemented, and how they might be improved. In effect, a conceptual model provides a means for investigating a problem situation systematically. Each activity in the model can be compared with current practice to identify differences that suggest opportunities for improvement. For example, in considering activity A6: stock cashpoints, it may emerge that some cashpoints occasionally run out of money. This would then lead to a consideration of how to overcome or reduce this problem—such as increasing the cash stock in the machines affected, adjusting restocking schedules, or simply diverting customers to other outlets.

This investigation of a model also has a validation role, in that missing activities may be revealed or some activities redefined. Another type of validation (and refinement) that can be performed is through the use of an alternative technique that views the same situation from a different perspective. In this case, one strong option is to check the model produced against Beer's Viable Systems Model (VSM) [2] [16] [17] [17] [22] [21]. A viable system is one that is robust against internal or external malfunction and/or disturbance; it has the ability to respond and adapt to unexpected stimuli, allowing it to accommodate change. Clearly, autonomic systems have the same objectives.

The VSM identifies five necessary and sufficient subsystems that together maintain overall system viability: S1: operations; S2: coordination; S3: control; S4: intelligence; and S5: policy. S1 subsystems are an exact match for autonomic elements, as they have an operational part (managed component) overseen by a management part (autonomic manager) [27]. When describing an organization, subsystems S2-S5 are usually considered to be management activities, implemented by individuals in the organization. For organisms, S1 to S3 are automatic. With autonomic computing, the goal is to develop systems that support all of these activities to some extent. Further information on the use of VSM to assess SSM models may be found in [8].

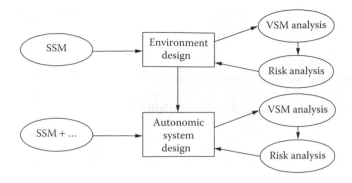

FIGURE 2.4
A Systems-oriented autonomic design process.

Yet another way to validate and refine a conceptual model in a way that is sympathetic to autonomic computing is through risk analysis [13] [14]. Although some risks have already been identified in the environmental constraints in the root definition, there are many more potential problems to consider. For example, these include customers forgetting to take away their cash or being robbed at the cashpoint. The conceptual model can be extended as necessary to cover such threats, thereby ensuring that they are taken into account in the design of the cashpoint and in organizing its operating environment.

Conceptual models can also be used to support further analysis toward the development of specific implementations of change, such as the creation or enhancement of information systems [25] [29] or computing systems [5]. They can also be used to create other types of model, such as dataflow diagrams [3], process models [4], and object models [5] [6] [15]. These developments facilitate the integration of SSM into existing software engineering processes.

Figure 2.4 suggests the basis of an overall approach to the development of an autonomic system design. Here the environment is identified and documented using SSM, the models of which are further validated and refined through VSM and risk analyses. The resulting models then form the basis of the autonomic system design, which can be taken to lower levels of detail through SSM and other modeling techniques. Again, the resulting models can be checked and enhanced through VSM and risk analyses.

The discussion so far has argued for a top-down approach to the development of autonomic systems because of the need to understand and document the environment in which such systems operate. This approach can also ensure that the environment itself is in an appropriate form to accommodate an autonomic system. Specifically, this means that the environment should be organized as a viable system, sharing many of the properties expected of the autonomic computing system that it hosts. The next section looks at the general requirements for autonomic systems and further emphasizes the interdependence between an autonomic system and its environment.

2.3   Generic Requirements for Autonomic Systems

IBM's eight defining characteristics [18] are essentially the main detailed requirements of autonomic systems. However, these apply only to the highest level of autonomic maturity and so effectively represent a long-term vision of how autonomic systems might evolve. In the near-to-medium term it seems useful to identify requirements that apply to all systems that aspire to be autonomic, regardless of their current progress along the maturity path.

From the eight defining characteristics, the main overall requirement seems to be that:

- *Autonomic systems should be robust;* that is, they should protect themselves from accidental and malicious harm; where damage does occur, they should attempt to recover.

In effect, this is an implicit requirement of all software, but from a user perspective, making the requirement explicit and giving it a high profile could lead to significant improvement in system design. It is an aspect of systems development that has been given particular attention in dependable, fault-tolerant system design [26], especially in the area of safety-critical applications, and its relevance to autonomic system development has been noted [28].

As indicated in the previous section, the environment is also a system, so it is important that it too be robust. Indeed, this argument can be taken one step further by recognizing the software development process as yet another system, with robustness sought there as well.

A second general engineering requirement that the autonomic concept brings to the fore is that:

- *Autonomic systems should be easy to use;* that is, they should take initiative in making necessary or desirable adjustments to their operation in response to changing circumstances, reducing the effort expected of users as much as possible.

"Plug and play" has now become so commonplace that most users expect computers to simply offer a menu choice when any new piece of equipment is connected to a USB port. Ideally, the same philosophy should be extended to interaction with users, a network, or other autonomic system. Again 'ease of use' would be a useful concept to extend to the business processes in the environment of an autonomic system and could also be an influence on the design of software development processes. This last possibility seems particularly important to IBM, which sees the development of autonomic systems as a way of dealing with growing complexity in system design, development, and evolution [18].

Yet another general requirement implicit in the basic interpretation of the autonomic concept is that:

- *Autonomic systems should be proactive;* that is, they should take action in support of system objectives without direct user control.

This is likely to mean that autonomic systems are implemented as concurrent programs. A certain amount can be achieved with a traditional sequential software structure, but it is limiting and leads to design compromises. For example, some existing applications when activated will check the Internet for updates. If any are found, the user is invited to install them. Unfortunately, most users will have started the application for another purpose and won't want to break away to deal with the update. Ideally, following the "easy-to-use" requirement, an application should be updated with minimal user inconvenience, which usually means making the change in parallel with executing the current version of the application or updating offline, at a different time.

Where there is independent behavior, there are additional general requirements to address. In particular:

- *Autonomic systems should be transparent;* that is, they should be able to report on their activity and provide explanations, as required.

Explanation is obviously important at the current stage of autonomic system evolution because of understandable anxiety about delegating decisions that may lead to system failure or data loss. Automatic updates are a particular concern because the new version of an application may be worse than the version it replaces. Thus, prior to an installation, the implications for each user need to be known, in terms of both the benefits that the change will bring and the consequences of not performing the update. This need for explanation will always be required and indeed seems likely to become more important as autonomic systems are given more autonomy and make greater use of artificial intelligence techniques.

In general, there is an inherent risk in delegating any responsibility, so a consequential requirement is that:

- *Autonomic systems should have flexible delegation arrangements;* that is, while they may be able to take full responsibility for particular operations, their actual degree of involvement will be agreed with the user.

In this way, control is given back to the user, with the autonomic system taking the role of a "good servant." Thus, the autonomic system will follow orders but be able to warn of problems and make suggestions for action. For example, instead of installing application updates as they appear, there could be different policies covering different types of change: for routine application updates there might be a built-in delay to see if others experience problems before making the change; but security updates would be installed immediately, as the associated threat is usually worse than the update risk associated involved.

Inevitably, no matter how much care is taken, an autonomic system will, from time to time, take action that is problematic. So, yet another consequential requirement is that:

- *Autonomic systems should be reversible;* that is, as far as possible, they should be able to undo their actions.

One typical scenario, for example, is that faults are found in a new version of an application just after it has been installed, perhaps through the use of an evaluation test suite. In such circumstances it should be straightforward to revert to the previous version. A more difficult situation to handle is where reversal is required after an application has been in use for some time and there are data compatibility issues between the two versions involved. In general, reversal can be achieved only if this requirement is taken into account in designing both the application and its change process. It is even more important that this requirement be considered when an autonomic system can create new implicit versions of itself through self-modification.

2.4 Conclusion

Although autonomic computing was proposed as a way of handling the growing complexity of some large computing systems, its associated properties are desirable in software of any size. Indeed, most software typically contains some aspects of self-management, and that use is growing through inclusion of features such as automated updates over the Internet. In effect, therefore, autonomic computing can be interpreted as a particular approach to software engineering. This chapter attempts to clarify that approach. It started by arguing for the use of a top-down analysis technique that sets the context for development before considering the details of the computing system itself. This is because of the importance of understanding and documenting the environment in which the autonomic system is used. Soft Systems Methodology, well known in the general systems field, was introduced as a suitable technique for performing such analysis, and was illustrated with the partial development of an environment model for an autonomic cashpoint. Integration with other well-established techniques was also proposed to further bring out the desirable properties of autonomic systems. In particular, it was suggested that a combination of risk management and assessment against the Viable System Model could improve the robustness of a design. The chapter finished by considering a set of general properties that autonomic computing is effectively promoting and which can be emphasized in a suitably defined software engineering process. It is believed that the use of such a process would help increase the autonomicity of computing systems in general, moving them toward the long-term vision for autonomic computing proposed by IBM. If successful, this approach has the potential to improve all computing systems, not only in terms of the quality of the software produced but also bringing benefits to the development process and the wider business context in which the system is used.

Although many research questions remain, especially in relation to the highest levels of autonomic behavior, it is certainly possible to develop a basic software engineering process for autonomic systems that builds on existing research results and established techniques. This chapter has suggested a top-level framework for such a process.

Acknowledgments

This work was undertaken through the Centre for Software Process Technologies, which is supported by the EU Programme for Peace and Reconciliation in Northern Ireland and the Border Region of Ireland (PEACE II).

References

1. Beck, K. Extreme Programming Explained, Boston: Addison–Wesley, 1999.
2. Beer, S. The Viable System Model: Its Provenance, Development, Methodology and Pathology. Journal of the Operational Research Society 35: 7–26, 1984.
3. Bustard, D. W., Oakes, R., and Heslin, E. Support for the Integrated Use of Conceptual and Dataflow Models in Requirements Specification. In Proceedings of Colloquium on Requirements for Software Intensive Systems, 37–44, 1993, Defense Research Agency (DRA) Malvern.
4. Bustard, D. W., and Lundy, P. J. Enhancing Soft Systems Analysis with Formal Modelling. In Proceedings of Requirements Engineering Symposium, 164–171. York, UK: IEEE Computer Society, 1995.
5. Bustard, D. W., Dobbin T. J., and Carey, B. Integrating Soft Systems and Object Oriented Analysis. In Proceedings of International Conference on Requirements Engineering, 52–59. Colorado Springs: IEEE Computer Society, 1996.
6. Bustard, D. W., He, Z., and Wilkie, F. G. (2000) Linking Soft Systems and Use-Case Modelling Through Scenarios. Interacting with Computers 13: 97–110, 2000.
7. Bustard, D. W., and Keenan, F. Strategies for Systems Analysis: Groundwork for Process Tailoring. In Proceedings of International Conference of Computer Based Systems, 357–362, IEEE Computer Society, 2005.
8. Bustard, D. W., Sterritt, R., Taleb-Bendiab, A., Laws, A., Randles, M., and Keenan, F. Towards a Systemic Approach to Autonomic Systems Engineering.

In Proceedings of International Conference on the Engineering of Computer Based Systems, 465–472, IEEE Computer Society, 2005.

9. Checkland, P. Systems Thinking, Systems Practice (with 30-year retrospective). John Wiley & Sons: Chichester, 1999.

10. Checkland, P., and Scholes, J. Soft Systems Methodology in Action. John Wiley & Sons: Chichester, 1990.

11. Cockburn, A. Agile Software Development, Boston: Pearson Education, 2002.

12. Gilb, T. Principles of Software Engineering Management, Addison Wesley, 1988.

13. Greer, D., Bustard, D. W., and Sunazuka, T., Effecting and Measuring Risk Reduction in Software Development, NEC R&D Journal, 40(3): 378–383.

14. Greer, D., and Bustard, D. W. Collaborative Risk Management, In Proceedings of International Conference on Systems, Man and Cybernetics, 406–410, Tunisia: IEEE Computer Society, 2002.

15. Guo, M., Wu, Z., and Stowell, F. A. Information Systems Specifications Within the Framework of Client-Led Design. In Bustard, D. W., Kawalek, P., and Norris, M. T. (eds.) Systems Modelling for Business Process Improvement, 199–212. Artech House, 2000.

16. Herring, C., and Kaplan, S. The Viable Systems Architecture. In Proceedings of Hawaii International Conference on System Sciences. Hawaii, 2001.

17. Herring, C. Viable Software: The Intelligent Control Paradigm for Adaptable and Adaptive Architecture. PhD Thesis, University of Queensland, Brisbane, Australia, 2002.

18. Horn, P. Autonomic Computing: IBM Perspective on the State of Information Technology. IBM T. J. Watson Labs., NY, 15 October 2001. Presented at AGENDA 2001, Scottsdale, AR (www.research.ibm.com/autonomic/).

19. IBM. An Architectural Blueprint for Autonomic Computing (3rd ed.), White Paper, 2005.

20. Kephart, J. O. Research Challenges of Autonomic Computing. In Proceedings of International Conference on Software Engineering (ICSE), 15–22: St. Louis, Missouri: IEEE Computer Society, 2005.

21. Laws, A. G., Taleb-Bendiab, A., and Wade, S. J., From Wetware to Software: A Cybernetic Perspective of Self-Adaptive Software In Proceedings of Second International Workshop on Self-Adaptive Software, 257–280, Berlin: Springer-Verlag, 2003.

22. Laws, A. G., Taleb-Bendiab, A., and Wade, S. J. Towards a Viable Reference Architecture for Multi-Agent Supported Holonic Manufacturing Systems, Journal of Applied Systems Studies, 2(1), 2001.

23. Lin, P., MacAuther, A., and Leaney, J. Defining Autonomic Computing: A Sofware Engineering Perspective. In Proceedings of Australian Software Engineering Conference (ASWEC), 88–97. Brisbane: IEEE Computer Society, 2005.

24. Mingers, J., and Taylor, S. The Use of Soft Systems Methodology in Practice. Journal of Operational Research 43(4):321–332, 1992.

25. Mingers, J. An Idea Ahead of its Time: The History and Development of Soft Systems Methodology. Systemist 24(2):113–139, 2002.

26. Nelson, V. P. Fault Tolerant Computing: Fundamental Concepts. Computer, 23(7): 19–25, 1990.

27. Sterritt, R., and Bustard, D. W. Towards an Autonomic Computing Environment. In Proceedings of 1st International Workshop on Autonomic Computing

Systems at 14th International Conference on Database and Expert Systems Applications, 694–698. Prague, Czech Republic: IEEE Computer Society, 2003.

28. Sterritt, R., and Bustard, D. W. Autonomic Computing—A Means of Achieving Dependability? In Proceedings of International Conference on the Engineering of Computer Based Systems, 247–251, Huntsville, Alabama: IEEE Computer Society, 2003.

29. Stowell, F. A. Information Systems Provision: The Contributions of SSM. McGraw-Hill: London, 1995.

30. Wilson, B. Systems: Concepts, Methodologies and Applications. John Wiley & Sons: Chichester, 1990.

31. Wilson, B. Soft Systems Methodology: Conceptual Model Building and Its Contribution. John Wiley & Sons: Chichester, 2001.

3

Autonomic Computing: A System-Wide Perspective

Robbert van Renesse and Kenneth P. Birman

CONTENTS

Autonomic computing promises computer systems that are self-configuring and self-managing, adapting to the environment in which they run and to the way they are used. In this chapter, we will argue that such a system cannot be built simply by composing autonomic components, but that a system-wide monitoring and control infrastructure is required as well. This infrastructure needs to be autonomic itself. We demonstrate this using a datacenter architecture.

3.1   Introduction

Computer systems develop organically. A computer system usually starts as a simple clean system intended for a well-defined environment and applications. However, in order to deal with growth and new demands, storage, computing, and networking components are added, replaced, and removed from the system, while new applications are installed and existing applications are upgraded. Some changes to the system are intended to enhance its functionality, but result in loss of performance or other undesired secondary effects. In order to improve performance or reliability, resources are added or replaced. The particulars of such development cannot be anticipated; it just happens the way it does.

Organic development seldom leads to simple or optimal systems. On the contrary, an organically developing system usually becomes increasingly complex and difficult to manage. Configurations of individual components become outdated. Originally intended for a different system, they are no longer optimal for the new environment. System documentation also becomes outdated, and it is difficult for new system managers or application developers to learn how and why a system works. While the original system was designed with clear guidelines, an organically developed system appears ad hoc to a newcomer.

The autonomic computing effort aims to make systems self-configuring and self-managing. However, for the most part the focus has been on how to make system *components* self-configuring and self-managing. Each such component has its own feedback loop for adaptation, and its own policies for how to react to changes in its environment. In an organic system, many such components may be composed in unanticipated ways, and the composition may change over time in unanticipated ways. This may lead to a variety of problems, some of which we now address.

First, the components, lacking global control, may oscillate. Consider, for example, a machine that runs a file system and an application that uses the file system. Both the file system and the application are adaptive—they both can increase performance by using more memory, but release memory that is unused. Now apply a load to the application. In order to increase request throughput, the application tries to allocate more memory. This memory is taken from the file system, which as a result cannot keep up with the throughput. Due to decreased throughput, the application no longer requires as much memory and releases it. The file system can now increase its throughput. In theory, an autonomic allocation scheme exists so that these components may converge to an optimal allocation of memory between them. However, if the components release the same amount of memory each time, they are more likely to oscillate and provide sub-optimal throughput. While with two components and one shared resource such a situation is relatively easy to diagnose and correct, more components and resources make doing so increasingly hard.

Second, even if all components of a system are autonomic, the configuration of the system as a whole is typically done manually. For example, some of the machines may be used to run a database, while other machines may be dedicated to run web servers or business logic. Such partitioning and specialization reduce complexity, but likely result in nonoptimal division of resources, particularly as a system changes over time. Perceived inefficiencies are usually solved by adding more hardware for specific tasks; for example, more database machines may be added. In a system spanning multiple time zones, performance-critical data or software licenses may need to be moved over time. Such changes often require that the configurations of many other machines have to be updated, and finding and tracking such dependencies is nontrivial.

Perhaps the most difficult problem is the one of scale. As an organic system grows in the number of resources, applications, and users, its behavior becomes increasingly complex, while individual components may become overloaded and turn into bottlenecks for performance. Few if any system administrators will have a complete overview of the system and its components, or how its components interact. This not only makes it more difficult to manage a system, but it also becomes harder to develop reliable applications. Installing new resources or applications may break the system in unexpected ways. To make matters even more complicated, such breakage may not be immediately apparent. By the time the malfunction is observed, it may no longer be clear which system change caused the problem.

In this chapter, we argue that a system-wide autonomic control mechanism is required. We describe a Scalable Monitoring and Control Infrastructure (SMCI) that acts as a system-wide feedback loop. The infrastructure allows administrators to view their system and zoom into specific regions with directed queries and to change the behavior of the system. Administrators can create global policies for how components should adapt. While it is still necessary for the individual components to adapt their behavior, the infrastructure guides such local adaptation in order to control system-wide adaptation. This way, internal oscillation may be all but eliminated, and the infrastructure can control how applications are assigned to hardware in order to maximize a global performance metric.

3.2 Scalable Monitoring and Control Infrastructure

An SMCI can be thought of as a database, describing the underlying managed system. For convenience, the database reflects an organizational hierarchy of the hardware in the system. We call the nodes in such a hierarchy "domains." For example, a building may form a "domain." Within a building, each machine room forms a domain. Each rack in the machine room forms a subdomain. Associated with each domain is a table with a record for each subdomain. The records reflect and control the subdomains. See Figure 3.1 for an example.

For example, in the table of the domain of a rack, there may be a record for each machine, specifying attributes like the type of central processing unit (CPU), the amount of memory, its peripherals, what applications it is running, and the load on the machine. Some attributes are read-only values, and can be updated only by the machine itself. An example is the number of applications that is running on the machine. Other attributes may be written by external

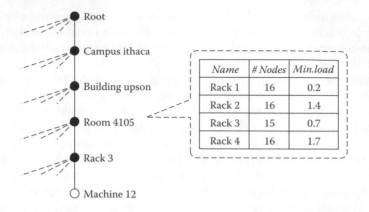

FIGURE 3.1
An example of an SMCI hierarchy. On the right is the domain table of the room 4015, which has four racks of machines. Each rack has three attributes: its name, the number of operational machines in the rack, and the minimum load across the machines.

sources in order to change the behavior of the component. For example, the machine record may contain attributes that govern relative priorities of the applications that it is running.

The tables look like database relations, and would typically be accessed using a query language like SQL. An SMCI will allow one-shot queries like: "How many machines are there?" as well as continuous standing queries like: "Inform me of updates to the number of machines." The latter query works in conjunction with a publish/subscribe or event notification mechanism and publishes updates to interested parties.

A unique feature of SMCIs as compared with ordinary relational databases is their ability to do "hierarchical aggregation queries." An ordinary aggregation query is limited to a single table, while a hierarchical query operates on an entire domain, comprising a tree of tables. For example, one may ask, "How many machines are there in room 4105?" This query adds up all the machines in all the racks in room 4105.

Note that unlike a database, an SMCI is not a true storage system. It gathers attributes from system components and can update those attributes, but it does not store them anywhere persistently. It may store them temporarily for caching purposes. In other words, the tables observed in an SMCI are virtual and are dynamically materialized as necessary in order to answer queries about the system. In order to do so, an SMCI does need to keep track of the organizational hierarchy of the system as well as the standing queries. In addition, an SMCI has to detect and reflect intentional reconfigurations as well as changes caused by machine crashes or network link failures.

An SMCI allows clients, which may include both administrators and running programs, to ask various questions about the system. Examples of such questions include: "Where is file foo?" or "Where is the closest copy of file foo?" "Which machines have a virus database with version less than 2.3.4?" "Which is the heaviest loaded cluster?" "Which machine in this cluster runs a DNS resolver?" "How many web servers are currently up and running?" "How many HTTP queries per second does the system receive as a whole?" "Which machines are subscribed to the topic bar?" And so on.

The real power of the system derives from the fact that the query results may be fed back into the system in order to control it. This creates an autonomic feedback loop for the system as a whole. It can drive the configuration of the system and have the system adapt automatically to variations in its environment and use, as well as recover from failures. For example, the number of replicas of a service may depend on the measured load on the service. The replica placement may depend on where the demand originates and the load on the machines. All this may be expressed in the SMCI.

In order for an SMCI to work well, it needs to be autonomic itself. Should an SMCI require significant babysitting compared with the system it is managing, it would cease to be useful. An SMCI should be robust and handle failures gracefully. It should be self-configuring to a large extent and provide accurate information using a minimum of network traffic. It should scale well, allowing for the hierarchy to be grown as new clusters, machine rooms, or buildings are added to the organizational structure.

Our choice of language deliberately suggests that a system would often have just one SMCI hierarchy, and indeed for most uses, we believe one to be enough. However, nothing prevents the user from instantiating multiple, side-by-side but decoupled SMCI systems, if doing so best matches the characteristics of a given environment.

A number of SMCIs have been developed, and below we give an overview of various such systems. All are based on hierarchies. Most use a single aggregation tree. Examples are Captain Cook [9], Astrolabe [10], Willow [11], DASIS [1], SOMO [13], TAG [5], and Ganglia [8, 6]. SDIMS [12] and Cone [2] use a tree per attribute. Below we give an example of two peer-to-peer SMCIs, one that uses a single tree and one that uses a tree per attribute.

3.2.1   Case Study: Astrolabe

The Astrolabe system [10] is a peer-to-peer SMCI, in that the functionality is implemented in the machines themselves and there are no centralized servers to implement the SMCI. Doing so improves the autonomic characteristics of the SMCI service. In Astrolabe, each machine maintains a copy of the table for each domain it is a member of. Thus it has a table for the root domain and every intermediate domain down to the leaf domain describing the machine itself. Assuming a reasonably balanced organizational hierarchy the amount of information that a machine stores is logarithmic in the number of machines.

The attributes of a leaf domain in Astrolabe are attributes of the machine corresponding to the leaf domain, and may be writable. Standing aggregation

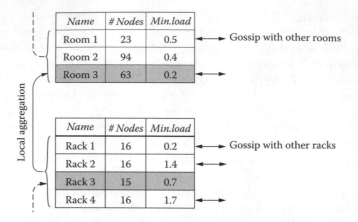

FIGURE 3.2
Data structure of a machine in rack 3 inside room 3 of some building. The machine generates the attributes of rack 3 and room 3 using local aggregation. The attributes of peer racks and rooms are learned through gossip.

queries generate the attributes of nonleaf domains (so-called internal domains) and are strictly read-only. The table of a domain is replicated on all the machines contained in that domain and kept consistent using a gossip protocol (Figure 3.2).

Using an organizational hierarchy based on Domain Name System (DNS) as an example, consider the leaf domain fox.cs.cornell.edu, and consider how this machine may generate the attributes of the cornell.edu subdomains cs, bio, chem, etc. The machine fox.cs.cornell.edu can generate the attributes of the cs.cornell.edu domain locally by aggregating the attributes in the rows of its copy of the domain table of cs.cornell.edu. In order to learn about the attributes of bio.cornell.edu and chem.cornell.edu, the machine has to communicate with machines in those domains.

In order to do so efficiently, Astrolabe utilizes its mechanisms to manage itself. Astrolabe has a standing query associated with each domain that selects a small number, say three, of machines from that domain. The selection criteria do not really matter for this discussion and can be changed by the system administrator. These selected machines are called the representatives of the domain. Peer representatives gossip with one another, meaning that they periodically select a random partner and exchange their versions of the domain table. The tables are merged pair-wise (Astrolabe uses a timestamp for each record in order to determine which records are most recent). Applying this procedure with random partners results in an efficient and highly robust dissemination of updates between representatives. The representatives gossip the information on within their own domains.

Thus, fox.cs.cornell.edu computes the attributes of the cs.cornell.edu domain locally and learns the attributes of bio.cornell.edu etc. through gossip. Applying this strategy recursively, it can then compute the attributes of cornell.edu locally and learn the attributes of the peer domains of cornell.edu through gossip. Using this information it can compute the attributes of the edu domain. Note, however, that there is no consistency guaranteed. At any point in time, fox.cs.cornell.edu and lion.cs.cornell.edu may have different values for the attributes in their common domains, but the gossip protocols disseminate updates quickly, leading to fast convergence.

3.2.2   Case Study: SDIMS

Another peer-to-peer approach to an SMCI is taken by SDIMS (Scalable Distributed Management Information System) [12]. SDIMS is based on Plaxton's scheme for finding nearby copies of objects [7] and also upon the Pastry Distributed Hash Table, although any Distributed Hash Table (DHT) that employs bit-correcting routing may be used.

In a DHT, there is a key space. Each machine has a key in this key space and maintains a routing table in order to route messages, addressed to keys, to the machines with the nearest-by keys. For each position in a key, a machine maintains the address of another machine. For example, suppose that keys consist of 8 bits and a particular machine has key 01001001. If this machine receives a message for key 01011100, the machine determines the common prefix, 010, and finds in its routing table the address of a machine with address 0101xxxx. It then forwards this message to that machine. This continues until the entire key has been matched or no machine is found in the routing table.

In SDIMS, the name of an attribute is hashed onto a key. The key induces a routing tree. This tree is then used to aggregate attributes of the same name on different machines. Unlike Astrolabe, SDIMS has a flexible separation of policy and mechanism that governs when aggregation happens. In SDIMS, one can request that aggregation happen on each read, in which case the query is routed from the requester to the root of the tree, down to the leaves in the tree, retrieving the values, back up the tree aggregating the values, and then down to the requester in order to report the result. For attributes that are read rarely, this is efficient. For attributes that are written rarely, SDIMS also provides a strategy that is triggered on each write, forwarding the update up the tree and down to subscribers to updates of the aggregate query. SDIMS provides various strategies in between these extremes, as well as a system that automatically tries to determine which is the best strategy based on access statistics.

In order to make Pastry suitable for use as an SMCI, several modifications were necessary. Most importantly, the original Pastry system does not support an organizational hierarchy, and cannot recover from partition failures.

3.2.3   Discussion

Because SDIMS utilizes a different tree for each attribute and optimizes aggregation for each attribute individually, it can support more attributes than Astrolabe. On the other hand, Astrolabe can support multi-attribute queries in a straightforward way. Astrolabe's gossip protocols are highly robust and work even in partially connected systems but are less efficient than SDIMS's protocols.

Both Astrolabe and SDIMS are self-configuring, self-managing protocols. Each requires that each node is given a unique name and an initial set of "contact nodes" in order to bootstrap the system. After that, failure detection and recovery and membership management in general are taken care of automatically.

3.3   Service-Oriented Architecture

Many large datacenters, including Google, eBay, Amazon, Citibank, etc., as well as at datacenters used in government, organize their systems as Service-Oriented Architecture (SOA). In [4], the authors introduce a common terminology for describing such systems. Below we will review this terminology and give an example of how a web-based retailer might create a self-configuring and self-managing datacenter using an SMCI.

A collection of servers, applications, and data at a site is called a *farm*. A collection of farms is called a *geoplex*.

A service may be either *cloned* or *partitioned*. Cloning means that the data is replicated onto a collection of nodes. Each node may provide its own storage (inefficient if there are many nodes), or use a shared disk or disk array. The collection of clones is called a *Reliable Array of Cloned Services*, or RACS. Partitioning means that the data is partitioned among a collection of nodes. Partitions may be replicated onto a few nodes, which then form a *pack*. The set of nodes that provide a packed-partitioned service is called a *Reliable Array of Partitioned Services*, or RAPS (Figure 3.3).

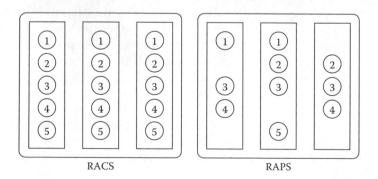

FIGURE 3.3
Example of RACS and RAPS architectures. In RACS, each object is replicated on all three machines. In RAPS, an object is replicated only on a subset of the machines, allowing more objects to be stored and offering greater flexibility in availability guarantees.

While a RAPS is functionally superior to a RACS, a RACS is easier to build and maintain, and therefore many datacenters try to maximize the use of the RACS design. RACS are good for read-only, stateless services such as are often found in the front end of a datacenter, while RAPS are better for update-heavy states, as found in the storage back end. A system built as a RAPS in which each service is a RACS potentially combines the best of both options but can be complex to implement.

SOA principles are being applied in most if not all large datacenters. All these systems have developed as described in the introduction, in an organic way, starting with a few machines in a single room. Now they are geoplexes with thousands to hundreds of thousands of machines, with hundreds or thousands of applications deployed. Literally billions of users depend on the services that these systems provide.

The configuration and management of such systems is a daunting task. Problems that result in black-outs or brown-outs need to be resolved within seconds. Automation is consequently highly desired.

The state of the art is that each service has its own management console. Yet the deployed services depend on one another in subtle or not so subtle ways. If one service is misbehaving, this may be caused by the failure of another service. Obtaining a global view of what is going on and quickly locating the source of a problem is not possible with the monitoring tools currently provided. Also, while such services may have some support for self-configuration and self-management built in, doing so on a global scale is not supported.

An SMCI can be used to solve such problems. It has a global presence and can monitor arbitrary sensory data available in the system. Subsequently, the SMCI can provide a global view by installing an appropriate aggregation query. While useful to a system administrator, such data can also be fed back into the system to help automate the control of resource allocation or drive actuators, rendering the system self-configuring and self-managing. We will demonstrate this for a web-based retailer.

3.4 An Example

A web-based retailer's datacenter typically consists of three tiers. The front end consists of web servers and handles HTTP requests from remote clients. The web servers issue parallel calls to services implemented by applications running in the middle tier. A web page typically requires calls into various services for the various sections of the page. Examples of services include an HTTP Session Service, product search and browsing, various services that describe a product (pictures, technical data, availability, pricing, etc.), a service that recommends related products, and many more. The services may need

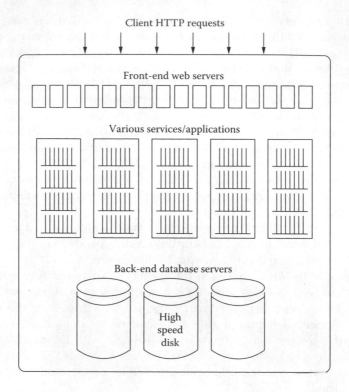

FIGURE 3.4
A datacenter is typically organized as three tiers.

to call on one another. Long-term storage is handled by the back-end tier (see Figure 3.4).

The front-end servers typically are stateless. For every incoming request they simply call into the second tier. A RACS architecture is an obvious choice: each web server is essentially identical and maintains little data. The third tier's storage reflects the actual assets held by the retailer and also deals with credit card information and transactions, and therefore needs to be highly reliable and consistent. A RACS architecture would not scale, as updates would have to be applied consistently at each server. Consequently, the third tier is typically implemented as a RAPS. For the middle tier, a choice between RACS and RAPS can be made on a per-service basis.

The description so far is a logical depiction of a datacenter. Physically, the datacenter consists of a collection of machines, disks, networks, and often, load balancers that distribute requests among nodes of a service. Both the logical and physical aspects of a datacenter are in flux. Physically, machines die, are added, or are replaced, while the network architecture may also change frequently. Logically, many datacenters deploy several new applications per week and retire others. As a result, the configuration of a datacenter must be updated continuously, and more frequently as a datacenter grows.

The web servers in the front end need to know the layout of a web page, which services it should invoke for the various sections, and how it should contact these services so as to balance the load across the nodes comprised by the service, while avoiding nodes that are faulty or overloaded. The services in the middle tier depend on other services in the middle and third tier, and also need to know how to contact those services. The number of machines assigned to a service depends on the load distribution among the machines. Which machines are assigned to which applications depends upon data placement, and it may be necessary to move data when the system is reconfigured. We will now show how an SMCI can be used in such an environment.

A simple SMCI organization for this system could be as follows. The root of the tree would describe a geoplex, with a child node for each farm. Each farm would have a child node for each cluster of machines. For each cluster, there would be a child node for each machine in the cluster. Each machine would be able to report what applications it is running, as well as various information about these applications.

3.4.1 Using a RACS

As an example, consider the Session Service which keeps track of sessions with web clients. When an HTTP request arrives at a web server, the web server contacts the Session Service in order to obtain information about the current session. In order to do so, the web server has to find a machine that runs the Session Service (called a Session Server henceforth) and knows about the session. If the Session Server is organized as a RACS, any machine will do. If it is organized as a RAPS, only some of the machines in the RAPS know about the session. For now we assume the Session Server is organized as a RACS.

The system has permanently installed an aggregation query in the SMCI that counts the number of Session Servers in an SMCI domain (1 for the leaf domains that represent the machines that run the Session Server). Starting in the root domain, the web server would first locate the child node corresponding to its own farm. Then it would locate a child node (a cluster) that has more than Session Server. If there is more than one such cluster, it could pick one at random, weighted by the number of Session Servers for fairness. Finally, it would pick one of the Session Servers at random from the cluster. Additionally, the Session Servers could report their load. Using an aggregation function that reports the minimum load in a domain, web servers could locate the least loaded Session Server.

3.4.2 Using a RAPS

Should the Session Service be implemented as a RAPS, then there are various options in order to locate one of the Session Servers in the pack for the given session identifier. One possibility is that the Session Servers themselves maintain such a map. A request from a web server is first sent to an arbitrary Session Server and then forwarded to the correct one. This delay can be avoided by using a Bloom filter [3]. A Bloom filter is a concise representation of a set in the form of a bitmap.

Each Session Server would report in a Bloom filter the set of sessions it is responsible for. An aggregation function would simply "bitwise or" these filters together in order to create a Bloom filter for each domain. As a result, each domain would report what sessions it is responsible for. Using this information, a web server can quickly find a Session Server in the pack responsible for a particular session.

So far we have described how the web servers can use the SMCI in order to find Session Servers, and this is part of how the datacenter manages itself. As another example of self-management, the Session Service can use the SMCI to manage itself. We will focus on how Session Servers in a RAPS architecture choose which sessions they are responsible for. This will depend on load distribution. Complicating matters, machines may be added or removed. We also try to maintain an invariant of k replicas for each session—no less, but also no more.

Each Session Server can keep track of all the machines in the Session Service simply by walking the SMCI domain tree. Using the Session Service membership, it is possible to apply a deterministic function that determines which machines are responsible for which sessions. For example, a session could be managed by the k machines with the lowest values of $\mathcal{H}(\text{machine ID, session ID})$, where $\mathcal{H}$ is a hash function. These machines would be responsible for running a replication protocol to keep the replicas consistent with one another.

3.5   Conclusion

A large autonomic computing system cannot be composed from autonomic components. The configuration of the components would not self-adapt, the composition could oscillate, and individual components might become bottlenecks. In order to have system-wide autonomy, a system-wide feedback loop is necessary. We described the concept of a Scalable Monitoring and Control Infrastructure and how it may be applied in an autonomic Service Oriented Architecture.

Acknowledgment

The authors were supported by grants from the Autonomic Computing effort at Intel Corporation, from DARPA's Self Regenerative Systems program, and from NSF through its TRUST Science and Technology Center.

References

1. K. Albrecht, R. Arnold, and R. Wattenhofer. Join and leave in peer-to-peer systems: The DASIS approach. Technical Report 427, Dept. of Computer Science, ETH Zurich, November 2003.
2. R. Bhagwan, G. Varghese, and G. M. Voelker. Cone: Augmenting DHTs to support distributed resource discovery. Technical Report CS2003-0755, UC, San Diego, July 2003.
3. B. Bloom. Space/time tradeoffs in hash coding with allowable errors. *CACM*, 13(7):422–426, July 1970.
4. B. Devlin, J. Gray, B. Laing, and G. Spix. Scalability terminology: Farms, clones, partitions, packs, racs and raps. Technical Report MSR-TR-99-85, Microsoft Research, 1999.
5. S. R. Madden, M. J. Franklin, J. M. Hellerstein, and W. Hong. TAG: a Tiny AGgregation service for ad-hoc sensor networks. In *Proceedings of the 5th Symposium on Operating Systems Design and Implementation*, Boston, MA, December 2002. USENIX.
6. M. L. Massie, B. N. Chun, and D. E. Culler. The Ganglia distributed monitoring system: Design, implementation, and experience. *Parallel Computing*, 30(7), July 2004.
7. C. G. Plaxton, R. Rajaraman, and A. W. Richa. Accessing nearby copies of replicated objects in a distributed environment. In *ACM Symposium on Parallel Algorithms and Architectures*, pages 311–320, 1997.
8. F. D. Sacerdoti, M. J. Katz, M. L. Massie, and D. E. Culler. Wide area cluster monitoring with Ganglia. In *Proceedings of the IEEE Cluster 2003 Conference*, Hong Kong, 2003.
9. R. van Renesse. Scalable and secure resource location. In *Proceedings of the 33rd Annual Hawaii International Conference on System Sciences*, Los Alamitos, CA, January 2000. IEEE, IEEE Computer Society Press.
10. R. van Renesse, K. P. Birman, and W. Vogels. Astrolabe: A robust and scalable technology for distributed system monitoring, management, and data mining. *ACM Transactions on Computer Systems*, 21(3), May 2003.
11. R. van Renesse and A. Bozdog. Willow: DHT, aggregation, and publish/subscribe in one protocol. In *Proceedings of the 3rd International Workshop on Peer-To-Peer Systems*, San Diego, CA, February 2004.
12. P. Yalagandula and M. Dahlin. A Scalable Distributed Information Management System. In *Proceedings of the '04 Symposium on Communications Architectures & Protocols*, Portland, OR, August 2004. ACM SIGCOMM.
13. Z. Zhang, S.-M. Shi, and J. Zhu. SOMO: Self-Organized Metadata Overlay for resource management in P2P DHT. In *Proceedings of the Second International Workshop on Peer-to-Peer Systems (IPTPS'03)*, Berkeley, CA, February 2003.

4

Autonomic Grid Computing: Concepts, Requirements, and Infrastructure

Manish Parashar

CONTENTS

Emerging pervasive wide-area Grid computing environments are enabling a new generation of applications that are based on seamless aggregation and interactions of resources, services, and information. However, Grid applications and the underlying Grid computing environment are inherently large, heterogeneous, and dynamic, resulting in significant development, configuration, and management challenges. Addressing these challenges has led researchers to consider autonomic self-managing solutions, which are based on strategies used by biological systems to deal with similar challenges. This chapter motivates and introduces autonomic Grid computing. It also introduces Project AutoMate and describes its key components. The goal of Project Automate is to investigate conceptual models and implementation architectures that can enable the development and execution of such self-managing Grid applications. Two application scenarios enabled by AutoMate are described.

4.1 Introduction

The Grid vision [38] has been described as a world in which computational power (resources, services, data) is as readily available as electrical power and other utilities, in which computational services make this power available to users with differing levels of expertise in diverse areas, and in which these services can interact to perform specified tasks efficiently and securely with minimal human intervention. Driven by revolutions in science and business and fueled by exponential advances in computing, communication, and storage technologies, Grid computing is rapidly emerging as the dominant paradigm for wide-area distributed computing. Its goal is to provide a service-oriented infrastructure that leverages standardized protocols and services to enable pervasive access to, and coordinated sharing of, geographically distributed hardware, software, and information resources. The Grid community and the Global Grid Forum are investing considerable effort in developing and deploying standard protocols and services that enable seamless and secure discovery of, access to, and interactions among resources, services, and applications. This potential for seamless aggregation, integration, and interactions has also made it possible for scientists and engineers to conceive a new generation of applications that enable realistic investigation of complex scientific and engineering problems.

However, the Grid computing environment is inherently large, heterogeneous, and dynamic, globally aggregating large numbers of independent computing and communication resources, data stores, instruments, and sensor networks. Furthermore, emerging applications are similarly complex and highly dynamic in their behaviors and interactions. Together, these characteristics result in application development, configuration, and management complexities that break current paradigms based on passive components and static behaviors and compositions. Clearly, there is a need for a fundamental change in how these applications are developed and managed. This has led researchers to consider alternative programming paradigms and management techniques that are based on strategies used by biological systems to deal with complexity, dynamism, heterogeneity, and uncertainty. The approach, referred to as *autonomic computing*, aims at realizing computing systems and applications capable of managing themselves with minimal human intervention [9].

This chapter has two objectives. The first is to investigate the challenges presented by Grid environments and to motivate self-management as a means for addressing these challenges. The second is to introduce Project AutoMate, which investigates autonomic solutions to deal with the challenges of complexity, dynamism, heterogeneity, and uncertainty in Grid environments. The overall goal of Project AutoMate is to develop conceptual models and implementation architectures that can enable the development and execution of such self-managing Grid applications. Specifically, it investigates programming models, frameworks, and middleware services that support definition of autonomic elements; the development of autonomic applications as dynamic and opportunistic compositions of these autonomic elements; and the policy-, content-, and context-driven execution and management of these applications.

In this chapter we first investigate the characteristics of Grid environments and the requirements of developing and managing Grid applications, review current research in Grid computing, and introduce autonomic Grid computing as a holistic approach that can address these requirements. We then introduce AutoMate and its key components and describe their underlying conceptual models and implementations. Specifically we describe the Accord programming system, the Rudder decentralized coordination framework, and the Meteor content-based middleware providing support for content-based routing, discovery, and associative messaging. We also present two autonomic Grid application scenarios enabled by AutoMate. The first application investigates the autonomic optimization of an oil reservoir by enabling a systematic exploration of a broader set of scenarios, to identify optimal well locations based on current operating conditions. The second application investigates the autonomic simulation and management of a forest file based on static and dynamic environment and vegetation conditions.

The rest of this chapter is organized as follows. Section 2 outlines the challenges and requirements of Grid computing. Section 3 presents autonomic Grid computing. Section 4 introduces Project AutoMate, presents its overall architecture, and describes its key components, i.e., the Accord programming framework, the Rudder decentralized coordination framework, and the Meteor content-based middleware. Section 5 presents the two illustrative Grid applications enabled by AutoMate. Section 6 presents a conclusion.

4.2   Grid Computing: Challenges and Requirements

The goal of the Grid concept is to enable a new generation of applications combining intellectual and physical resources that span many disciplines and organizations, providing vastly more effective solutions to important scientific, engineering, business, and government problems. These new applications must be built on seamless and secure discovery of, access to, and interactions among resources, services, and applications owned by many different organizations.

Attaining these goals requires implementation and conceptual models [37]. Implementation models address the virtualization of organizations, which leads to Grids, the creation and management of virtual organizations as goal-driven compositions of organizations, and the instantiation of virtual machines as the execution environment for an application. Conceptual models define abstract machines that support programming models and systems to enable application development. Grid software systems typically provide capabilities for: (i) creating a transient "virtual organization" or virtual resource configuration, (ii) creating virtual machines composed from the resource configuration of the virtual organization, (iii) creating application programs to execute on the virtual machines, and (iv) executing and managing application execution. Most Grid software systems implicitly or explicitly incorporate a programming model, which in turn assumes an underlying abstract machine with specific execution behaviors, including assumptions about reliability, failure modes, etc. As a result, failure to realize these assumptions by the implementation models will result in brittle applications. The stronger the assumptions made, the greater the requirements for the Grid infrastructure to realize these assumptions and consequently its resulting complexity. In this section we first highlight the characteristics and challenges of Grid environments, and outline key requirements for programming Grid applications. We then introduce autonomic self-managing Grid applications that can address these challenges and requirements.

4.2.1   Characteristics of Grid Execution Environments and Applications

The key characteristics of Grid execution environments and applications are:

Heterogeneity: Grid environments aggregate large numbers of independent and geographically distributed computational and information resources, including supercomputers, workstation clusters, network elements, data storages, sensors, services, and Internet networks. Similarly, applications typically combine multiple independent and distributed software elements such as components, services, real-time data, experiments, and data sources.

Dynamism: The Grid computation, communication, and information environment is continuously changing during the lifetime of an application. This includes the availability and state of resources, services, and data. Applications similarly have dynamic runtime behaviors in that the organization and interactions of the components/services can change.

Uncertainty: Uncertainty in Grid environments is caused by multiple factors, including (1) dynamism, which introduces unpredictable and changing behaviors that can be detected and resolved only at runtime, (2) failures, which have an increasing probability of occurrence and frequencies as system/application scales increase, and (3) incomplete knowledge of global system state, which is intrinsic to large decentralized and asynchronous distributed environments.

Security: A key attribute of Grids is flexible and secure hardware/software resource sharing across organization boundaries, which makes security (authentication, authorization, and access control) and trust critical challenges in these environments.

4.2.2   Requirements of Grid Applications and Environments

The characteristics listed above impose requirements on the programming and runtime systems for Grid applications. Grid programming systems must be able to specify applications which can detect and dynamically respond during execution to changes in both the state of execution environment and the state and requirements of the application. This requirement suggests that: (1) Grid applications should be composed from discrete, self-managing components which incorporate separate specifications for all of functional, non-functional, and interaction-coordination behaviors. (2) The specifications of computational (functional) behaviors, interaction and coordination behaviors, and nonfunctional behaviors (e.g., performance, fault detection and recovery, etc.) should be separated so that their combinations are composable. (3) The interface definitions of these components should be separated from their implementations to enable heterogeneous components to interact and to enable dynamic selection of components.

Given these features of a programming system, a Grid application requiring a given set of computational behaviors may be integrated with different interaction and coordination models or languages (and vice versa) and different specifications for nonfunctional behaviors such as fault recovery and quality of service (QoS) to address the dynamism, heterogeneity, and uncertainty of applications and the environment.

Furthermore, the underlying core enabling services must address the unreliability and uncertainty of the execution environment, and support decentralized behaviors, dynamic and opportunistic discovery and compositions, and asynchronous and decoupled interactions.

4.2.3   Grid Computing Research

Grid computing research efforts over the last decade can be broadly divided into efforts addressing the realization of virtual organizations and those addressing the development of Grid applications. The former set of efforts has focused on the definition and implementation of the core services that enable the specification, construction, operation, and management of virtual organizations and instantiation of virtual machines that are the execution environments of Grid applications. Services include (1) security services to enable the establishment of secure relationships between a large number of dynamically created subjects and across a range of administrative domains, each with its own local security policy, (2) resource discovery services to enable discovery of hardware, software, and information resources across the Grid, (3) resource management services to provide uniform and scalable mechanisms for naming and locating remote resources, support the initial registration/discovery and ongoing monitoring of resources, and incorporate these resources into applications, (4) job management services to enable the creation, scheduling, deletion, suspension, resumption, and synchronization of jobs, (5) data management services to enable accessing, managing, and transferring of data, and provide support for replica management and data filtering. Efforts in this class include Globus [6], Unicore [44], Condor [43], and Legion [8]. Other efforts in this class include the development of common APIs, toolkits, and portals that provide high-level uniform and pervasive access to these services. These efforts include the Grid Application Toolkit (GAT) [1], DVC [41] and the Commodity Grid Kits (CoG Kits) [18]. These systems often incorporate programming models or capabilities for utilizing programs written in some distributed programming model. For example, Legion implements an object-oriented programming model, while Globus provides a capability for executing programs utilizing message passing.

The second class of research efforts, which is also the focus of this paper, deals with the formulation, programming, and management of Grid applications. These efforts build on the Grid implementation services and focus on programming models, languages, tools and frameworks, and application runtime environments. Research efforts in this class include GrADS [3], GridRPC [33], GridMPI [11], Harness [32], Satin/IBIS [36] [35], XCAT [7] [17], Alua [45], G2 [15], J-Grid [30], Triana [42], and ICENI [5].

These systems have essentially built on, combined, and extended existing models for parallel and distributed computing. For example, GridRPC extends the traditional RPC model to address system dynamism. It builds on Grid system services to combine resource discovery, authentication/authorization, resource allocation, and task scheduling with remote invocations. Similarly, Harness and GridMPI build on the message passing parallel computing model, and Satin supports divide-and-conquer parallelism on top of the IBIS communication system. GrADS builds on the object model and uses reconfigurable object and performance contracts to address Grid dynamics, while XCAT and Alua extend the component based model. G2, J-Grid, Triana, and ICENI build on various service based models. G2 builds on .Net [34], J-Grid builds on Jini [13], and current implementations of Tirana and ICENI build on JXTA [14]. While this is natural, it also implies that these systems implicitly inherit the assumptions and abstractions that underlie the programming models of the systems upon which they are based and thus in turn inherit their capabilities and limitations.

4.2.4   Self-Managing Applications on the Grid

As outlined above, the inherent scale, complexity, heterogeneity, and dynamism of emerging Grid environments — and the resulting uncertainty — present application programming and runtime management complexities that break current paradigms. This is primarily because the programming models and system and the abstract machine underlying these models make strong assumptions about common knowledge, static behaviors, and system guarantees that can no longer be realized by Grid virtual machines and which are not true for Grid applications. Addressing these challenges requires re-defining the programming and runtime systems to address the separations outlined above. Specifically, it requires (1) static (defined at the time of instantiation) application requirements and system and application behaviors to be relaxed, (2) the behaviors of elements and applications to be sensitive to the dynamic state of the system and the changing requirements of the application and be able to adapt to these changes at runtime, (3) required common knowledge to be expressed semantically (ontology and taxonomy) rather than in terms of names, addresses, and identifiers, and (4) the core enabling middleware services (e.g., discovery, coordination, messaging) to be driven by such a semantic knowledge. Furthermore, the implementations of these services must be resilient and must scalably support asynchronous and decoupled behaviors.

4.3   The Autonomic Computing Paradigm

An autonomic computing paradigm, modeled after the autonomic nervous system, must have a mechanism whereby changes in its essential variables can trigger changes in the behavior of the computing system such that the system is brought back into equilibrium with respect to the environment. This state of stable equilibrium is a necessary condition for the survivability of the organism. In the case of an autonomic computing system, we can think of survivability as the system's ability to protect itself, recover from faults, reconfigure as required by changes in the environment, and always maintain its operations at near optimal performance. Its equilibrium is impacted by both the internal environment (e.g., excessive memory/utilization of the central processing unit (CPU)) and the external environment (e.g., protection from an external attack).

An autonomic system requires: (a) sensor channels to sense the changes in the internal and external environments, and (b) motor channels to react to and counter the effects of the changes in the environment by changing the system and maintaining equilibrium. The changes sensed by the sensor channels have to be analyzed to determine if any of the essential variables have gone out of their viability limits. If so, it has to trigger some kind of planning to determine what changes to inject into the current behavior of the system such that it returns to the equilibrium state within the new environment. This planning would require knowledge to select the right behavior from a large set of possible behaviors to counter the change. Finally, the motor neurons execute the selected change. 'Sensing', 'Analyzing', 'Planning', 'Knowledge', and 'Execution' are in fact the keywords used to identify an autonomic computing system [10]. We use these concepts to present the architecture of an autonomic element and autonomic applications and systems.

4.3.1 Autonomic Grid Computing: A Holistic Approach

As motivated above, the emerging complexity in Grid computing systems, services, and applications requires the system/software architectures to be adaptive in all their attributes and functionalities (performance, security, fault tolerance, configurability, maintainability, etc.). We have been successful in designing and implementing specialized computing systems and applications that are adaptive. However, the design of general purpose dynamically programmable computing systems and applications that can address the emerging needs and requirements remains a challenge. For example, distributed (and parallel) computing has evolved and matured to provide specialized solutions to satisfy very stringent requirements in isolation, such as security, dependability, reliability, availability, performance, throughput, efficiency, pervasive/amorphous, automation, reasoning, etc. However, in the case of emerging systems and applications, the specific requirements, objectives, and choice of specific solutions (algorithms, behaviors, interactions, etc.) depend on runtime state, context, and content and are not known a priori. The goal of autonomic computing is to use appropriate solutions based on current state/context/content and on specified policies.

The computer evolution has gone through many generations, starting from single-process, single-computer systems, to multiple processes running on multiple geographically dispersed heterogeneous computers that could span several continents (e.g., Grid). The approaches for designing the corresponding computing systems and applications have been evolutionary and ad hoc. Initially, the designers of such systems were concerned mainly about performance and focused intensive research on parallel processing and high-performance computer architectures and applications to address this requirement. As the scale and distribution of computer systems and applications evolved, the reliability and availability of the systems and applications became the major concern. This in turn has led to separate research in fault tolerance and reliability, and to systems and applications that were ultra-reliable and resilient, but not high performance. In a similar way, ultra-secure computing systems and applications have been developed to meet security requirements in isolation.

This ad hoc approach has resulted in the successful design and development of specialized computing systems and applications that can optimize a few of the attributes or functionalities of computing systems and applications. However, as we highlighted above, the emerging systems and applications and their contexts are dynamic. Consequently, their requirements will change during their lifetimes and may include high performance, fault tolerance, security, availability, configurability, etc. Consequently, what is needed is a new computing architecture and programming paradigm that takes a holistic approach to the design and development of computing systems and applications. Autonomic computing provides such an approach by enabling the design and development of systems/applications that can adapt themselves to meet requirements of performance, fault tolerance, reliability, security, etc., without manual intervention. Every element in an autonomic system or application consists of two main modules: the functional unit that performs the required services and functionality, and the management/control unit that monitors the state and context of the element, analyzes its current requirements (performance, fault tolerance, security, etc.), and adapts to satisfy the requirement(s).

4.4 Project AutoMate: Enabling Self-Managing Grid Applications

Project AutoMate [39] investigates autonomic solutions that are based on the strategies used by biological systems to deal with similar challenges of complexity, dynamism, heterogeneity, and uncertainty. The goal is to realize systems and applications that are capable of managing (i.e., configuring, adapting, optimizing, protecting, healing) themselves. Project AutoMate aims at developing conceptual models and implementation architectures that can enable the development and execution of such self-managing Grid applications. Specifically, it investigates programming models, frameworks and middleware services that support the definition of autonomic elements, the development of autonomic applications as the dynamic and opportunistic composition of these autonomic elements, and the policy-, content-, and context-driven definition, execution, and management of these applications.

A schematic overview of AutoMate is presented in Figure 4.1. Components of AutoMate include the Accord [27, 28] programming system, the

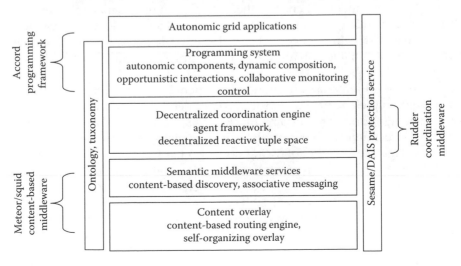

FIGURE 4.1
A schematic overview of AutoMate.

Rudder [21, 20, 19] decentralized coordination framework, and the Meteor [12] content-based middleware providing support for content-based routing, discovery, and associative messaging. Project AutoMate additionally includes the Sesame [47] context-based access control infrastructure, the DAIS [46] cooperative-protection services, and the Discover collaboratory [29] services for collaborative monitoring, interaction, and control, which are not described here.

4.4.1   Accord: A Programming Framework for Autonomic Applications

The Accord programming system [27, 28] addresses Grid programming challenges by extending existing programming systems to enable autonomic Grid applications. Accord realizes three fundamental separations: (1) a separation of computations from coordination, and interactions, (2) a separation of non-functional aspects (e.g., resource requirements, performance) from functional behaviors, and (3) a separation of policy and mechanism—policies in the form of rules are used to orchestrate a repertoire of mechanisms to achieve context-aware adaptive runtime computational behaviors and coordination and interaction relationships based on functional, performance, and QoS requirements.

Accord builds on the separation of composition aspects of elements (object, components, and services) from their computational behaviors that underlie the object-, component-, and service-based paradigms, and extends it to enable the computational behaviors of elements as well as their organizations, interactions, and coordination to be managed at runtime using high-level rules. The Accord system (1) defines autonomic elements that encapsulate functional and nonfunctional specifications, rules, and mechanisms for self-management, (2) enables the formulation of self-managing applications as dynamic compositions of autonomic elements, and (3) provides a runtime infrastructure for correct and efficient rule execution to enforce self-management behaviors in response to changing requirements and execution context.

Autonomic Elements: An autonomic element in Accord extends traditional programming elements (i.e., objects, components, and services) to define a self-managing software building block with specified interfaces and explicit context dependencies. It encapsulates rules, constraints, and mechanisms for self-management via the following ports: (1) the functional port, defining a valid set of behaviors in terms of an input-output set; (2) the control port, managing the set of sensors and actuators (sensors are interfaces that provide information about the element while actuators are interfaces for modifying its state); and (3) the operational port, defining the interfaces to formulate, dynamically inject, and manage rules that are used to control the runtime behavior of the element and the interactions among elements and between elements and their environments, and the coordination of elements within an application.

The control and operational ports enhance element interfaces to export information about their behavior and adaptability to system and application dynamics. An autonomic element also embeds an element manager that is delegated to manage its execution. The element manager monitors the state of the element and its context and controls the execution of rules. Element managers may cooperate with each other to satisfy application objectives.

Rules in Accord: Accord defines two classes of rules: (1) behavioral rules that control the runtime functional behaviors of an autonomic element (e.g., the dynamic selection of algorithms, data representation, input/output format used by the element), and (2) interaction rules that control the interactions between elements and between elements and their environment, and the coordination within an autonomic application (e.g., communication mechanism, composition and coordination of the elements). Rules are executed by collecting element and system runtime information using sensors, evaluating conditions, and then invoking corresponding actuators specified in the control port. Behavioral rules are executed by the element manager associated with a single element without affecting other elements. Interaction rules define interactions among elements. For each interaction pattern, a set of interaction rules are defined and dynamically injected into the interacting elements. The coordinated execution of these rules results in the desired interaction and coordination behaviors between the elements.

Autonomic Composition in Accord: Dynamic composition enables relationships between elements to be established and modified at runtime. Operationally, dynamic composition consists of a composition plan or workflow generation and execution. Plans may be created at runtime, possibly based on dynamically defined objectives, policies, requirements, and context. Plan execution involves discovering elements, configuring them, and defining interaction relationships and mechanisms. This may result in elements being added, replaced, or removed, or interaction relationships changed.

In Accord, composition plans may be generated using the Accord Composition Engine or other approaches. Plan execution is achieved by a control network of element managers. A composition relationship between two elements is defined by the control structure (e.g., loop, branch) and/or the communication mechanism used (such as RPC, shared space). A composition manager translates this into a suite of interaction rules, which are then injected into corresponding element managers. Element managers execute the rules to establish coordination and communication relationships among these elements. Rules can be similarly used to add or delete elements. Accord defines a library of rule sets for common control and the communication relationships between elements.

Rule Execution in Accord: A three-phase rule execution model [24] is used by element managers to ensure consistent and efficient parallel rule execution. The three phases of rule execution are batch condition inquiry, condition evaluation and conflict resolution/reconciliation, and batch action invocation. During the batch condition inquiry phase, each element manager queries, in parallel, all the sensors used by the rules, gets their current values, and then generates the precondition. During the next phase, condition evaluation for all the rules is performed in parallel. Rule conflicts are detected at runtime when rule execution will change the precondition (defined as sensor-actuator conflicts), or the same actuator will be invoked with different values (defined as actuator-actuator conflicts). Sensor-actuator conflicts are resolved by disabling those rules that will change the precondition. Actuator-actuator conflicts are resolved through relaxing the precondition according to user-defined strategies until no actuator will be invoked with different values. Further, the framework also provides mechanisms for reconciliation among manager instances, which is required to ensure consistent adaptations, since each processing node may independently propose different adaptation behaviors based on its local state and context. After conflict resolution and reconciliation, the postcondition, consisting of a set of actuators and their new values, is generated. The postcondition is enforced by appropriately invoking the actuators in parallel during the batch action invocation phase. Note that the rule execution model presented here focuses on correct and efficient rule execution and provides mechanisms to detect and resolve conflicts at runtime. However, the correctness of rules and conflict resolution strategies are the users' responsibility.

4.4.2 Rudder Coordination Framework

Rudder [21, 20] is a scalable coordination middleware for supporting self-managing applications in decentralized distributed environments. The goal of Rudder is to provide the core capabilities for supporting autonomic compositions, adaptations, and optimizations. Rudder consists of two key components: (1) COMET, a fully decentralized coordination substrate that enables flexible and scalable coordination among agents and autonomic elements, and (2) an agent framework composed of software agents and agent interaction and negotiation protocols. The adaptiveness and sociableness of software agents provides an effective mechanism for managing individual autonomic elements and their relationships in an adaptive manner. This mechanism enables appropriate application behaviors to be dynamically negotiated and enacted by adapting classical machine learning, control, and optimization models and theories. The COMET substrate provides the core messaging and eventing services for connecting agent networks and scalably supporting various agent interactions, such as mutual exclusion, consensus, and negotiation. Rudder effectively supports the Accord programming framework and enables autonomic self-managing applications.

The COMET Substrate: COMET [19] provides a global virtual shared coordination space associatively accessible by all peer agents, and the access is independent of the physical location of the tuples or identifiers of the host. The virtual coordination space builds on an associative messaging substrate and implements a distributed hash table, where the index space is directly generated from the semantic information space (ontology) used by the coordinating entities. COMET also supports dynamically constructed, transient spaces to enable context locality to be explicitly exploited for improved performance.

COMET consists of layered abstractions prompted by a fundamental separation of communication and coordination concerns. It provides an associative communication abstraction and guarantees that content-based query messages, specified using flexible content descriptors, are fully served with bounded costs. This layer essentially maps the virtual information space in a deterministic way to the dynamic set of currently available peer nodes in the system, while maintaining content locality. The COMET coordination abstraction extends the traditional data-driven coordination model with event-based reactivity to changes in system state and data access operations. It defines a reactive tuple abstraction, which consists of additional components: a *condition* that associates *reaction* to events, and a *guard* that specifies how and when the reaction will be executed (e.g., immediately, once). The *condition* is evaluated on an access event. If it evaluates to true, the corresponding reaction is executed. The COMET coordination abstraction provides the basic Linda-like primitives, such as Out, In, and Rd. These basic operations operate on regular as well as reactive tuples and retain their Linda semantics.

The Agent framework: The Rudder agent framework [20, 21] is composed of a dynamic network of software agents existing at different levels, ranging from individual system/application elements to the overall system/application. Agents monitor the element states, manage element behaviors and dependencies, coordinate element interactions, and cooperate to manage overall system/application behaviors. An agent is a processing unit that performs actions based on rules, which are dynamically defined to satisfy system/application requirements. Further, agents use profiles which are used to identify and describe elements, interact with them, and control them. A profile consists of a set of (functional and nonfunctional) attributes and operators, which are semantically defined using an application-specific ontology. The framework additionally defines a set of protocols for agent coordination and application/system management. Discovery protocols support the registering, unregistering, and discovery of system/application elements. Control protocols allow the agents to query element states, control their behaviors, and orchestrate their interactions. These protocols include negotiation, notification, and mutual exclusion. The agent coordination protocols are scalably and robustly implemented in using the abstractions and services provided by COMET. COMET builds on an associative communication middleware, Meteor, which is described below.

4.4.3   Meteor: A Content-Based Middleware

Meteor [12] is a scalable content-based middleware infrastructure that provides services for content routing, content discovery, and associative interactions. The Meteor stack consists of three key components: (1) a self-organizing content overlay, (2) a content-based routing engine and discovery service (Squid), and (3) the Associative Rendezvous Messaging Substrate (ARMS).

The Meteor overlay is composed of Rendezvous Peer (RP) nodes, which may be any node on the Grid (e.g., gateways, access points, message relay nodes, servers, end-user computers). RP nodes can join or leave the network at any time. The content overlay provides a single operation, *lookup(identifier)*, which requires an exact content identifier (e.g., name). Given an identifier, this operation locates the peer node where the content should be stored or fetched.

Squid [40] is the Meteor content-based routing engine and decentralized information discovery service. It supports flexible content-based routing and complex queries containing partial keywords, wildcards, and ranges. Squid guarantees that all existing data elements that match a query will be found. The key innovation of Squid is the use of a locality-preserving and dimension-reducing indexing scheme, based on the Hilbert Space Filling Curve (SFC), which effectively maps the multidimensional information space to the peer identifier space. Keywords can be common words or values of globally defined attributes, depending on the nature of the application that uses Squid, and are based on common ontologies and taxonomies.

The ARMS layer [12] implements the Associative Rendezvous (AR) interaction paradigm. AR is a paradigm for content-based decoupled interactions with programmable reactive behaviors. Rendezvous-based interactions provide a mechanism for decoupling senders and receivers, in both space and time. Such decoupled asynchronous interactions are naturally suited for large, distributed, and highly dynamic systems such as pervasive Grid environments. AR extends the conventional name/identifier-based rendezvous in two ways. First, it uses flexible combinations of keywords (i.e., keyword, partial keyword, wildcards, and ranges) from a semantic information space, instead of opaque identifiers (names, addresses) that have to be globally known. Interactions are based on content described by these keywords. Second, it enables the reactive behaviors at the rendezvous points to be encapsulated within messages, therefore increasing flexibility and enabling multiple interaction semantics (e.g., broadcast multicast, notification, publisher/subscriber, mobility, etc.).

4.4.4   Current Status

The core components of AutoMate have been prototyped and are currently being used to enable self-managing applications in science and engineering. Current prototypes of Accord include an object-based prototype (DIOS++) [26], a component-based prototype based on the DoE Common Component Architecture (Accord-CCA) [25], and a service-based prototype based on the WS-Resource specifications (Accord-WS) [22]. Current prototypes of Rudder and Meteor build on the JXTA [14] platform, use existing Grid middleware services, and have been deployed on several wide-area systems, including PlanetLab. Current applications include autonomic oil reservoir optimizations [31, 2], autonomic forest fire management [16], self-managing scientific simulations [25], autonomic runtime application management [22, 4], advanced feature-based visualization techniques [23], and enabling sensor-based pervasive applications [12]. The first two applications are briefly described below. Further information about AutoMate and its components and applications can be obtained from http://automate.rutgers.edu.

4.5   Autonomic Grid Applications

4.5.1   Autonomic Oil Reservoir Optimization

One of the fundamental problems in oil reservoir production is determining the optimal locations of the oil production and injection wells. However, the selection of appropriate optimization algorithms, the runtime configuration and invocation of these algorithms, and the dynamic optimization of the reservoir remain challenging problems. In this research we use AutoMate to support autonomic aggregations, compositions, and interactions and enable an autonomic self-optimizing reservoir application. The application consists of: (1) sophisticated reservoir simulation components that encapsulate complex mathematical models of the physical interaction in the subsurface and execute on distributed computing systems on the Grid; (2) Grid services that provide secure and coordinated access to the resources required by the simulations; (3) distributed data archives that store historical, experimental, and observed data; (4) sensors embedded in the instrumented oil field providing real-time data about the current state of the oil field; (5) external services that provide data relevant to optimization of oil production or of the economic profit such as current weather information or current prices; and (6) the actions of scientists, engineers, and other experts in the field, the laboratory, and management offices.

The main components of the autonomic reservoir framework [31, 2] are (i) instances of distributed multi-model, multi-block reservoir simulation components, (ii) optimization services based on the Very Fast Simulated Annealing (VFSA) [31] and Simultaneous Perturbation Stochastic Approximation (SPSA) [2], (iii) economic modeling services, (iv) real-time services providing current economic data (e.g., oil prices), and (v) archives of data that have already been computed, and (vi) experts (scientists, engineers) connected via pervasive collaborative portals.

The overall oil production process is autonomic in that the peers involved automatically detect suboptimal oil production behaviors at runtime and orchestrate interactions among themselves to correct this behavior. Further, the detection and optimization process is achieved using policies and constraints that minimize human intervention. Policies are used to discover, select, configure, and invoke appropriate optimization services to determine optimal well locations. For example, the choice of optimization service depends on the size and nature of the reservoir. The SPSA algorithm is suited for larger reservoirs with relatively smooth characteristics. In case of reservoirs with many randomly distributed maxima and minima, the VFSA algorithm can

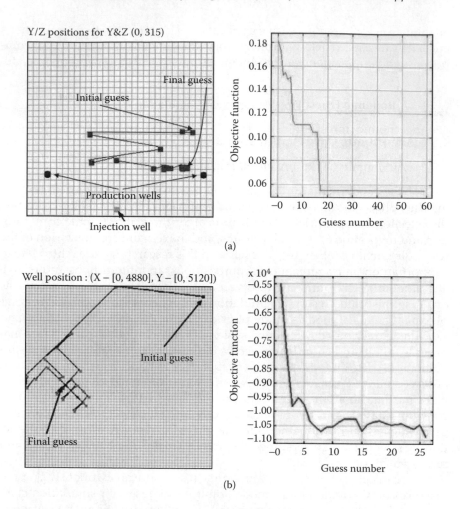

FIGURE 4.2
Autonomic optimization of the well placement problem using (a) VFSA algorithm (b) SPSA algorithm.

be employed during the initial optimization phase. Once convergence slows down, VFSA can be replaced by SPSA. Similarly, policies can also be used to manage the behavior of the reservoir simulator, or may be defined to enable various optimizers to execute concurrently on dynamically acquired Grid resources, and select the best well location among these based on some metric (e.g., estimated revenue, time, or cost of completion).

Figure 4.2 illustrates the optimization of well locations using the VFSA and SPSA optimization algorithms for two different scenarios. The well position plots (on the left in 4.2(a) and (b)) show the oil field and the positions of the wells. Black circles represent fixed injection wells and a gray square at the bottom of the plot is a fixed production well. The plots also show the sequence of guesses for the position of the other production well returned by the optimization service (shown by the lines connecting the light squares), and the corresponding normalized cost value (plots on the right in 4.2(a) and (b)).

4.5.2 Autonomic Forest Fire Management Simulation

The autonomic forest fire simulation, composed of *DSM* (Data Space Manager), *CRM* (Computational Resource Manager), *Rothermel, WindModel*, and *GUI elements*, predicts the speed, direction, and intensity of the fire front as the fire propagates using static and dynamic environment and vegetation conditions. *DSM* partitions the forest represented by a 2D data space into subspaces based on current system resources information provided by *CRM*. Under the circumstance of load imbalance, *DSM* repartitions the data space. *Rothermel* generates processes to simulate the fire spread on each subspace in parallel based on current wind direction and intensity simulated by the *Wind-Model*, until no *burning* cells remain. Experts interact with the above elements using the *GUI* element.

We use the *Rothermel, DSM*, and *CRM* as examples to illustrate the definition of the Accord functional, control, and operational ports, as shown in Figure 4.3. *Rothermel*, for example, provides *getSpaceState* to expose space information as part of its **Functional Port**, and provides the sensor *getDirection* to get the fire spread direction and the actuator *setCellState* to modify the state

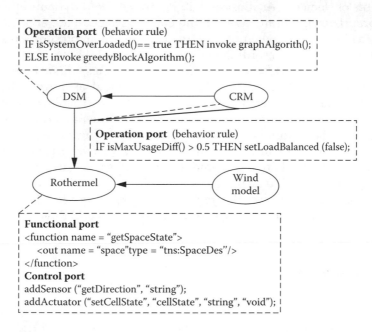

FIGURE 4.3
Examples of the port definition and rules.

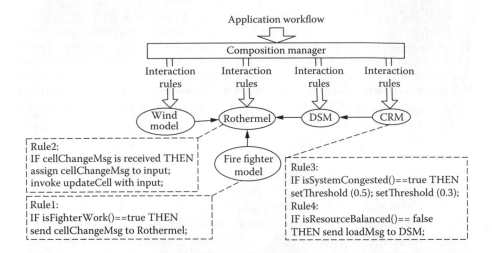

FIGURE 4.4
Add a new component, *Fire Fighter Model*, and change the interaction relationship between *CRM* and *DSM*.

of a specified cell as part of its **Control Port**. The *DSM* and *CRM* receive rules to manage their runtime behaviors through the **Operation Port**.

Behavior rules can be defined at compile time or at runtime and injected into corresponding element managers to dynamically manage the computational behaviors of elements. As illustrated in Figure 4.3, *DSM* dynamically selects an appropriate algorithm based on the current system load, and *CRM* will detect load imbalance when the maximal difference among resource usage exceeds the threshold according to the behavior rules shown.

The application workflow is decomposed by the *Composition Manager* into interaction rules, which are injected into individual elements. Therefore, addition, deletion, and replacement of elements can be achieved using corresponding interaction rules. For example, a new element, *Fire Fighter Model*, modeling the behaviors of the fire fighters, is added to the application, as shown in Figure 4.4, by inserting Rule1 into *Fire Fighter Model* and Rule2 into *Rothermel*. Similarly, changing an interaction relationship can be achieved by replacing the existing interaction rules with new rules. As shown in Figure 4.4, *CRM* dynamically decreases the frequency of notifications to *DSM* when the communication network is congested based on Rule3 and Rule4.

4.6   Summary and Conclusion

The increasing complexity, heterogeneity, and dynamism of emerging wide-area pervasive Grid environments and applications have necessitated the development of alternate development and management approaches. In this chapter, we first presented the autonomic Grid computing paradigm as one approach for addressing these challenges. The paradigm investigates solutions that are based on the strategies used by biological systems to deal with similar challenges, and aims at realizing systems and applications that are capable of managing (i.e., configuring, adapting, optimizing, protecting, and healing) themselves. We then presented Project AutoMate and described its key components. The overall goal of Project AutoMate is to investigate conceptual models and implementation architectures that can enable the development and execution of such self-managing Grid applications. Specifically, it investigates programming models, frameworks, and middleware services that support the definition of autonomic elements, the development of autonomic applications as the dynamic and opportunistic composition of these autonomic elements, and the policy-, content-, and context-driven definition, execution, and management of these applications. Two case-study applications—autonomic oil reservoir optimization and autonomic forest fire management—enabled by AutoMate were also presented.

Acknowledgments

The author would like to acknowledge the contributions of J.C. Browne and S. Hariri to the concepts and material presented in this chapter. The research presented in this paper is supported in part by the National Science Foundation via grants numbers ACI 9984357, EIA 0103674, EIA 0120934, ANI 0335244, CNS 0305495, CNS 0426354 and IIS 0430826, and by the DOE SciDAC CPES FSP (DE-FG02-06ER54857).

References

1. G. Allen, K. Davis, K.N. Dolkas, N.D. Doulamis, T. Goodale, T. Kielmann, A. Merzky, J. Nabrzyski, J. Pukacki, T. Radke, M. Russell, E. Seidel, J. Shalf, and I. Taylor. Enabling applications on the grid: A Gridlab overview. *International Journal of High Performance Computing Applications: Special issue on Grid Computing: Infrastructure and Applications*, page to appear, 2003.

2. W. Bangerth, V. Matossian, M. Parashar, H. Klie, and M. F. Wheeler. An autonomic reservoir framework for the stochastic optimization of well placement. *Cluster Computing: The Journal of Networks, Software Tools, and Applications, Kluwer Academic Publishers*, 8(4), 2005.

3. F. Berman, A. Chien, K. Cooper, J. Dongarra, I. Foster, D. Gannon, L. Johnsson, K. Kennedy, C. Kesselman, J. Mellor-Crummey, D. Reed, L. Torczon, and R. Wolski. The grads project: Software support for high-level grid application development. *International Journal of High Performance Computing Applications*, 15(4): 327–344, 2001.

4. S. Chandra, M. Parashar, J. Yang, Y. Zhang, and S. Hariri. Investigating autonomic runtime management strategies for samr applications. *International Journal of Parallel Programming*, 33(2-3):247–259, 2005.

5. N. Furmento, J. Hau, W. Lee, S. Newhouse, and J. Darlington. Implementations of a service-oriented architecture on top of jini, jxta and ogsa. In *Proceedings of UK e-Science All Hands Meeting*, 2003.

6. The globus alliance. http://www.globus.org.

7. M. Govindaraju, S. Krishnan, K. Chiu, A. Slominski, D. Gannon, and R. Bramley. Xcat 2.0: A component-based programming model for grid web services. Technical Report-TR562, Dept. of Computer Science, Indiana Univ., June 2002.

8. A. S. Grimshaw and W. A. Wulf. The legion vision of a worldwide virtual computer. *Communications of the ACM*, 40(1): 39–45, 1997.

9. S. Hariri, B. Khargharia, H. Chen, J. Yang, Y. Zhang, M. Parashar, and H. Liu. The autonomic computing paradigm. *Cluster Computing: The Journal of Networks, Software Tools, and Applications*, 9(1): 5–17, 2006.

10. P. Horn. Autonomic Computing: IBM's perspective on the State of Information Technology. http://www.research.ibm.com/autonomic/, Oct 2001. IBM Corp.

11. Y. Ishikawa, M. Matsuda, T. Kudoh, H. Tezuka, and S. Sekiguchi. The design of a latency-aware mpi communication library. In *Proceedings of SWOPP03*, 2003.

12. N. Jiang, C. Schmidt, V. Matossian, and M. Parashar. Content-based decoupled interactions in pervasive grid environments. In *First Workshop on Broadband Advanced Sensor Networks, BaseNets'04*, San Jose, California, October 2004.

13. Jini network technology. http://wwws.sun.com/software/jini

14. Project jxta. http://www.jxta.org, 2001.

15. W. Kelly, P. Roe, and J. Sumitomo. G2: A grid middleware for cycle donation using .net. In *Proceedings of The 2002 International Conference on Parallel and Distributed Processing Techniques and Applications*, 2002.

16. B. Khargharia, S. Hariri, and M. Parashar. vgrid: A framework for building autonomic applications. In *1st International Workshop on Heterogeneous and Adaptive Computing–Challenges of Large Applications in Distributed Environments (CLADE 2003)*, 19–26, Seattle, WA, USA, 2003. Computer Society Press.

17. S. Krishnan and D. Gannon. Xcat3: A framework for cca components as ogsa services. In *Proceedings of HIPS 2004, 9th International Workshop on High-Level Parallel Programming Models and Supportive Environments*, 2004.

18. G. von Laszewski, I. Foster, and J. Gawor. Cog kits: A bridge between commodity distributed computing and high-performance grids. In *ACM 2000 Conference on Java Grande*, 97–106, San Francisco, CA USA, 2000. ACM Press.

19. Z. Li and M. Parashar. Comet: A scalable coordination space in decentralized distributed environments. In *Proceedings of the 2nd International Workshop on Hot Topics in Peer-to-Peer Systems (HOT-P2P 2005)*, 104–111. IEEE Computer Society Press, 2005.

20. Z. Li and M. Parashar. Enabling dynamic composition and coordination of autonomic applications using the rudder agent framework. *The Knowledge Engineering Review*, 2006.

21. Z. Li and M. Parashar. Rudder: An agent-based infrastructure for autonomic composition of grid applications. *Multiagent and Grid System—An International Journal*, 2006.

22. H. Liu, V. Bhat, M. Parashar, and S. Klasky. An autonomic service architecture for self-managing grid applications. In *Proceedings of the 6th IEEE/ACM International Workshop on Grid Computing (Grid 2005)*, Seattle, WA, 2005.

23. H. Liu, L. Jiang, M. Parashar, and D. Silver. Rule-based Visualization in the Discover Computational Steering Collaboratory. *Journal of Future Generation Computer System, Special Issue on Engineering Autonomic Systems*, 21(1):53–59, 2005.

24. H. Liu and M. Parashar. A Framework for Rule-Based Autonomic Management of Parallel Scientific Applications. In *The 2nd IEEE International Conference on Autonomic Computing (ICAC-05)*, Seattle, WA, 2005.

25. H. Liu and M. Parashar. Enabling Self-management of Component-based High-Performance Scientific Applications. In *The 14th IEEE International Symposium on High Performance Distributed Computing (HPDC-14)*, 59–68, Research Triangle Park, NC, 2005.

26. H. Liu and M. Parashar. Rule-based Monitoring and Steering of Distributed Scientific Applications. *International Journal of High Performance Computing and Networking (IJHPCN)*, 3(4):272–282, 2005.

27. H. Liu and M. Parashar. Accord: A programming framework for autonomic applications. *IEEE Transactions on Systems, Man and Cybernetics, Special Issue on Engineering Autonomic Systems*, 36(3), 2006.

28. H. Liu, M. Parashar, and S. Hariri. A Component-based Programming Framework for Autonomic Applications. In *the 1st IEEE International Conference on Autonomic Computing (ICAC-04)*, 10–17, New York, 2004.

29. V. Mann, V. Matossian, R. Muralidhar, and M. Parashar. DISCOVER: An environment for Web-based interaction and steering of high-performance scientific applications. *Concurrency and Computation: Practice and Experience*, 13(8–9):737–754, 2001.

30. J. Mathe, K. Kuntner, S. Pota, and Z. Juhasz. The use of jini technology in distributed and grid multimedia systems. In *MIPRO 2003, Hypermedia and Grid Systems*, 148–151, Opatija, Croatia, 2003.

31. V. Matossian, V. Bhat, M. Parashar, M. Peszynska, M. Sen, P. Stoffa, and M. F. Wheeler. Autonomic oil reservoir optimization on the grid. *Concurrency and Computation: Practice and Experience, John Wiley and Sons*, 17(1):1–26, 2005.

32. M. Migliardi and V. Sunderam. The harness metacomputing framework. In *Proceedings of Ninth SIAM Conference on Parallel Processing for Scientific Computing*, San Antonio, TX, 1999. SIAM.

33. H. Nakada, S. Matsuoka, K. Seymour, J. Dongarra, C. Lee, and H. Casanova. Gridrpc: A remote procedure call api for grid computing, 2003.

34. Microsoft .net. http://www.microsoft.com/net/.

35. R. V. van Nieuwpoort, J. Maassen, G. Wrzesinska, R. Hofman, C. Jacobs, T. Kielmann, and H. E. Bal. Ibis: A flexible and efficient java-based grid programming environment. *Concurrency & Computation: Practice & Experience*, 17(7-8):1079–1107, 2005.

36. R. V. van Nieuwpoort, J. Maassen, G. Wrzesinska, T. Kielmann, and H. E. Bal. Satin: Simple and efficient java-based grid programming. *Journal of Parallel and Distributed Computing Practices*, 2004.

37. M. Parashar and J. C. Browne. Conceptual and implementation models for the Grid. *Proceedings of the IEEE, Special Issue on Grid Computing*, 93(3):653–668, 2005.

38. M. Parashar and C. A. Lee. Scanning the issue—Grid computing: An evolving vision. *Proceedings of the IEEE, Special Issue on Grid Computing*, 93(3):479–484, 2005.

39. M. Parashar, H. Liu, Z. Li, V. Matossian, C. Schmidt, G. Zhang, and S. Hariri. Automate: Enabling autonomic grid applications. *Cluster Computing: The Journal*

of Networks, Software Tools, and Applications, Special Issue on Autonomic Computing, 9(2):161–174, 2006.

40. C. Schmidt and M. Parashar. Enabling flexible queries with guarantees in p2p systems. *IEEE Internet Computing*, 8(3):19–26, 2004.

41. N. Taesombut and A. Chien. Distributed virtual computer (dvc): Simplifying the development of high performance grid applications. In *Workshop on Grids and Advanced Networks (GAN '04), IEEE Cluster Computing and the Grid (CCGrid2004) Conference*, Chicago, 2004.

42. I. Taylor, M. Shields, I. Wang, and R. Philp. Distributed p2p computing within triana: A galaxy visualization test case. In *International Parallel and Distributed Processing Symposium (IPDPS'03)*, Nice, France, 2003. IEEE Computer Society Press.

43. D. Thain, T. Tannenbaum, and M. Livny. *Condor and the Grid*. John Wiley & Sons Inc., 2002.

44. Unicore forum. http://www.unicore.org

45. C. Ururahy and N. Rodriguez. Programming and coordinating grid environments and applications. *Concurrency and Computation: Practice and Experience*, 16(5), 2004.

46. G. Zhang and M. Parashar. Corporative defense against ddos attacks. *Journal of Research and Practice in Information Technology (JRPIT)*, 38(1):66–84, 2006.

47. G. Zhang and M. Parashar. Sesame: Scalable, environment sensitive access management engine. *Cluster Computing: The Journal of Networks, Software Tools, and Applications*, 9(1):19–27, 2006.

5

Architecture Overview for Autonomic Computing

John W. Sweitzer and Christine Draper

CONTENTS

IBM

This chapter introduces an architecture for autonomic computing by detailing the four main aspects of the architecture (process definition, resource definition, technical reference architecture, and application patterns), by presenting the fundamental functions of the *autonomic manager* and *touchpoint* building blocks, and by summarizing an initial set of application patterns commonly found in an autonomic computing system.

5.1   Architecture Overview

Figure 5.1 shows the four main aspects of an architecture for autonomic computing, consisting of:

- The **process definition architecture**. This aspect describes the business processes that are fully or partially automated in an autonomic computing system, along with the organizational structure that establishes the deployment context.

- The **resource definition architecture**. This aspect describes the resource types that are to be managed in an autonomic computing system.

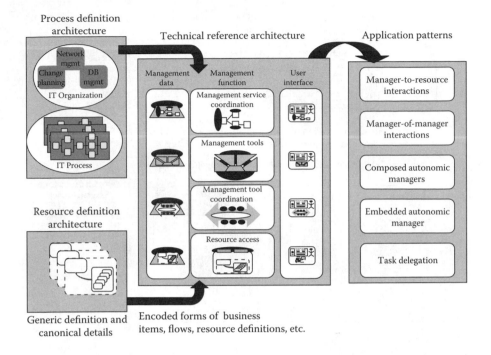

FIGURE 5.1
Aspects of an autonomic computing architecture.

- The **technical reference architecture**. This aspect provides a set of building blocks for describing how IT management system elements may be integrated together to support the services delivered by an IT organization.
- The **application patterns**. This aspect provides templates for combining building blocks into specific situations commonly found in real deployments.

5.1.1   Process Definition Architecture

IT organizations define the work to be performed by IT professionals as a collection of best practices and processes such as those defined in the IT Infrastructure Library (ITIL, from the Office of Government Commerce in the United Kingdom). An example of an IT management process is the *change management process* outlined in Figure 5.2 that choreographs the activities typically involved in changing IT systems. An activity is a coordinated set of tasks (basic units of work) that are performed within IT organizations. The high level activities defined for such processes characterize some of the work an autonomic computing system must perform to make the system more self managing.

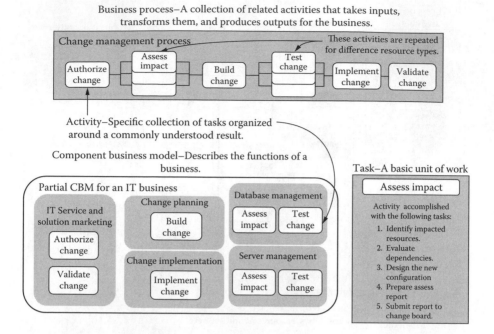

FIGURE 5.2
IT management process example.

The organizational context for the architecture is provided by *a component business model* (CBM) for IT, see [1]. A CBM is a model of the organizational capabilities or functions that are found within a business of a particular type. As shown in Figure 5.2, components for the IT business can be dedicated to specific resource types (for example, database management and server management), or can be responsible for cross-resource functions such as change planning or change implementation. The definition of a component includes details for the activities that component is typically responsible for delivering.

IT professionals within the components perform the tasks needed to complete the activities for that component. From the IT professional's perspective, tasks are the basic units of work that are assigned to them to perform. In an autonomic computing system, IT professionals can delegate tasks to technologies in the system, so the system can manage itself.

5.1.2 Resource Definition Architecture

The resource definition architecture provides a common "vocabulary" for describing the resources to be managed, in terms of their types, properties and relationships. Several resource models already exist in the industry, such

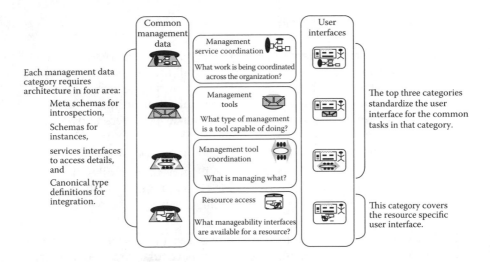

FIGURE 5.3
Technical reference architecture.

as DMTF Common Information Model [2] and domain-specific models like JSR 77 for J2EE [3]. The autonomic computing architecture embraces these existing models wherever possible, but more canonical definitions about relationships, properties, and capabilities across resource types are required to deliver system wide autonomic computing.

5.1.3 Technical Reference Architecture

Figure 5.3 shows the four key functional areas (horizontal) covered by the technical reference architecture and the two overlay areas (vertical) that are pervasive across the functional areas.

A building block is defined for each area to represent the realization of functions associated with that area. For example, the autonomic manager building block is the architectural building block that defines the management tools' functional area.

The four functional areas (described in Sections 1.3.1 to 1.3.4) are:

- Management Service Coordination
- Management Tools
- Management Tool Coordination
- Resource Access

The first three of these can be thought of as defining the managing system; the final one, resource access, can be thought of as defining the managed system.

The two vertical overlay areas are:

- Common Management Data — defines the *knowledge types* relevant to a functional area. The corresponding building block is a *knowledge source*, which enables the knowledge (such as symptoms, policies, change requests, release plans, topology information, historical logs, and metrics) to be stored, queried, and retrieved. These knowledge types include metadata so the autonomic computing system can manage more of itself with self-describing building blocks.
- User Interfaces — defines the capabilities needed to allow IT professionals to interact effectively with the autonomic systems. The corresponding building block is a *manual manager*.

These two areas overlay the four functional areas to specify the common management data and user interfaces for a particular area. As shown in Figure 5.3, the common management data area and the user interface area include a specialized building block for each of the four functional areas.

5.1.3.1   Management Service Coordination

The *management service coordination* area includes the capabilities involved in coordinating the work of an IT organization and automating the flow of work across the organization. That is, this area focuses on the capabilities that integrate the activities of the organization so that it can be more productive. This includes capabilities such as work flow engines and work queues. The content of the management service coordination area is driven by the activities defined by the process definition architecture.

The building block for this area is the *process flow module*.

5.1.3.2   Management Tools

The *management tools* area includes the capabilities that assist in automating the work required to perform management tasks. This area focuses on the software that automates the detailed work performed by IT professionals, who can be more productive when they can delegate some of their assigned tasks to the system. The content of the management tools area is driven by the tasks defined by the process definition architecture rather than the activities (which define the content of the management service coordination area).

An *autonomic manager* is the building block that automates tasks. This building block is the centerpiece of the autonomic computing architecture, and is described in more detail in Section 5.2.

5.1.3.3   Management Tool Coordination

The *management tool coordination* area deals with understanding what management tools are managing what resources. The capabilities provided in this area include functions such as assigning resources to a management tool, discovering what resources a management tool is managing, and determining the type of management that a tool is performing on a particular collection of resources.

What is being managed can be described in many different forms. It could be a single resource, a collection of homogeneous or heterogeneous resources, a business system, or a service.

The building block associated with this area is the *enterprise service bus*, which assists in integrating other building blocks (for example, autonomic managers and touchpoints) by directing the messages and interactions among these building blocks. Enterprise service bus (ESB) is a concept in services-oriented architecture for integrating applications [6].

5.1.3.4 Resource Access

The *resource access* area deals with the capabilities involved in accessing and controlling the *managed resources* within the system. *Managed resources* include resource types such as a server, storage unit, database, application server, service, application, or other entity. Capabilities include manageability interfaces, resource registries, and metadata descriptions of resource capabilities.

A *touchpoint* is the building block that implements the sensor and effector behavior (see Section 2.1.1 for more details) for one or more of a managed resource's manageability mechanisms. It also provides a standard manageability interface. Deployed managed resources are accessed and controlled through these manageability interfaces.

5.1.4 Application Patterns

Application patterns provide a "template" for applying the reference architecture to solve specific problems. The patterns described in Section 5.3 are chosen to illustrate some of the ways in which the building blocks described in the technical reference architecture for autonomic computing can be applied in specific contexts to address particular problem statements.

5.2 Autonomic Managers — the Centerpiece of the Architecture

The autonomic manager building block is the centerpiece of the architecture because it is responsible for automating the IT management functions that make a system "self-managing." Each autonomic manager automates one or more management functions, and externalizes this function according to the behavior defined by management interfaces. The four primary functions of an autonomic manager are shown in Figure 5.4 and detailed next:

- The *monitor* function collects, aggregates, correlates and filters details from managed resources, through the touchpoint sensor interface, until it recognizes a *symptom* that needs to be analyzed. For example, a monitor function might recognize an "increased

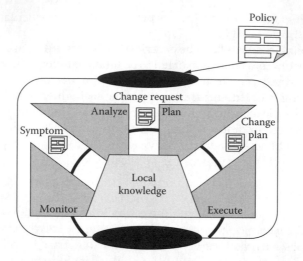

FIGURE 5.4
Autonomic manager building block details.

transaction time" symptom based on response time metrics collected in real time from the system. The details observed by the monitor function can include topology information, events, metrics, configuration property settings and so on. Logically, the symptom is passed to the analyze function.

- The *analyze* function provides the mechanisms to observe and analyze symptoms to determine if some change needs to be made. For example, symptoms about "increased transaction time" and "unavailable servers" might be analyzed to determine that more servers are needed to avoid violating a "response time" policy. In this case, a *change request* for "three more servers to be assigned to the degraded business application" might be generated to avoid a "response time" violation and to reduce the growing workload backlog. The change request is logically passed to the plan function. After passing a change request to the plan function, the analyze function can watch for an expected change (for example, the growing backlog is reduced) to validate that the implementation of the change request resolved the original symptoms.

- The *plan* function creates or selects a procedure to change managed resources. The plan function generates the appropriate *change plan*, which represents a desired set of changes to be performed on the managed resource, and logically passes that change plan to the execute function. The details of the change plan may be a simple command for a single managed resource or it may be complex work flow that changes hundreds of managed resources.

- The *execute* function provides the mechanism to schedule and perform the necessary changes to the system. Once an autonomic manager has generated a change plan that corresponds to a change request, some *actions* may need to be taken to modify the state of one or more managed resources. The actions are performed on the managed resource through the touchpoint effector interface. In addition, part of the execution of the change plan could involve updating the knowledge that is used by the autonomic manager.

When these functions can be automated, an intelligent *control loop* is formed that automates a set of tasks commonly performed by professionals in an IT organization. Autonomic managers use *policies* (goals or objectives) to govern how the four functions are accomplished, so the policies governing the professionals' work need to be available to the autonomic manager in an encoded form.

The capabilities of an autonomic manager may be extended by specific knowledge types. For example, the monitor function can be extended by providing new symptom definitions, which tell an autonomic manager how to detect a condition in a resource that might require attention or action.

To facilitate the management of a management system (for example, to enable decisions to be made about which resources should be managed by which managers), metadata must be available about what sorts of resources autonomic managers are capable of managing. This information comes in two parts: what types of management an autonomic manager can perform (availability, performance, capacity, etc.) and what resources it can actually manage (databases, applications, routers, servers, etc.).

5.2.1 Interacting with Resources

Traditionally, management tools observe and control resources through a variety of manageability interfaces such as log files, events, commands, application programming interfaces (APIs) and configuration files. These mechanisms typically vary by resource type and vendor. The diverse mechanisms and the lack of coordinated content across the mechanisms are major contributors to the complexity involved in managing IT infrastructures. The architecture for autonomic computing introduces the *touchpoint* building block (detailed in Figure 5.5) to specify a more consistent manageability interface.

The coarse-grained resource that is being managed is referred to as the *managed resource* and the fine-grained elements of that managed resource are referred to as the *hosted resources*. A managed resource is an entity such as a hardware server, storage unit, database, application server, or application. A relational database managed resource will have constituent hosted resources such as a database service, database, tables, indexes and so on; a hardware server managed resource will have constituent hosted resources such as CPU, disks, memory, and so on.

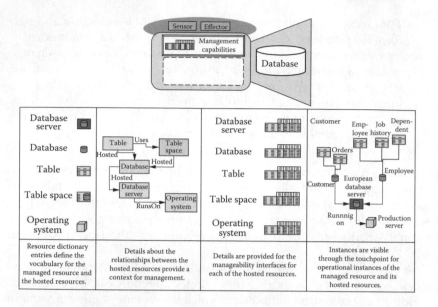

FIGURE 5.5
Touchpoint building block details.

5.2.1.1 Sensor and Effector Interaction Styles

An autonomic manager observes and controls manageable resources, whether these are managed resources or hosted resources, through a *manageability interface* specified as a sensor and effector interface. The *interaction styles* used to observe details about the manageable resources are aggregated into a *sensor*, and include:

- **Request-Response**: Operations that expose information about the current state of a manageable resource, such as properties and relationships.
- **Send-Notification**: A set of management events that occur when the manageable resource undergoes state changes that merit reporting.

The interaction styles used to change the behavior or state of the manageable resources are aggregated into an *effector*, and include:

- **Perform-Operation**: Operations that allow the behavior or state of the manageable resource to be changed in some way.
- **Solicit-Response**: Operations that are implemented by autonomic managers to allow the manageable resource to make requests from its manager.

A touchpoint building block for a managed resource exposes a sensor and effector interface for each hosted resource.

5.2.1.2 Manageability Capabilities

The sensors and effectors in the architecture need to be linked together. For example, a configuration change that occurs through the effector should be reflected as a configuration change notification through the sensor interface. A *manageability capability* refers to a logical collection of manageable resource state information and operations. Some examples of manageability capabilities are:

- **Identification**: state information and operations used to identify an instance of a manageable resource
- **Metrics**: state information and operations for measurements of a manageable resource, such as throughput or utilization
- **Configuration**: state information and operations for the configurable attributes of a manageable resource

For each manageability capability, an autonomic manager should be able to access metadata about the manageability capability (for example, to discover properties that configure a manageable resource, or information that specifies which resources can be hosted by the manageable resource) so the resource is self-describing. Using the standard sensors and effectors, the autonomic manager should then be able to determine how to observe a change that it made to the resource, or to initiate a change in response to observed state.

5.2.1.3 Consistent Content

The interaction styles and the manageability capabilities present a consistent structure for the interface to the manageable resources, but the architecture further requires that the content for these resources also be defined consistently. The material in the lower section of Figure 5.5 summarizes how this is accomplished.

First, a dictionary for the manageable resources needs to define a standard set of terms ("dictionary entries") that are used throughout the system for those manageable resources. Additionally, the dictionary entries include details about the relationships among the manageable resources, so that the topology of the IT infrastructure can be traversed using common algorithms.

Further dictionary information describes the capabilities of the specific manageability interfaces used to access the resources.

Standardizing the mechanisms for describing and interacting with all elements of the IT resource stack (hardware, operating system, database, middleware, application) is one of the necessary steps to enable autonomic managers to work at multiple levels in the IT infrastructure.

5.2.2   Incremental Delivery of Autonomic Managers

The architecture allows for autonomic managers that perform only a subset of the control loop functions either because:

- The implementation supports only a subset; or
- The IT professional who is responsible for the deployed autonomic manager has activated only a subset of the function that the autonomic manager is capable of performing. This can happen when the professional's job responsibility does not include all the functions or when the professional decides not to take advantage of all the functions the autonomic manager offers.

This enables the architecture to deal with the practical reality that the control loop functions need to be delivered and deployed incrementally. The control loop detailed in Figure 5.4 defines the architecture for an autonomic manager that enables a system to manage itself. Figure 5.6 illustrates using some combination of the monitor, analyze, plan, and execute functions to define the incremental capability of an autonomic manager from a coarse-grain perspective.

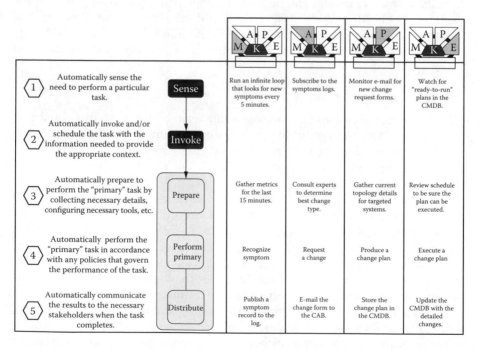

FIGURE 5.6
Architecture for partial autonomic manager.

The monitor, analyze, plan and execute functions of the control loop correspond to tasks performed by IT professionals. For example, an autonomic manager that only implements tasks for the monitor function can perform the primary task *recognize symptoms*. So, instead of an operator watching console tools to observe details that signal a symptom such as "increased response time" (that is, to perform the primary task of a monitor function) he or she can delegate this work to a partial autonomic manager that implements the monitor function.

For an IT professional to be able to fully delegate a task, the autonomic manager must do more than just automate the primary tasks. Autonomic managers must be able to do the five parts enumerated in Figure 5.6. Using the "recognize symptom" task as an example, the following five parts need to be automated for the operator to not be involved in any aspect of the task:

(1) Automatically *sense* the need to recognize a symptom.

(2) Automatically *invoke* the function that knows how to recognize the symptom, with the appropriate context.

(3) Automatically *prepare* to recognize the symptom by collecting and organizing the necessary details and tools.

(4) Automatically *perform* the work to actually recognize the symptom in accordance with any objectives or constraints (i.e., policies) that have been established to govern the performance of the task.

(5) Automatically *communicate* the results (that is, the recognized symptoms) to the necessary stakeholders when the task completes.

Technologies that automate any of these five parts assist IT professionals because the system eliminates the need for the professional to do that part of the work. The ability for an IT professional to *fully delegate* a complete task requires all five parts shown in Figure 5.6 to be automated in a reliable manner. In this case, the automated primary task can be performed without human involvement. The ability to fully delegate primary tasks is a criterion for distinguishing autonomic systems from simple automation.

5.3 Application Patterns for Autonomic Systems

This section contains patterns for applying the reference architecture to solve specific problems. Each of the application patterns is described in two parts:

Problem: This states the context of the problem to be solved, a summary of the problem itself (highlighted in bold), and a description of the underlying issues that could constrain the solution to the problem.

Solution: This consists of a summary of the solution (highlighted in bold), a solution description, the results or benefits that can be realized, and any specific limitations.

5.3.1   Pattern 1a: Use of Enterprise Service Bus for Manager-to-Resource Interactions

5.3.1.1   Problem

The simplest autonomic manager pattern involves a single manager managing a single resource. In autonomic systems, there is a need for multiple managers, each providing a different management discipline, to interact with the same resources.

How can multiple autonomic managers manage the same resource, without driving an $N \times M$ increase in configuration complexity?

Autonomic managers that perform specific functions such as change management, problem remediation, backup and recovery, security management, etc., need to interact with the same resources. This can lead to an exponential explosion in configuration complexity caused by the $N \times M$ interactions among managers and resources.

> Whenever a new resource is added, not only does the resource need to be configured, but all managers that are to manage the new resource need to be informed.

> Whenever a new manager is added, not only does the manager need to be configured, but all of the resources that are to be managed by that manager need to be informed.

Figure 5.7 illustrates this for the configuration of notification-style interactions, although this complexity applies to other interaction styles too.

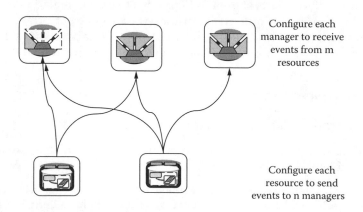

FIGURE 5.7
$N \times M$ interactions among managers and resources.

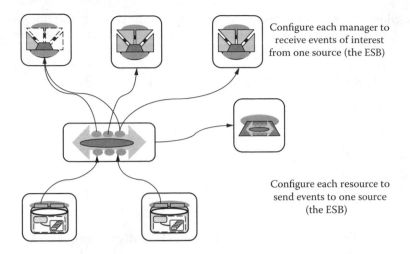

Configure each manager to receive events of interest from one source (the ESB)

Configure each resource to send events to one source (the ESB)

FIGURE 5.8
Using the ESB to manage notifications from resources to managers.

5.3.1.2 Solution

Configuration of manager-to-resource interactions should be externalized into functions within an Enterprise Service Bus (ESB) [6].

The interactions among autonomic managers and resources should be constructed so that the resources and managers are unaware of the details of the multiplicity. As shown in Figure 5.8, these details should be handled by patterns within the ESB, such as publish-subscribe and message routing, (see [5]).

To support the appropriate configuration of the interactions, each autonomic manager provides a description of its capabilities for managing resources, and each resource provides a description of its capabilities to be managed. This includes interaction styles and manageability capabilities. For example, a resource indicates that it can produce a particular set of events and uses the send-notification interaction style to communicate those events; an autonomic manager indicates that it can accept events using the send-notification interaction style and that it is interested in a particular set of events.

Finally, the management system makes available information about the existing management relationships among managers and resources.

Use of the ESB to externalize the configuration of manager-to-resource interactions facilitates the dynamic reconfiguration of the management system in response to change, and makes the configuration of the management system more visible and manageable.

An example of using this pattern might be monitoring the deployment of updates to applications. An operating system resource generates a notification whenever an application is updated. The notification causes the configuration compliance manager to check that the change is approved, and the problem manager responsible for the resource to update its symptom definitions for the new level of the resource.

This pattern does not resolve conflicts between managers. Patterns for conflict resolution are an advanced topic not covered in this chapter.

5.3.2   Pattern 1b: Shared Resource Data Among Managers

5.3.2.1   Problem

As described in Pattern 1a: Use of Enterprise Service Bus for Manager-to-Resource Interactions, multiple autonomic managers may manage a single resource and need to share data about that resource.

How can data about resources be shared efficiently among multiple autonomic managers?

Information about manageable resources is held in multiple sources, including knowledge sources, autonomic managers and the managed resource's touchpoint. Each source has partial and potentially out-of-date knowledge about the resource. An autonomic manager needs to understand the currency of the available information and be able to obtain further information as needed.

5.3.2.2   Solution

A federated Configuration Management Database (CMDB) provides efficient access to information about resources that may be held in multiple sources.

Important information about resource instances and their relationships is stored in a knowledge source, known as the CMDB. This information includes the location of other sources of information for that resource, including the authoritative source(s).

The interface to the CMDB may provide the ability to transparently collate information from the federated sources to satisfy data requests, as shown in Figure 5.9. This may take into account locality of the data, the requester and the requester's currency requirements.

This pattern allows resource data to be accessed by a requester, without requiring the requester to understand the complexity of how and where that data was acquired.

An example of using federated resource data would be a change deployment manager populating the CMDB with information about the solution it has deployed, its component resources and dependencies. This relationship information might then be used by problem determination for problem diagnosis; or by incident management to perform impact assessment.

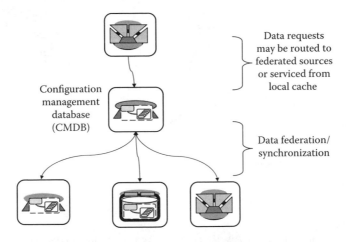

Data requests may be routed to federated sources or serviced from local cache

Configuration management database (CMDB)

Data federation/ synchronization

FIGURE 5.9
Use of a CMDB to federate access to resource information.

5.3.3 Pattern 2: Manager-of-Manager Interactions

5.3.3.1 Problem

Autonomic managers within an autonomic computing system need to be managed in a coordinated way to deliver a self-managing system.

How are the elements of an autonomic management system managed?

One of the potential pitfalls of any management system is that the management technology itself introduces even more complexity into the environment, and that "managing the management system" becomes as onerous a task as the manual management of the resources it was meant to replace.

To avoid this pitfall, the elements of an autonomic management system need to be managed in a straightforward and consistent way.

5.3.3.2 Solution

Autonomic managers can manage other autonomic managers using the same sensor and effector interfaces that are used to manage resources.

As illustrated in Figure 5.10, autonomic managers, in a manner similar to touchpoints, provide sensor and effector manageability interfaces for other autonomic managers and manual managers to use. This directly follows from the discussion in Section 5.1, because the answer to "what is an autonomic manager managing?" can be "another autonomic manager." The four interaction styles for the sensor and effector interfaces, as well as many of the manageability capabilities, can be used between autonomic managers in the same fashion as between an autonomic manager and a manageable resource. Autonomic managers have unique manageability capabilities that

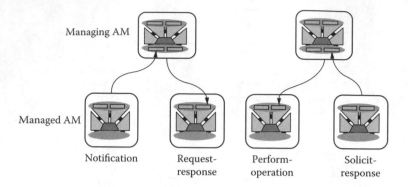

Managing AM

Managed AM

Notification   Request-   Perform-   Solicit-
                   response   operation   response

FIGURE 5.10
Manager-of-manager interactions use the same sensor and effector interfaces as resources.

allow the manager of an autonomic manager to assign manageable resources to the managed autonomic manager, to introspect about the type of management that the managed autonomic manager performs and to set the policies that govern how the managed autonomic manager accomplishes its mission.

An example of using this pattern is a business transaction workload manager that delegates management of a tier of resources to a local manager, which optimizes performance within that tier. The global manager assigns to the local managers the resources they are to manage with response time goals.

5.3.4   Pattern 3a: Composed Autonomic Managers

5.3.4.1   *Problem*

Section 5.2 describes the partial autonomic manager building block, which automates a subset of the functions that form a control loop. These partial autonomic managers need to be composed together to form a complete control loop.

How can multiple partial autonomic managers be integrated together to provide a complete control loop?

Partial autonomic managers are a useful building block because they allow customers to acquire and implement autonomic technology in an incremental manner and to choose implementations (potentially from different vendors) that best suit their organization's needs.

As an organization seeks to increase the autonomic maturity of its management systems, these incremental "islands" of automation need to be connected together to automate a larger context. Without structuring the connections, the resulting system may be complex to implement and manage, as illustrated in Figure 5.11.

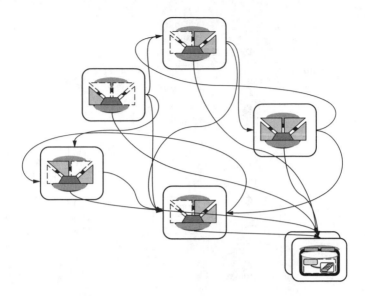

FIGURE 5.11
Arbitrary composition of autonomic managers.

What is needed is an architected way to compose the autonomic managers (this pattern) and to manage the federation of the data shared among the managers (Pattern 1b: Shared resource data among managers).

5.3.4.2 Solution

Partial autonomic managers should be composed to form more complete control loops, using canonical interfaces and data types.

The partial autonomic managers are provided with canonical functional interfaces for the tasks corresponding to the control loop functions that they perform. For example, a partial autonomic manager performing an analyze function might provide a "diagnose problem" interface that consumes a canonical symptom. In cases in which these interfaces are not provided natively by the implementation, they could be implemented using standard application integration patterns such as adapters [5]. These canonical interfaces can then be used to compose multiple partial autonomic managers into more complete control loops.

Use of this composition pattern provides a structured way to compose partial autonomic managers into complete control loops as shown in Figure 5.12. Patterns developed to manage the interaction of full autonomic managers can now be applied to the composite loops. This section shows a simple situation in which only one composed control loop manages each resource. Section 5.3.1 describes a pattern for handling cases in which a resource is managed by multiple control loops.

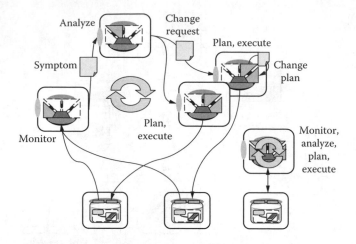

FIGURE 5.12
Composing partial autonomic managers.

Also depicted in Figure 5.12, routing decisions between the two partial managers that provide plan and execute functions are made statically in the manager performing the analysis function. Section 5.3.5 describes a pattern that externalizes this routing.

An example of using this pattern would be to compose an autonomic manager that monitors SLAs for workload management with one that is capable of provisioning a new server when the existing capacity is insufficient to meet demand.

5.3.5   Pattern 3b: Use of ESB for Composing Autonomic Managers

5.3.5.1   *Problem*

The "Composing autonomic managers" pattern (Section 5.3.4) describes how partial autonomic managers can be integrated together to form a more complete control loop. However, the composition of managers is either static (e.g., built into the manager or the adapter for the manager) or it requires manager-specific configuration.

How can the composition of autonomic managers be made readily reconfigurable?

Static configuration of the composition relationships between autonomic managers makes it more difficult for the IT organization to respond to changes and problems. Inconsistent manager-specific configuration mechanisms increase the complexity associated with managing the management system, resulting in increased effort and cost, slower time-to-value and more configuration errors.

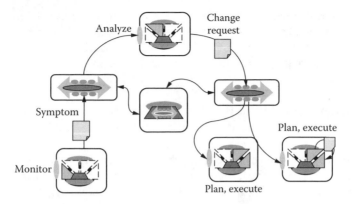

FIGURE 5.13
Using the ESB to configure interactions between autonomic managers.

What is needed is a way to separate the routing decisions from the auto-nomic manager implementations, allowing a single consistent configuration mechanism.

5.3.5.2 Solution

Configuration of manager-to-manager communications should be external-ized into a function within the ESB.

The communication between two autonomic managers should be constructed so that the invoking manager is unaware of the details of the invoked manager. As shown in Figure 5.13, these details should be handled by patterns within the ESB, such as mediation and content-based routing (see [5] and [6]).

To support the appropriate configuration of the communications, each au-tonomic manager should provide a description of its functional capabilities, as described in Section 5.2. The management system maintains information about the actual configuration of the autonomic managers and the resources they are managing.

Use of the ESB to externalize the configuration of manager-to-manager interactions facilitates the dynamic reconfiguration of the management sys-tem in response to change, and also makes the configuration of the manage-ment system more visible and manageable. Use of the ESB also facilitates non-intrusive monitoring of the management system, by monitoring the key manager-to-manager events as they flow across the ESB.

5.3.6 Pattern 4: Embedded Autonomic Manager

5.3.6.1 Problem

Resource providers build logic into the runtime of their managed resources to perform management tasks that would otherwise have to be performed by IT professionals.

Hence, some portions of control loops are implemented in the run-time environment of a managed resource. When any of the details for the control loop are visible, this pattern is used so that the control loop can be configured through the manageability interface of the managed resource (for example, a disk drive).

How can the management function delivered in the runtime of a resource participate in the autonomic computing system?

Resource providers reduce the total cost of ownership by building management functions into the resource. The resource provider is attempting to eliminate the need for the IT professionals to deal with some aspect of managing the resource.

The details of these implementations are specific to the resources, because the implementation must work within the resource's runtime. The engineers of these embedded management capabilities are challenged to incorporate sophisticated management functions without impacting the performance of the resource.

5.3.6.2   Solution

The embedded management function is exposed as an autonomic manager manageable resource through the touchpoint for the managed resource.

The logic embedded in the resource exposes its functionality as an "autonomic manager" manageable resource through the touchpoint for the resource itself. A critical characteristic of this pattern is that the "embedded management function" is exposed to the rest of the system as an autonomic manager, just like any other management function in the system. Using the structure for a touchpoint that was shown in Figure 5.5, the basic steps for introducing a manageable resource for an autonomic manager are:

(1) Introduce a manageable resource for the autonomic manager into the manageable resource dictionary;

(2) Declare a *has component* relationship between the managed resource and the autonomic manager manageable resource; and

(3) Define the manageability interface for the autonomic manager manageable resource so that it exposes the management function that it performs through the sensor and effector interfaces.

Because this manageable resource is an autonomic manager, it can have other interfaces that are unique to an autonomic manager. These details need to be discoverable through the sensor and effector interfaces for the autonomic manager manageable resource.

Figure 5.14 shows how an embedded autonomic manager is represented as a manageable resource that is visible to autonomic manager *A*. When the

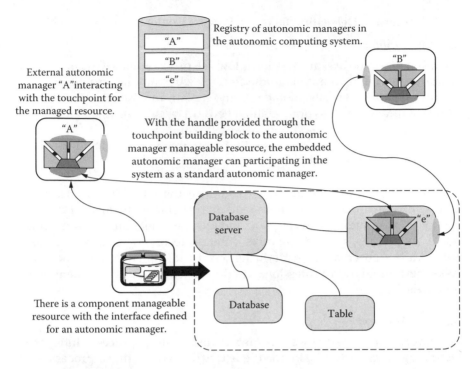

FIGURE 5.14
Sketch for "Embedded Autonomic Manager" pattern.

managed resource is assigned to autonomic manager *A*, autonomic manager *A* discovers that the managed resource has a constituent "autonomic manager" manageable resource. The details about how to access this embedded autonomic manager can be placed into a registry for autonomic managers so that other managers in the system can be configured to work with the embedded autonomic manager. In the figure, this situation is shown by the link between autonomic manager *B* and the embedded autonomic manager. Because an addressable "handle" is available for the autonomic manager manageable resource, that embedded manager can be treated just like any other autonomic manager in the system and can participate in any of the patterns that involve more than one autonomic manager, such as composed partial autonomic managers.

Note that the way embedded autonomic managers interact with the managed resource runtime is not within the scope of the architecture, because these implementations are specific to the managed resource.

An example of using this pattern is a database optimizer that monitors typical database transactions and creates or recommends appropriate search indexes to optimize overall database performance.

5.3.7   Pattern 5: Delegating Tasks in a Process Context

5.3.7.1   *Problem*

Process integration and automation allows IT professionals to delegate tasks to automation and to automate the flow of work among professionals. In many IT organizations, the monitor, analyze, plan, and execute functions that form a control loop cross organizational boundaries.

How can IT management processes that span organizational boundaries be incrementally automated?

The goal of process automation is to reduce the effort that professionals invest in coordinating routine systems management. In many IT organizations, the monitor, analyze, plan, and execute functions cross organizational boundaries, so, in many cases different technologies and procedures are used. This observation results in the need to link together the work (that is, tasks) performed by IT professionals in different business departments in the organization.

5.3.7.2   *Solution*

IT management processes are integrated using process integration technologies. Selected tasks for the activities within those processes are incrementally delegated to autonomic managers.

The workflows integrate across organizational boundaries by linking together the work performed by each business department.

Within each workflow, activities that implement the monitor, analyze, plan and execute functions of a control loop can be identified. The tasks for these activities can be incrementally delegated to autonomic managers, as described in Section 5.2.2. This results in the implementation of a control loop that crosses organization boundaries and that is integrated via workflow, as illustrated in Figure 5.15. As the number of automated tasks increases and the IT organization gains trust in the autonomic capabilities, the conditions under which manual intervention (e.g., approval steps) is required can be reduced by changing the policies that control the workflow.

By clearly defining the interfaces that a particular business department provides and by making the integration point occur at organizational boundaries, automation decisions can be made within an organization without the need for cross-organizational agreements, provided that the required interfaces are maintained and cross-organizational policies are upheld.

ESB integration patterns can be used to mediate the service interface invocations between organizations, and to manage the binding to autonomic managers within a flow.

By using the integration pattern described here, an IT organization can incrementally introduce autonomic capabilities into their management systems,

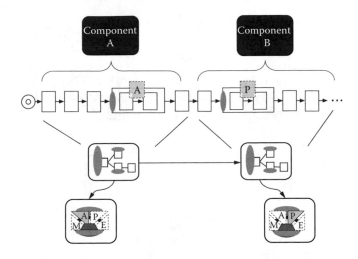

FIGURE 5.15
Automation of tasks for the activities in IT processes by partial autonomic managers.

and can do so in a way that allows different parts of the organization to retain a level of control over the degree of automation and the selection of automation technology.

An example of using this pattern is automating the software deployment task in the "Implement Change" activity of the change management process in Figure 5.2, using a provisioning application that implements the planning and execute functions of an autonomic manager.

5.3.8 Pattern 6: Process Integration through Shared Data

5.3.8.1 Problem

A variation of the process integration scenario just described involves an autonomic manager that provides a full control loop that is implemented independently of the management processes.

How can an independent autonomic manager be integrated into IT management processes?

For IT management processes to work effectively, key information (such as planned and actual changes) must be kept current, even if changes are initiated by a completely independent autonomic manager. Without this information, the IT professionals cannot understand the overall performance and state of the system. Worse, actions initiated manually could be erroneous or conflict with actions taken by the autonomic manager, because the data (e.g., policies) on which those actions are based is inaccurate.

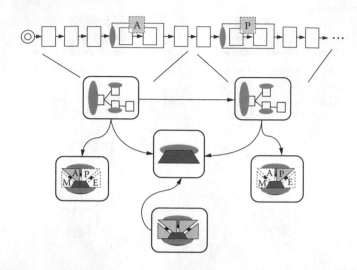

FIGURE 5.16
Integrating autonomic managers into IT processes using shared data.

5.3.8.2   Solution

An independent autonomic manager can be integrated into IT management processes by sharing data.

Autonomic managers that perform functions that need to be visible to the IT organization through management processes contribute the necessary data (e.g., detected symptoms or changes performed) to allow the reporting and oversight processes in the IT organization to function correctly as illustrated in Figure 5.16. This can be achieved by using application integration techniques to extract the necessary information from the autonomic managers and to populate the knowledge sources with that information. This data sharing allows autonomic managers to be integrated into IT management processes. Conversely, it allows process-level control loops that, over time, have become completely automated to be removed from the normal management workflow, becoming autonomic managers.

An example of using this pattern would be for an independent autonomic manager that detects and fixes a configuration problem to create an incident report, describing the problem and how it was fixed.

5.4   Summary

The material presented in this chapter demonstrates how a relatively small number of building blocks can be used to address many of the patterns that exist in an autonomic computing system. Many patterns deal with different ways in which the work performed by IT professionals can be performed automatically through the use of technology managing technology.

Reaching back to the analogy of the human autonomic nervous system that was used to name this area of computing, we posit that effective implementations of the autonomic computing architecture will allow portions of the work currently performed by IT professionals to be performed in their "subconscious." The autonomic nervous system handles numerous complex functions without our direct attention. However, we do have the option to become involved if and when we desire. This is why the "delegation" metaphor is used to describe the act of an IT professional taking advantage of technologies that automate some of the tasks they perform. Autonomic computing systems, like automated systems, need to earn the trust of their users and afford them the level of control that they desire.

To be successful, the future development and implementation of architecture for autonomic computing must keep the following objectives in mind:

(1) Embrace existing management architectures so that deployment of the autonomic computing architecture does not require the IT infrastructure to be re-tooled before value can be delivered.

(2) Exploit accepted programming techniques (adapters, enterprise service bus, flow composition, etc.) to integrate capabilities in the system to avoid unique management technologies that introduce additional complexity.

(3) Emphasize the importance of standardizing key content definitions (for example, common relationship types, common symptom types, etc.) that support management processes, so that a high level of integration is accomplished, beyond what is common for many existing management architectures.

(4) Adopt rigorous metadata descriptions so that the building blocks are self-describing. This makes it easier to manage an autonomic system, because building blocks like autonomic managers can discover how to manage new resources using metadata from a touchpoint building block.

(5) Simplify the business of managing diverse technology by standardizing interfaces and data across resource types.

(6) Reduce the need for IT professionals to be familiar with all of the system's complexity to deliver IT management services (that is, hide the complexity).

(7) Accommodate a wide range of organizational configurations (small and large, centralized and decentralized) and the process variability that occurs in different organizations.

(8) Simplify management activities by promoting the adoption of standard approaches to management (e.g., ITIL) that reduce unnecessary diversity.

In other words, the architecture must adapt to existing approaches while also establishing patterns that simplify the overall business of managing complex IT infrastructures.

5.5   Acknowledgments

The architecture presented in this chapter is derived from many collaborative interactions with individuals inside and outside IBM. Without candid and engaging discussions, this architecture for autonomic computing might still be about the nervous system. Our IBM colleagues who actively participated in this effort include Peter Brittenham, Nick Butler, Randy George, Brent A. Miller, Thomas Studwell, Richard Szulewski, and Mark Weitzel.

Without the input of people who actually manage complex systems, architectures such as this one could stray from reality. Since the introduction of Autonomic Computing, the following highly qualified practitioners have helped guide the evolution of the architecture by participating in both exploratory proposals and specific deliverables: Dave Searle, Tony Callister, Wolfgang Diefenbach, Kevin Hurdle, Hans van der Kruijf, Guy Liégeois, Paul McAvoy, Jon Sutcliffe. Much of the new material presented in this chapter about delegation and process integration is based on detailed discussions with these individuals.

References

1. IBM Corporation, *Component Business Modeling*, http://www-1.ibm.com/services/us/bcs/html/bcs_componentmodeling.html
2. DMTF Common Information Model (CIM) www.dmtf.org/standards/cim/
3. JSR 77: J2EE Management http://www.jcp.org/en/jsr/detail?id = 77
4. Jonathan Adams, Srinivas Koushik, Guru Vasudeva and George Galambos, *Patterns for e-business: A Strategy for Reuse*, 2001, IBM Press, ISBN: 1-931182-02-7
5. Gregor Hohpe and Bobby Woolf, *Enterprise Integration Patterns — Designing, Building and Deploying Messaging Systems*, 2004, Reading, MA: Addison-Wesley, ISBN: 0-321200-68-3
6. Martin Keen, et al, *Patterns: Implementing an SOA using an Enterprise Service Bus*, 2004, IBM Redbooks, http://www.redbooks.ibm.com/
7. IBM Corporation, *Autonomic Computing*, http://www.ibm.com/autonomic

Part II

Achieving *Self-**
Properties — Approaches
and Infrastructures

6

A Taxonomy for Self-* Properties in Decentralized Autonomic Computing

Tom De Wolf and Tom Holvoet

CONTENTS

6.1 Introduction

Decentralization is a characteristic of many modern computing systems and implies an increased complexity in managing the system behavior. Autonomic computing (AC) [12] is essential to keep such systems manageable. The problem is that many decentralized systems make central or global control impossible. For example, the information needed to make decisions cannot be gathered centrally (e.g., ad-hoc networks). In such systems AC is only possible when decentralized entities autonomously interact and coordinate with each other to maintain the required *self-** properties. We denote this kind of AC as *Decentralized Autonomic Computing* (DAC).

Typically, most *self-** properties in AC are the responsibility of a single autonomous entity (or manager) which possibly controls a hierarchy of autonomic managers. However, in DAC, many *self-** properties are achieved collectively, i.e., by a group of autonomous entities that coordinate in a peer-to-peer fashion. The lack of global or central control implies the need for new techniques to design and verify *self-** properties. Many properties of the system behavior can be categorized as *self-**, but not all *self-** properties can be designed or verified with the same technique. There is a need for more structure and guidance in this chaos of *self-** properties.

The contribution of this chapter is twofold. Firstly, a taxonomy of *self-** properties in DAC is described. Secondly, based on that taxonomy, guidance to design and verify *self-** properties is given. The engineer can thus situate the required *self-** properties in the taxonomy and is guided in designing and verifying those properties. The chapter is structured as follows. Section 6.2 introduces two case studies that are used throughout the chapter. Then, Section 6.3 explains DAC and its implications on engineering *self-** properties. After that, Section 6.4 describes a taxonomy that structures the *self-** properties, and Section 6.5 gives guidelines to design and verify them. Finally, Section 6.6 concludes this chapter.

6.2 Case Studies

Automated Guided Vehicles (AGVs). In an industrial research project [6], our group develops a decentralized autonomic solution for AGV transportation systems. Multiple transport vehicles, called AGVs, are used and guided by a computer system (on the AGV itself or elsewhere). The vehicles get their energy from a battery, and they move incoming packets (i.e., loads, materials, and/or products) in a warehouse to their destination(s). The AGV problem is very dynamic: many layouts are possible, packets can arrive at any moment, AGVs are constantly moving and can fail, obstacles can appear, etc. The screen-shot of our simulator[1] in Figure 6.1 shows one example of a simple factory layout: locations where packets must be picked up are located at the left, destinations are located at the right, and the network in-between consists of stations and connecting segments on which the AGVs can move. Packets are depicted as rectangular boxes, and some AGVs hold a packet.

The industrial project [6] has shown that a central server to control all AGVs is not flexible enough for a very dynamic and large AGV system. The central server constantly monitors the warehouse to detect changes and reacts to them by steering each AGV. The central server becomes a bottleneck in the presence of frequent changes. Also, the system is not flexible for deployment,

---

[1] http://www.cs.kuleuven.be/~distrinet/agentwise/agvsimulator/

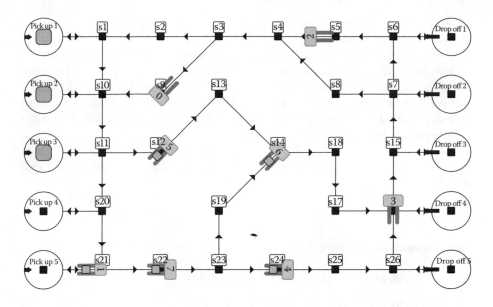

FIGURE 6.1
A screenshot of an AGV simulator showing an example factory layout.

i.e., manual optimization is needed for each new warehouse. Therefore, a decentralized autonomic solution in which the AGVs adapt without any central or external control is promising.

Mobile Ad-Hoc Network Management. A mobile ad-hoc network (MANET) is a local area network or other small network of portable and mobile devices, in which some of the network devices are connected to the network only while in some close proximity to the rest of the network (see Figure 6.2). The devices or nodes are free to move randomly and organize

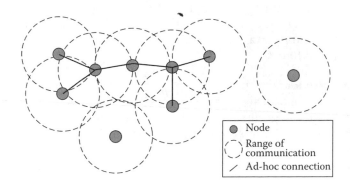

FIGURE 6.2
A mobile ad-hoc network where two nodes are not connected to the network.

themselves arbitrarily; thus, the network's wireless topology may change rapidly and unpredictably.

The goal of managing such a network is to maintain the connectivity between the nodes. This management task cannot be achieved by introducing a central controlling entity because the information needed for making management decisions is completely decentralized and distributed. Nodes can leave, join, and fail at every moment. In such a dynamic context it is impossible to aggregate enough information about a network centrally. By the time the information would have been aggregated, something has already changed which makes the information obsolete. Therefore, a decentralized solution is needed.

6.3 Decentralized Autonomic Computing

In this section we introduce the term *decentralized autonomic computing* and the implications of decentralization on engineering *self-** properties.

6.3.1 What Is Decentralized Autonomic Computing?

Decentralized systems are systems for which the control is decentralized by nature. The system behavior cannot be controlled by a central entity. A system that is decentralized is commonly composed of many entities which work together to form a stable structure. In such systems, autonomic computing is essential to keep the system behavior manageable.

Typically, each *self-** property in AC is the responsibility of a single autonomous entity (or manager) which possibly controls a hierarchy of autonomic managers to achieve the *self-** property. However, in DAC, the main implication of decentralization is that no central or global control is possible. Many *self-** properties (e.g., self-configuration, self-optimization, self-healing, and self-protecting) are achieved by a group of autonomous entities that coordinate in a peer-to-peer fashion. Therefore we introduce the term *decentralized autonomic computing*:

> *Decentralized autonomic computing (DAC)* is achieved when a system is constructed as a group of locally interacting autonomous entities that cooperate in order to adaptively maintain the desired system-wide behavior without any external or central control.

Such a system is also called a self-organizing emergent system [4]. Self-organization is achieved when the behavior is constructed and maintained adaptively without external control. A system exhibits emergence when there is coherent system-wide or macroscopic behavior that dynamically arises from the local interactions between the individual entities at the microscopic level. The individual entities are not explicitly aware of the resulting macroscopic behavior; they only follow their local rules. The term 'macroscopic' refers to the dynamics of the system as a whole, and the term 'microscopic' refers to the dynamics of the individual entities within the system. In other words, the *self-** properties are so-called macroscopic properties. It is, however, not mandatory that all *self-** properties become macroscopic properties. For example, self-protection can still be achieved locally by a single entity if the protection of the system is controlled by a single entry point. If, however, an attack on the system has to be countered by a defensive group attack on the intruders, then self-protection becomes a macroscopic property.

In the AGV case, a solution is considered a DAC solution when there is no central server that commands all AGVs. Yet, the system as a whole autonomously achieves the system behavior. Consider the *self-** property that an efficient throughput is to be maintained over time. In a decentralized solution, the throughput has to be achieved by the group of AGVs that coordinate with each other to dispatch and transport incoming packets to their destination. Each individual AGV is not aware of the global throughput that is achieved. Still, it emerges from the interactions between the autonomous AGVs.

In ad-hoc networks, decentralized control is the only possible way to manage the system behavior because all necessary information to make appropriate decisions is inherently distributed and decentralized. A *self-** property is to adaptively maintain a certain degree of connectivity within the network so that communication between different nodes remains possible. Because it is impossible to impose a central controller, a group of autonomous entities residing on the nodes need to coordinate their local movement in order to maintain an acceptable global connectivity. Again, an individual node is not aware of the overall connectivity of the network. Still, this connectivity should emerge from the interactions between the nodes.

6.3.2   Implications on Engineering *Self-** Properties

Engineering a decentralized solution to achieve *self-** properties as macroscopic properties has a number of implications. The design has to find new decentralized techniques to coordinate the behavior between the autonomous entities. Also, verification should use techniques that address macroscopic issues.

Design. Designing a decentralized autonomic system implies explicitly considering a way to coordinate the group of autonomous entities to achieve the desired macroscopic *self-** properties. In fact, more emphasis should be given to the interaction between entities than to the internal processing of each autonomous entity. The problem-solving power resides in the coordinated interaction instead of inside an individual autonomous entity.

For this purpose there is a need to identify possible coordination mechanisms that each allow the emergence of specific types of macroscopic properties. Using that set of mechanisms as a guide enables one to engineer more easily the required *self-** properties. A number of coordination mechanisms exist from the community on engineering self-organizing and emergent applications [2]. Section 6.5 elaborates on them by indicating for which kind of *self-** properties each mechanism is useful.

Verification. It is hard to give guarantees about the required *self-** properties in DAC because of the lack of any central or global control. As a consequence, the system is a complex interacting system for which it is practically infeasible to give formal verifications [19]. However, such systems are acceptable in an industrial context only if the required behavior has been verified. Therefore, traditional (formal) verification methods have to be complemented by usable metrics and (more empiric) verification methods for macroscopic *self-** properties. In Section 6.5 a number of metrics and verification approaches are given by indicating for which kind of *self-** properties they can be used.

6.4   A Taxonomy of *Self-** Properties

Many *self-** properties are very different in nature, and this has two implications. Firstly, not every *self-** property can be designed with the same coordination mechanism. Secondly, not every *self-** property can be verified with the same verification method. There is a need for more structure and guidance in the chaos of *self-** properties, coordination mechanisms, and verification methods in DAC.

A good approach to give more structure and guidance is to categorize the different types of *self-** properties and to indicate for which type of *self-** properties a certain design or verification technique is useful. Such a taxonomy guide is described in this section and Section 6.5 and originated from case studies described in literature and from our own experience in using the described design and verification techniques. Section 6.4.1 describes the structure used to categorize *self-** properties. After that, Section 6.4.2 schematically illustrates this taxonomy structure with examples from the AGV and ad-hoc network cases. Note that we do not claim that this taxonomy is complete—further additions and adjustments are to be expected. At this moment, a category is introduced only when there are also some guidelines for design or verification available at the moment of writing. As such, for example, self-protection is a category, but self-healing, self-configuring, and self-optimizing are not yet.

To use this taxonomy guide, an engineer should situate required *self-** properties in the taxonomy structure and then go to Section 6.5 to find guidelines on how such a property can be designed and verified.

6.4.1 Categorization

To distinguish the different *self-** properties, two main categories are used. Firstly, Section 6.4.1.1 elaborates on categories based on the characteristics of the *self-** properties that do not directly denote a specific functionality. Secondly, Section 6.4.1.2 gives an overview of categories based on a specific type of functionality achieved by the *self-** property.

6.4.1.1 Characteristics of Self- Properties*

Macroscopic versus Microscopic. A *self-** property is considered 'macroscopic' if it is a property of the system as a whole, i.e., the property is achieved by the coordinated interactions between the entities of the system. 'Microscopic' means that the property is achieved without global coordination with other autonomous entities. In other words, it can be solved by a single autonomous entity that decides what to do based on its own state and observations of its immediate vicinity or context, i.e., its locality. An important issue to decide if a property is solved locally or globally is the way 'locality' is defined in the system. Typically, local properties refer to properties of a single entity in the system and its immediate vicinity. For example, an AGV has to recharge its battery at regular intervals. The AGV can monitor its own battery level and move to a charging station at regular intervals without having to coordinate. To achieve this, the AGV knows the location of the charging station or is able to observe the station in its vicinity.

Self-* properties can be macroscopic in one system while being microscopic in another. This depends solely on how the *self-** property can be solved. For example, self-protection of an ad-hoc network can be solved locally when a single node acts as an entry point for a network. This node authorizes new nodes by giving them the credentials (e.g., an encrypted key) needed to communicate in the network. However, when protection consists of a group-coordinated counterattack against intruders, then this is no longer a local property. For example, consider a military ad-hoc network on unmanned aerial vehicles (UAVs) that has to explore an unknown territory. The vehicles coordinate and adjust their positions in order to stay invisible and thus protect themselves from the enemy. Meanwhile, they have to maintain the connectivity in the network to communicate about newly detected enemies.

Ongoing versus one-shot. Most *self-** properties are so-called ongoing properties. They require that the system adaptively maintains a certain property over a longer period in time. The requirement that an ad-hoc network has to self-optimize a certain degree of connectivity is an example of an ongoing property.

In contrast, 'one-shot' properties are properties for which the behavior to achieve is triggered, and after a finite time the required effects are achieved. There is no ongoing evolution. For example, in the AGV case the system has to dispatch new packets to available AGVs. This property is triggered by an incoming packet and ends when that packet is assigned to a certain AGV. If, however, the system constantly has to reassign packets because another AGV has become more suited (e.g., shorter distance to reach), then this is also an ongoing property.

Time/History Dependent versus Time/History Independent. "Time/history dependent" means that the property cannot be measured at one moment in time. It is measured over a history of the behavior, i.e., a certain time period. For example, the throughput of an AGV system is expressed as the number of packets that are transported within a certain period in time (e.g., within an hour).

"Time/history independent" means that the property does not depend on historical information. Such a property can be measured at every moment in time. For example, in ad-hoc networks the current utilization load (i.e., used bandwidth for all connections) can be measured at each moment in time.

Continuous or Smooth Evolution. Another characteristic to distinguish *self-** properties is concerned with their evolution. Assume that the evolution of the property can be measured and plotted in a graph. If this graph has a gradual evolution without sudden jumps, the property can be considered to have a 'smooth' or 'continuous' evolution. Typically, in stochastic or nondeterministic systems, many properties are not smooth or continuous at all.

Adaptation-Related. These properties are mostly expressed in terms of a change in the system and how a property can cope with it. These changes can be size changes (i.e., scalability) or other dynamic events (i.e., adaptability, robustness for optimizing a certain property, self-optimizing, self-configuring, self-healing, etc.). The robustness or stability of the degree of connectivity in a mobile ad-hoc network, with respect to the failure of nodes, is an example of an adaptation-related property.

6.4.1.2 *Functional Categories of Self-* Properties*

Spatial versus Nonspatial. Some *self-** properties require that a certain spatial structure be constructed and maintained in the face of changes. For such spatial properties, there is an extra distinction that can be made:

- *Specific shapes*: Specific spatial shapes are constructed or self-organized by the system. For example, in an ad-hoc network a requirement could be that a certain topology should be maintained. Another example is the paths that are followed and constructed for routing AGVs.

- *Distribution-alike*: A certain spatial distribution has to be achieved and maintained. For example, in the AGV case, a requirement could be that the distribution of AGVs on the factory floor should cover the floor as much as possible. More concretely if a number of equally important regions are identified, then an equal distribution of AGVs among those regions is required.

Resource Allocation. Certain *self-** properties require that the system autonomously manage a limited resource by deciding which entity in the system can use or be assigned to the resource. 'Resource' should be interpreted in the broadest sense. For example, the available bandwidth in an ad-hoc network is limited, and a fair use policy should be enforced between the services active on the network. Another way of interpreting 'resource' involves dispatching or assigning tasks. For example, in the AGV case, incoming packets are assigned to the available AGVs.

Group Formation. Groups or teams can also be formed autonomously. Therefore, the teams can be adapted and restructured based on specific changes that occur. Also, clustering of items or data is a kind of group formation. For example, in ad-hoc networks, nodes can cluster based on common interest or service needs.

Role-Based Organizations. In several multi-agent systems (MASs) a requirement is to enforce a so-called organization. An organization of a multi-agent system comprises roles and their interrelations. A role clusters types of behavior into a meaningful unit that contributes to the group's overall goal (e.g., master role and slave role). Role interrelations (e.g., hierarchical or class-membership relations) that provide communication paths among agents and role interdependencies (e.g., temporal or resource dependency relations) can be exploited for effective agent coordination and communication. As such, roles and their interrelations impose a formal structure on a MAS in order to implement coherent teams. Enforcing such an organization autonomously involves making sure that all required roles are assigned and active when required and that only the allowed number of agents execute a certain role.

Self-Protection. Self-protection requires that the system detect, identify, and protect itself against various types of attacks to maintain overall system security and integrity. More than simply responding to failures or running periodic checks for symptoms, an autonomous system will need to remain on alert, anticipate threats, and take necessary action. In a decentralized autonomic system, this can even imply that a certain attack needs to be countered by a coordinated group defense (e.g., UAVs in military defense operations).

6.4.2   Overview and Examples of *Self-** Properties

To make the structure from Section 6.4.1 more concrete, Table 6.1 gives a schematic overview that indicates for a number of example *self-** properties with which categories they can be related (X indicates strong relation, (X) indicates a possible relation). Each example is described in what follows. In general, many of the example properties are *macroscopic, adaptation-related*, and can be considered to evolve *smoothly* if a good average is taken over time or over the system as a whole.

For the AGV case, some required *self-** properties are:

- *A1: Average waiting time until dispatch of packet.* The time that a packet has to wait before it is assigned to an AGV should be self-optimized to a minimum. Clearly, this is a *time-dependent resource allocation* property. The average waiting time can change over time and the

TABLE 6.1

Example Properties and Their Place in the Taxonomy

	A1	A2	A3	A4	M1	M2	M3	M4
Smooth	(X)	(X)	(X)		(X)	(X)		
Ongoing	X	X	X		X	X	X	(X)
One-shot								(X)
Time-dependent	X		X			X		
Time-independent		X			X		X	
Adaptation	X	X	X	X	X	X	X	X
Macro	X	X	X	X	X	X	X	(X)
Micro								(X)
Resource allocation	X			(X)			X	
Group formation		X			(X)			
Organizations							X	
Protection								X
Spatial		X			(X)	X		
— Shapes						X		
— Distribution		X						

system has to self-optimize by constantly adapting the assignments (e.g., when another AGV becomes more suitable). As such, it is an *ongoing* property.

- *A2: Average distribution of AGVs.* On a factory floor there are a number of 'important' regions for which it is desirable to constantly have an AGV available. Therefore, requiring that the system self-organize a *spatial distribution* of AGVs between those regions at each moment in time (i.e., *time-independent and ongoing*) can accommodate this. As such, the property can also be considered as the *formation of groups* of AGVs for each region.

- *A3: Throughput of packets.* Throughput is the number of packets that are transported in a certain time span (e.g., an hour). This *time-dependent* property has to be achieved and maintained (*ongoing*) by the system.

- *A4: Scalability w.r.t. number of incoming packets.* An AGV system must be able to handle and *adapt* to an increase in the number of incoming packets. The *resource allocation* of packets to AGVs should react efficiently.

For a MANET, the examples are:

- *M1: Average fraction of nodes in largest network component.* At each moment in time (*time-independent*) the system has to maintain (*ongoing*) a network structure in which as much as possible nodes are part of the largest connected network component or *group*. Because the spatial movement of the nodes significantly influences this property, it is closely related to a *spatial* property.

- *M2: Average network packet routing efficiency.* In a network the communication packets have to be routed from source to destination as efficiently as possible. Therefore, self-optimization of routes, which are specific *spatial structures*, is required. The efficiency of a route can be calculated only after the route has been completed. This is therefore a history- or *time-dependent* property of which the average overall routes can evolve over time (*ongoing*).

- *M3: Organization with limited number of server roles.* For an ad-hoc network, a requirement could be to *adaptively* assign the role or task of 'server' to nodes (i.e., *resource allocation*) in order to accommodate a good service success. However, at each moment in time (*time-independent*), the system should maintain (*ongoing*) only a limited number of nodes that execute the server role to limit server replication and synchronization overhead between servers. Thus, a specific *organization* is to be enforced.

- *M4: Self-protection against malicious nodes.* The network should *self-protect* against malicious nodes that want to enter the network. When such protection can be handled at a single entry point which gives each node authorization, then this is a *micro* property. However, if such nodes can be identified and eliminated only by a coordinated defensive group attack of the other nodes, then this is a *macro* property. Most intrusions can be handled by triggering a *one-shot* behavior of the system. However, the system constantly has to be alert, so the degree of protection over time (*ongoing*) can be important.

6.5   Guidelines for Design and Verification

To design and verify *self-** properties in DAC, multiple techniques are available. Some techniques are very straightforward and easy to apply for an engineer. Other techniques require a substantial amount of expertise to use them. The engineer has to choose an appropriate technique based on the guidelines given in this section and the available expertise. The guidelines indicate which technique can be useful for which categories of *self-** properties (see section 6.4).

6.5.1   Design: Coordination Mechanisms

To achieve the required *self-** properties for a decentralized system, a number of issues have to be coordinated between the autonomous entities. This section and Table 6.2 give an overview of possible coordination mechanisms and guidelines that indicate for which categories of *self-** properties the mechanisms can be useful. Most coordination mechanisms are useful for achieving *ongoing* properties at the *macroscopic* level that have to *adapt* to changing circumstances.

TABLE 6.2

This Table Indicates Which Coordination Mechanism Can Be Appropriate for Which Type of *Self-** Property

	Pheromones	Co-fields	Markets	Tags	Tokens
Smooth					
Ongoing	X	X	X	X	X
One-shot					
Time-dependent			(X)		
Time-independent					
Adaptation	X	X	X	X	X
Macro	X	X	X	X	X
Micro					
Resource allocation			X		X
Group formation				X	
Organizations					X
Protection				X	X
Spatial	X	X			
— Shapes	X	X			
— Distribution		X	X		

Digital Pheromone Coordination. Inspired by nature, many emergent systems communicate through *stigmergy*. Stigmergy is a method of communication in which the individual entities of the system communicate with each other by modifying their local environment. Stigmergy was first observed in nature—ants communicate with each other by laying down pheromone trails. This pheromone idea can be translated into a digital pheromone coordination mechanism to design certain macroscopic *self-** properties [3]. Typically pheromones are used for paths which entities of the system can form and follow. Therefore, *spatial properties* such as *routing* and the optimization of those routes can be achieved. For example, in networks the routing of network packets could be done by a pheromone solution. This mechanism is inherently *adaptive* because old and not reinforced pheromones (i.e., old information) gradually disappear or evaporate (i.e., an *ongoing truth-maintenance* mechanism).

Gradient-Field or Co-Field Coordination. Another stigmergic coordination mechanism is a so-called computational field or *co-field* [14]. The key idea of co-fields can be described as follows. Autonomous entities spread out computational fields through the environment. The field forms a gradient map which conveys useful context information for the entities' coordination tasks. The coordination policy is realized by letting autonomous entities move following the waveform of these fields, e.g., uphill or downhill. Environmental dynamics and movement of entities induce changes in the fields' surface, composing a feedback cycle that influences how entities move. This feedback cycle lets the system *adaptively self-organize*. This coordination mechanism is promising to guide spatial movement of entities and thus construct *spatial shapes* and enforce *spatial distributions*.

Market-Based Coordination. Market-based methods view computational systems as virtual marketplaces in which economic entities interact by buying and selling. Although decision making by these entities is only local, economic theory provides means to generate and predict macroscopic properties. Typically this coordination mechanism is used to achieve efficient *resource allocation* [1]. An economic market for scarce resources is constructed. For example, consider a network in which autonomous entities have to find resources to complete an assigned task by moving to suitable hosts. These resources could include access to the central processing unit (CPU), network and disk interfaces, and data storage. A required *self-** property is that the entities be allocated efficiently. There is the danger that entities will exploit network resources and cause unwanted resource contention. Agents need the ability to coordinate at a high level to distribute their load evenly across the network and over time. Market-based control can address each of these points. An autonomous entity that arrives at a host will use electronic currency to purchase the resources necessary to complete its task. To coordinate this, information is needed about the resource usage at hosts. A market provides usage information through prices. High prices connote congestion. This gives agents incentive to distribute themselves evenly in the network by choosing another host or waiting for lower prices, i.e., less congestion. Thus, prices support coordination for both spatial and temporal load balancing. In other words, *spatial distribution* and *time-dependent* allocation properties can be supported.

Tag-Based Coordination. Tags [10] are observable labels, markings, or social cues. Agents can observe tags of others and maybe even adjust or add tags to other agents. Coordination emerges because agents may discriminate based on tags. Therefore, a number of macroscopic *self-** properties can be achieved. For example, when tags on an agent can be added/adjusted only by other agents, a kind of trust and security can be enforced by excluding agents that are tagged as badly behaving agents (i.e., *self-protection*). Another property that can be achieved is *group formation* (or tribes) [8]. Assume that agents interact preferentially with those sharing the same tag. Groups are formed around similar tags. Using and adjusting the right tags can serve to achieve desired group cooperations.

Token-Based Coordination. Tokens are objects that are passed around in a group of autonomous entities and encapsulate anything that needs to be shared by the group, including information, tasks, and resources [21]. For each type of token, there are a limited number of instances, and the agent holding a token has exclusive control over whatever is represented by that token. Hence, tokens provide a type of access control (*self-protection*). Agents either keep tokens or pass them to team-mates. Such a coordination mechanism can be used for a number of *self-** properties. For example, *task and resource allocation* is possible. An agent holding a resource token has exclusive access to the resource represented by that token and passes the token to transfer access to that resource. Another example of a *self-** property is management and construction of *role-based organizations*. A token then represents a certain

TABLE 6.3

This Table Indicates Which Verification Methods Can Be Appropriate for Which Type of *Self-** Property

	V1	V2	V3	V4	V5	V6	M1	M2	M3	M4
Smooth			X	X	X		(X)		X	(X)
Ongoing			X	X	X	X	X	(X)	X	X
One-shot	X									
Time-dependent			X	(X)	(X)					
Time-independent			X	X	X					X
Adaptation			X	X	X	X		X		
Macro	(X)		X	X	X	X			X	X
Micro	X	X	X							
Resource allocation			X							
Group formation			X							X
Organizations			X							X
Protection	(X)	(X)	X							
Spatial			X					X		X
—Shapes			X							X
—Distribution			X					X		
Other			X						(X)	

role in the group, and by fixing the number of tokens the number of entities in a certain role can be limited. The organizational structure can thus be enforced and roles can be passed on when dynamic changes imply a better allocation.

6.5.2 Verification: Methods and Metrics

Verification methods allow to verify if the system has achieved its requirements. Because there are many different types of *self-** properties, each type of property can require its own verification method. Therefore, a complete verification process involves applying multiple verification methods. Table 6.3 gives a schematic overview of the described guidelines. This table is a tool for the engineer to quickly find a (set of) verification methods to try based on the categories from Section 6.4 to which a required property is related. In what follows, each verification method or metric is discussed in detail.

Method V1: Unit-Based and Integration Testing. In traditional software engineering processes the verification consists of unit-tests and integration tests [11]. A unit test is a piece of code used to verify if a particular unit of the system is working properly. The goal of unit testing is to isolate each part of the system and show that the individual parts are correct. Integration testing is the phase of testing in which individual units are combined and tested as a group. It involves testing if the postconditions are achieved by a chain of units when the system is triggered.

By definition, a unit test applies only to the functionality of a single unit. Therefore, it will not catch performance problems and other system-wide issues. Macroscopic *self-** properties that involve a set of complex interacting units cannot be verified with unit tests. Although integration testing is for issues that involve multiple units, such tests verify only postconditions at a certain moment in time. They do not test the evolution of a property, and most *self-** properties are ongoing properties. Unit testing and integration testing are suited only for *one-shot* properties which are *microscopic* in nature, i.e., properties of a single unit or limited set of units (*macro*). Such properties are not concerned with complex interactions in the system. An example is *self-protection* that is handled locally.

Method V2: Formal Proof. Formal methods refer to mathematical techniques for the specification and verification of software systems. Once a formal specification or model of the system behavior has been developed, that model is used as the basis for proving properties of the system behavior analytically.

However, constructing a formal model and correctness proof of a complex interacting computing system is infeasible. Wegner [19] proves this based on the fact that all possible behavior of an interaction model cannot be captured formally, and thus formally proving correctness of interactive models (e.g., DAC) is not merely difficult but impossible. This does not mean that formal methods are completely useless. A formal proof may still be feasible for properties that are quite simple and solved locally, i.e., *microscopic* properties.

Method V3: Statistical Experimental Verification. If formal verification of a *self-** property is impossible, the only alternative is to use experimental methods. Typically this means observing a number of experiments and processing the results using statistical techniques. Statistics is a type of data analysis which includes the planning, summarizing, and interpreting of observations of a system, possibly followed by predicting future events. The basic idea of statistics is that the system behavior can be represented by a sample of that behavior (i.e. experiments) when the sample is sufficiently large and when the parameter space of the system is covered as much as possible. Statistical theory provides methods for determining how large a sample is needed to get statistically significant results. Then statistics can be used as a tool to verify conformance to the requirements of the system behavior.

Most categories of measurable *self-** properties can be verified with this method. The method is especially well suited for observing the evolution of *long-term ongoing* behavior. However, because a sufficiently large number of experiments is needed, this can be a very expensive verification method.

Method V4: Equation-Based Macroscopic Verification. To achieve reliable and scientifically founded results, using mathematical methods is often a good idea. In the scientific computing community, numerical algorithms exist for analyzing the system dynamics [9]. Typically a *macroscopic self-** property is modeled as a variable in a (partial) differential equation which represents the evolution of the *self-** property. Numerical algorithms then analyze the evolution of that equation (illustrated in the left of Figure 6.3): the algorithm repeatedly evaluates the equation for certain initial values (X_i) for the property to obtain the value some time steps later (X_{i+1} in Figure 6.3). The evolution is thus analyzed on the fly by the algorithm.

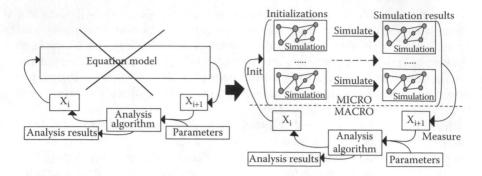

FIGURE 6.3
Equation-Free Accelerated Simulation guided by the analysis algorithm.

For example, a bifurcation algorithm allows analysis for qualitative changes in the behavior of the system as a result of a changing parameter. To illustrate, consider a mobile ad-hoc network. The communication range used by each node is an important parameter. For certain values of that parameter, the network can maintain itself as one connected network. For other values, the network can be constantly partitioned or completely disconnected. Bifurcation analysis allows dectection of which type of behavior is achieved for which communication ranges. This also allows checking for *adaptation-related* properties. For example, scalability of the AGV case with respect to the utilization load can be analyzed by using the maximum utilization load as a parameter to be varied from low to very high utilization. The bifurcation algorithm analyzes the evolution of the performance for each value.

To be able to apply this method, a first requirement is an equation for the system behavior. For *self-** properties that result from complex interactions between autonomous entities, this is as impossible as constructing a formal model. Also, equations typically have a smooth and continuous evolution. As a consequence, the algorithms used to analyze the behavior also rely on this. Therefore, only *self-** properties that have a *simple and continuous/smooth* evolution can be represented like this. If an equation is found, then the *long-term ongoing* behavior can be verified scientifically. However, even if an equation can be derived, there is the risk of having results that differ significantly from the real system behavior (e.g., [20]). To be able to construct an equation, a number of simplifications are often required, while every detail in the behavior can have a significant impact on the macroscopic result.

Method V5: Equation-Free Macroscopic Verification. The problem with the equation-based macroscopic verification is the need to construct an equation. An alternative approach that solves this problem is called "equation-free macroscopic analysis" [13]. The observation is that most numerical algorithms have no need for the equation itself; they need only a routine that 'evaluates' the equation for a given value in order to return a value some time steps later. The idea is to replace the equation evaluations with realistic 'simulations' (see Figure 6.3). Simulations are initialized to reflect the given values (X_i) for the properties under study (init). After simulating for the requested time (simulate), the new values (X_{i+1}) are measured from the simulation state (measure). Therefore, numerical analysis algorithms can be applied to verify the system behavior (e.g., Newton steady state detection, bifurcation analysis, etc.). A challenge for this approach is finding macroscopic variables reflecting the evolution of the *macroscopic self-** properties and an initialization operator to initialize simulations based on these variables at a certain moment in time.

This approach can verify *long-term ongoing* behavior if the evolution of that behavior is rather *continuous and smooth*. If not, focusing on verifying trends in the average behavior can still be possible and useful. Because an initialization at each moment in time is necessary, the property has to be *time-independent*. For a *time-dependent* property, this method is possible only if that property can be translated into a (set of) time-independent properties from which the evolution of the original property can be calculated as a postanalysis step. In contrast to the equation-based approach, the complexity of the interaction causing the *self-** property is no longer an issue because no equation-based or formal model has to be constructed. A realistic simulation or even the system itself can be used in the verification. This still enables a bifurcation algorithm to verify *adaptation-related* properties such as scalability and robustness. For example, in an ad-hoc network a bifurcation parameter could be the degree of node failure. Using this in a bifurcation analysis allows verification of how much node failure the network can cope with. We refer to [5] for more details, results on an AGV transportation system, and a discussion of challenges and (dis)advantages.

Method V6: Time Series Analysis Based on Chaos Theory. Today we often make use of simple measures (e.g., averages over time) to characterize the behavior of a system. Although useful characterizations for static systems, they are less helpful if the system is nonlinear and constantly subjected to change [18]. We need more advanced analysis techniques. A lot of tools and techniques are available from chaos theory, which was developed to describe nonlinear behavior. Chaos theory offers [17]: a conceptual framework with which dynamic behavior can be characterized, and experimental techniques that can estimate these characteristics from time series of state variables in an observed system. For example, there exists a technique that determines the Lyapunov exponent of a time series which indicates how sensitive the evolution is to changes in its initial conditions [15]. This refers to how stable the behavior actually is, relative to changes in the system. Of course this can be useful for verifying the *adaptation-related* property of robustness. Other techniques allow measurement of the complexity of the behavior, i.e., Is it very straightforward or almost completely chaotic? Clearly, these are techniques concerned with statements about the *long-term ongoing* behavior.

Metrics. Many of the above verification methods require some kind of metric that reflects the *self-** property. Some guidelines for metrics are:

- *M1: Entropy*: A metric that is often used for macroscopic properties is the so-called Entropy [7]. This metric originates from Shannon Entropy [16] in information theory, and is defined as follows:

$$E = \frac{-\sum_n^N p_n \cdot \log p_n}{\log N} \tag{6.1}$$

  where p_n is the probability that state n out of all N states occurs. The denominator $\log N$ normalizes to the range [0, 1]. An entropy close to 1 indicates rather equal probabilities, while close to 0 indicates a nonequal situation. Entropy can be used for properties that can be translated into a number of states with their probabilities and where a measure of the equality or inequality is useful.

  This *continuous and possible smooth* evolving metric can be used to verify the *long term ongoing* evolution of a number of *self-** properties. Especially the *spatial distribution-alike* properties. For example, consider the AGV case where a number of equally important regions are defined on the factory floor. Assume that a required *self-** property is that an equal spatial distribution of the AGVs has to be maintained between those regions in order to have an AGV available at all times in each region. This can be measured by using the entropy metric. For each region there is the state that an AGV is in that region. Then p_n represents the probability of finding an AGV in a region. If these probabilities are equal, then E is close to 1 and an equal distribution is reached, otherwise E is closer to 0. This does not mean that all spatial properties can be expressed with an entropy metric.

- *M2: Lyapunov exponent*: This is a metric from chaos theory that measures how sensitive the *ongoing* system behavior is regarding changes in the initial conditions of the system [15]. Therefore, this is a good metric to measure the stability or robustness of a *self-** property (*adaptation-related*).

- *M3: Averages*: An average is in *many cases* a metric worth considering. Some macroscopic properties are only the average of a local property of entities in the system. Thus, an average can often serve to translate *microscopic into macroscopic properties*. Averages tend to evolve *smooth and continuous* over time (*ongoing*), which implies that for example the equation-free macroscopic verification method can be applied with this kind of metric.

- *M4: Distances*: To measure whether a *self-** property is present, the difference between the current state and the desired state can be measured. A 'distance' metric should be constructed that can be measured at each moment in time (*time-independent and ongoing*).

Consider the *self-** property to achieve a *specific spatial shape, a group structure, or even an organizational structure*. Measuring this by calculating the distance between the current structure and the desired structure can be a good metric. However, how this distance metric is constructed remains a challenge.

6.6 Conclusion

It is clear that autonomic computing is essential in decentralized systems to keep them manageable. A group of locally interacting autonomous entities coordinate to achieve the desired *self-** properties without external or central control. We denote this as decentralized autonomic computing. Because central or global control is impossible, advanced design (i.e., coordination mechanisms) and verification techniques are needed. To structure the many different *self-** properties and different design and verification techniques, this chapter has given engineers a taxonomy to situate their *self-** properties and to guide them in designing and verifying those properties with suitable coordination mechanisms, verification methods, and/or metrics.

Future work[2] includes additions and adjustments to the taxonomy guide presented in this chapter. Developing support for using certain verification approaches and coordination mechanisms is also necessary, e.g., tools, modeling languages, etc.[3]

References

1. Jonathan L. Bredin. *Market-Based Control of Mobile Agent Systems*. PhD thesis, Dartmouth College, Hanover, NH, June 2001.
2. S. Brueckner, G. Di Marzo Serugendo, A. Karageorgos, and R. Nagpal, editors. *Engineering Self-Organising Systems: Methodologies and Applications*, volume 3464 of *Lecture Notes in Computer Science*. Springer Verlag, 2005. Post-proceedings of the Second International Workshop on Engineering Self-Organising Applications (ESOA '04) @ AAMAS 2004, New York.
3. Sven Brueckner. *Return From the Ant—Synthetic Ecosystems for Manufacturing Control*. PhD thesis, Humboldt-Universitt, Berlin, 2000.
4. Tom De Wolf and Tom Holvoet. Emergence versus self-organization: different concepts but promising when combined. In *Engineering Self Organising Systems: Methodologies and Applications*, volume 3464 of *Lecture Notes in Computer Science*, 1–15. Springer Verlag, Berlin, 2005.

---

[2] Future work on this topic will be available at http://www.cs.kuleuven.be/~tomdw

[3] This work is supported by the K. U. Leuven research council as part of the concerted research action on Autonomic Computing for Decentralized Production Systems (project 3E040752)

5. Tom De Wolf, Giovanni Samaey, Tom Holvoet, and Dirk Roose. Decentralised Autonomic Computing: Analysing Self-Organising Emergent Behaviour using Advanced Numerical Methods. In *Proceedings of IEEE International Conference on Autonomic Computing (ICAC'05)*, Seattle, June 2005.

6. Egemin and DistriNet. Emc²: Egemin modular controls concept, IWT-funded project with: Egemin (http://www.egemin.be) and DistriNet (research group of K.U. Leuven). From March 1, 2004, to February 28, 2006.

7. S. Guerin and D. Kunkle. Emergence of constraint in self-organizing systems. *Journal of Nonlinear Dynamics, Psychology, and Life Sciences*, 8(2), 2004.

8. David Hales. Choose your tribe!—evolution at the next level in a peer-to-peer network. In *Proceedings of the 3rd Workshop on Engineering Self-Organising Applications (EOSA 2005)*, Utrecht, Netherlands, 2005.

9. Michael T. Heath. *Scientific Computing: An Introductory Survey, Second Edition*. McGraw-Hill, New York, 2nd ed., 2002.

10. J. Holland. The effect of labels (tags) on social interactions. Technical Report SFI Working Paper 93-10-064, Santa Fe Institute, Santa Fe, NM, 1993.

11. I. Jacobson, G. Booch, and J. Rumbaugh. *The unified software development process*. Addison-Wesley, 1999.

12. Jeffrey O. Kephart and David M. Chess. The vision of autonomic computing. *IEEE Computer Magazine*, 36(1):41–50, January 2003.

13. I. G. Kevrekidis, C. W. Gear, J. M. Hyman, P. G. Kevrekidis, O. Runborg, and C. Theodoropoulos. Equation-free, coarse-grained multiscale computation: enabling microscopic simulators to perform system-level analysis. *Comm. in Mathematical Sciences*, 1(4):715–762, 2003.

14. M. Mamei, F. Zambonelli, and L. Leonardi. Co-fields: A physically inspired approach to motion coordination. *IEEE Pervasive Computing*, 3(2), 2004.

15. Ulrich Nehmzow and Keith Walker. Is the behaviour of a mobile robot chaotic? In *Proceedings of the AISB'03 Symposium on Scientific Methods for Analysis of Agent-Environment Interaction*, 12–19, 2003.

16. C. E. Shannon. A mathematical theory of communication. *Bell System Technical Journal*, 27:379–423 and 623–656, 1948.

17. H. D. Van Parunak. Complexity theory in manufacturing engineering: Conceptual roles and research oportunities. Technical report, Industrial Technology Institute, 1993.

18. H. D. Van Parunak. The heartbeat of the factory: Understanding the dynamics of agile manufacturing enterprises. Technical report, Industrial Technology Institute, 1995.

19. Peter Wegner. Why interaction is more powerful than algorithms. *Communications of the ACM*, 40(5):80–91, May 1997.

20. W.G. Wilson. Resolving discrepancies between deterministic population models and individual-based simulations. *American Naturalist*, 151(2):116–134, 1998.

21. Yang Xu, Paul Scerri, Bin Yu, Steven Okamoto, Michael Lewis, and Katia Sycara. An integrated token-based algorithm for scalable coordination. In *Proceedings of the fourth international joint conference on Autonomous agents and multiagent systems (AAMAS)*, 407–414, Utrecht, Netherlands, 2005.

7

Exploiting Emergence in Autonomic Systems

Richard Anthony, Alun Butler, and Mohammed Ibrahim

CONTENTS

In the future self-* systems must be constructed to work with and exploit emergent properties. This will entail embracing features of naturally occurring emergent systems like redundancy, interaction, identity, loose coupling, feedback, failure, and composition.

Emergence is well known among entomologists, biologists, and bio-mathematicians; among complexity and chaos theorists; economists and game theorists. It has been identified (in a popular text on evolution) as early as 1923 [1], and has been recognized in computer science in fields as diverse as Cellular Automata [2], Computational Fluid Dynamics, and latterly Autonomic Computing [3]. Inspired by many naturally occurring cases and long recognized as an observable phenomenon, only recently has it been seriously regarded as a design paradigm for human crafted software artefacts.

Similarly, natural systems are the initial inspiration for the field of Autonomic Computing. To what extent are the two terms *autonomic* and *emergent* coupled? How would it be for a system to genuinely exhibit self-* properties without emergence? Consider two simple examples.

While we could describe a dual redundant server system (in which the shut down sequence of the active partner includes a signal to the passive partner to become active) as "self-healing," this hardly qualifies as a genuine autonomic system. Such a system lacks genuine redundancy and is tightly coupled to the defined location of its neighbor. By contrast consider a system of nodes, passively monitoring local neighbors (whoever they may be and by whatever means) and becoming active according to interacting policies — now that *is* autonomic. Yet in such a loosely-coupled system it is apparent that the node that will become active is undetermined at design time — in other words, the choice of active node and the pattern of that choice over time, is emergent.

Similarly an operating system that increments the system date every 24 hours may be termed self-configuring (it is configuring its own date!), yet in a manner that is essentially uninteresting. Consider a system that, on first power up, passively senses time signals from a series of sources (radio, mobile telephony signal, local network personal computers), gradually weighting them according to perceived authority, sets (in due course) its own system time and influences (and is influenced by) similarly configured systems to synchronize system times. In the latter, more clearly autonomic, case the exact time of cooperating systems is unlikely to be predictable — it will *emerge*. Similarly the nature of the stability of synchronization is emergent [4].

In the latter systems we recognise some of the advantages of the initial biological models that inspire the autonomic computing metaphor. Such systems demonstrate redundancy of operation (resisting local failure and message loss), locality of messaging (enhancing scalability). With some extra features (which we will discuss) such a system may become self-optimizing and even self-protecting, but each time with a reduction in the microscopic determinism (although macroscopically the emergent structure may appear stable to clients).

From an alternate direction, imagining emergent phenomenon without many agents and many interactions may seem somewhat unusual, but it is definitely possible — imagine the emergent repetitive structure of the Mandelbrot Set for example. These "simple" emergent systems become more possible still if we are prepared to exclude natural input phenomena as "agents," while still allowing their input. A suitably programmed sensor reacting to wind may produce "emergent" output (a regular resonance effect for example). We can characterize this output as emergent if it demonstrates structure (or, in some circumstances, a complete, unexpected absence of structure). Naturally, properly designed and working autonomic systems are intended to result in positive outcomes (the self-* properties), but emergence may well be a negative phenomenon — for example in the synchronization timing example above it is possible for the system to cascade into two or more dissimilar system times across the population [4], never reaching a consensus, or resonant feedback can mean sufficient stability is never achieved.

Thus while emergence and autonomic systems may exist independently, we suggest that together they realize the full suggestive power of both metaphor and phenomenon. Further we also propose that, to be effective, emergence must be actively engineered and actively exploited.

The body of this chapter is divided into two sections. In section A we discuss the nature of emergence, how this nature can impact on autonomic computer system design, and current work in the field. In section B we look at implementation issues, including describing an exemplar system built by the authors.

7.1 Emergence and Autonomics

This section discusses the various meanings and interpretations of emergence, particularly those properties, behavioral characteristics, and evolutionary aspects that apply to the engineering of self-managing systems. We then look at current work in the field, first reflecting on the impact of various biological and other metaphors for emergence, and then considering how these metaphors impact on the design and composition of realized autonomic systems.

7.1.1 Facets of Emergence

There are almost as many definitions of emergence as there are formalisms that claim emergence for their own. De Wolf et al. [5] list no less than eight references with plausible definitions before giving in to temptation and providing one of their own. Corning [6] discusses numerous definitions in an attempt to explore whether his redefinition (based on a holistic approach to emergence) yields a more complete view of reality. One of Corning's examples is illustrative. If one considers the parts of a car all in a heap, we have a disordered system without purpose. However, if the parts are assembled as intended, a functional car emerges from the chaos. Thus a car — or the property of *car-ness* — is an emergent phenomenon arising from a particular disposition of its elements.

The extent to which the car argument is convincing may depend somewhat on your point of view. Anyone who has owned a car with a "personality" may well have sympathy with emergent car-ness. The static nature of car-ness may be troublesome, but how much of the finished car can we remove before it ceases to exhibit car-ness? Necessary and sufficient conditions for emergence are slippery. While we recognize that a "we know it when we see it" approach lacks formalism, we nevertheless take a pragmatic approach to the meaning of emergence — if the term "emergence" is helpful and descriptive, we say it is emergent. Thus we withstand the allure of providing yet another definition, but simply describe some of the key examples associated with emergence, bearing in mind that none of them alone are necessary or sufficient to rigorously label any system with those properties as designed for "emergence."

Emergent systems are characterized by the whole being more than the sum of the parts [7] (for example a market economy). Many things follow from this, not least that emergent systems, to greater or lesser extent, are composed [8]. This means we include certain elements and exclude others when considering emergence, which implies that the system as a whole and perhaps the parts have identity; are identifiable (we meaningfully refer to both the individual birds and the flock as identifiable). Most natural systems incorporate a concept of death or failure (cells die, enterprises go bust) and we are familiar with systems that allow evolution, through direct feedback and reconfiguration (such as the stable gaps between the asteroid belts caused by the orbit of Jupiter) or by Lamarkian/Darwin-Mendelian means.

As described in the previous section, the parts of composition may be within the system (particularly when we are considering artefacts) or provided by nature (the wind sensor example). In either case it is often the interaction of these parts that provides us with insights into engineering emergence. Micro-level interactions often lead to macro-level behavior with little or no apparent centralized governance, with interactions being local to agents (flocking). The loose coupling and unrestricted temporal dimension of these interactions often means emergent systems have a highly dynamic aspect (the Great Red Spot on Jupiter, the ever changing detail of an ant colony). Within the context of software systems these prerequisites imply some practical considerations; for example that agents must have a means of announcing or advertising their presence at start up [9] so that interaction may commence.

Emergent systems often demonstrate indeterminate behavior. That is to say, while there may be evident structure, the detail of that structure is unpredictable (although the nature of this unpredictability may be fundamentally intractable or merely practically intractable).

Memory in emergent systems is problematic. Interactions between agents, being local and small, are often stateless. Agents may preserve state between interactions and base decisions on stored knowledge, but each agent's experience is necessarily parochial (think of what the ant "knows"). Experience may be encoded in system as a whole (through evolution for example), but in this sense we must recognize that such learning is itself an emergent phenomenon and is separate from the system already utilizing emergence. We will show that this distinction is important.

Each of these characteristics has the capacity to be exploited for beneficial engineering effect. If many agents are homogenous (or have the capacity to appear so), are loosely coupled, and interactions are stateless, it becomes possible to substantially increase the robustness of an engineered system. Each agent can act as a replacement for the others, so they are fault tolerant in the event of agent failure, are maintainable because they can be hot-swapped and have built-in redundancy. Decentralization of control removes a single point of failure. Locality of interaction enables us to build scaleable systems, where adding agents does not necessarily compromise the system efficiency (because the number of interactions grows in a linear rather than exponential fashion with the number of actors).

The composite nature of emergent systems can allow conjunctions of interacting agents without the necessity of foresight. This is easiest to see in nature — an ant colony can react to the growth in power or locality of a neighboring colony, or with other emergent systems (forests, weather) and still maintain identity and function. Surviving emergent systems are inherently self-stable.

If randomness is incorporated into the design of our emergent system, we break symmetry — in particular the temporal coupling between components. If failure (death) is allowed, or evolution with or without death, for the first time we have a stable system that can learn and improve. Individual death can be invisible to clients (the ant colony survives the ant).

7.1.2 The State of the Art

It is ironic that a primal motive force behind the autonomic movement is to manage complexity, while much of the literature surrounding emergence focuses on how complexity arises from simple interactions [2]. Thus literature spawned from diverse fields that nonetheless relate to emergence share two common themes, influence from natural metaphors and the stability arising thereof.

Perhaps the seminal example is flocking (such as in [10]) where stability emerges from the micro-level interactions of many agents. The flock is able to change direction and even stay together when passing around obstacles without internal collision — a global behavior that arises from simple exchanges between neighbors. Thus a new individual arises (a flock) from a composition of agents (the birds), but the second-by-second density, shape and direction of this emergent individual is non-deterministic. The flock is, none-the-less, a self-healing entity (able to survive flowing around obstacles for example).

The other key example is ant colonies (perhaps half the papers in our bibliography mention ants somewhere — it would be uninteresting to list which). Here we have a considerably more goal-oriented system, where message passing remains minimal, local, and simple, but global state is complex and interdependent — reflecting the sophistication of the system, for example ant trails supporting efficient food gathering which in turn supports larvae nurture. While both metaphors have proved fertile ground for autonomic design [9], [11], the sophistication of outcomes in the second demonstrates there need be few boundaries in the complexity of what can be achieved with such systems. Camorlinga et al [12] represent simple local interactions as core to "Emergent Thinking."

The generality of inspiration has led to the development of a number of interest groups and some loosely related fields. Some groups, like the Santa Fe Institute [13], are multi-disciplinary, while others express the unconstrained nature of the phenomenon in a broadness of subject area output (such as Institute for the Study of Complex Systems [14], and the Emergent Phenomena Research Group [15]). More tightly focused on autonomic computing and resultant systems are groups like Ozalp Babaoglu's team [16]. Babaoglu's projects are motivated and inspired by nature, but their interest in emergence is, well, emergent.[1]

Babaoglu's approach [17][18] acts as an exemplar of the relative design simplicity of a system in which emergence is built-in. He describes a collection of peers, each peer running two infinite loops, one active, one passive. The active thread waits a random time, selects a peer, and sends its state to that peer, receiving the peer's state by return and updates its own state based on that returned state. The passive thread simply waits for messages, sends its state in response to any message, and then updates its state based on the new state received from the peer in the message. Using this simple, loosely coupled interaction Babaoglu is able to demonstrate agents building predefined structures and stable distributed systems. Devils remain in the detail — e.g. what is the nature of the update-state call? — but it should be noted that systems like this need not be "learning" or "evolving" systems, they simply evince stability directly in behaviour. We identify emergent *behavior* as **first order emergence**, while the ability of such systems to *evolve* is **second order emergence**.

While natural systems are the springboard for much research into emergent phenomena, the seminal example of emergence is arguably a human artefact. In *Wealth of Nations*, Adam Smith [19] demonstrates how emergent market forces (the *invisible hand*) balances supply and demand. When supply is low, sustained demand causes the price to rise; this encourages suppliers to supply more (to take advantage of the high price) until demand and supply are again in balance. If demand falls, the price falls, discouraging production; again matching low demand to low supply.

We believe markets provide an excellent metaphor when engineering emergent autonomic systems and economic systems have been influential in the work of Kephart [20][21] and Eymann [22] (among many others) as a model of interaction and as a means of injecting decision support [23]. Markets provide an example of complex and goal-oriented systems functionally dependent on a single variable (money). They also illustrate how such artefacts can use composition in the real world. Economic agents are flexible beasts — an individual can form a company and employ other agents (people and companies) to do their bidding, but from the point of view of their trading partners, still remain a consistent entity.

In the design of autonomic computer systems problems remain. A precept of utility functions is that given any two baskets of goods (or "bads") it is possible for agents to order them in terms of preference. In economic models the basis for the preferences are exogenous to the system — they are simply a given. In realized autonomic systems this void may be filled by establishing policy driven functions — but such functions will fall short of the dynamical aspects of a true market. As autonomic systems are composed, the interaction of policy and preference are themselves required to be emergent. Market failure, the stepped nature of demand for computing resources (the supply

---

[1] It is a non-trivial point that emergence begets further emergence.

curve is essentially horizontal for a computing node for a range of quantities, before becoming vertical at a certain level of use) and contrarian local minima (such as Giffen goods, goods that become more demanded as prices rise — typically goods that are essential, but cheap), mean that emergent equilibria are not always achievable.

Studies of coupling and composition of emergent systems are at present relatively nascent. Fisher et al. [8] [24] present a call to action, arguing that such "unbounded" systems are inevitable and that interoperability based on local self-interest may provide profound benefits. Fisher is attempting to develop a suite of autonomic algorithms to address interoperability. Sadjadi [25] explains how dynamic composition can be achieved using aspect-oriented programming techniques, but simpler engineering solutions will also do the trick, particularly if the problem space is constrained and trust is high. Fromm [26] examines the dynamic composition of agents, cautioning that unconstrained emergence can result in huge monolithic individuals, or splits into non-viable parts.

Butler [27] identifies that inserting autonomic elements into composed loosely-coupled systems, such as any given Service Oriented Architecture, will inevitably lead to emergent behavior (whether good or bad). Butler suggests that active simulation, engineered into systems at design stage, provides a means for interacting agents to plumb new configuration without causing system failure.

Bush et al. [28] suggest a list of protocols (which could be considered to be design patterns) arising from natural examples that might be susceptible to tooling. For example "Fire Fly" where random flashes are timed by progressive excitation in each fire fly, leading to an emergent synchronization of flashes, could be used for contention avoidance in networks. Bush's list of protocols is only five items long, but a little thought suggests many more patterns with less biological antecedence, e.g. Trust, where if I trust you, I trust the people you trust — which could be applicable to resource sharing applications, or Wikipedia [29], where knowledge and description is incrementally improved by many agents participating as writers and readers — which could be applicable to sensor networks.

It is only in the final example above where we have entered the territory of second order emergence; the realm of machine learning. Learning is a "nice to have" in an autonomic system, not a "must have." Many emergent natural systems do exhibit this phenomenon and the applicability of genetic algorithms, neural nets, and many other techniques are set out in many sources, including [30]. However all these evolving systems share the common facet — a concept of death or of failure (cells die, animals expire, firms go bankrupt). Incorporating failure into live systems is a troubling problem, although Butler [27] has suggested a plausible means by which this could be accomplished, by establishing a state (computer dreaming) which is a shared super-state of all other states of the system, but which is idempotent.

Self-healing, self-stabilizing systems have an inherently secure architecture (in terms of survivability), but Yurcik et al. [31] have pointed out that cluster interaction generated emergent security problems (such as local bandwidth spikes mirroring denial-of-service attacks). This suggests that once part of a system is composed of emergent autonomic behaviors, further (emergent/autonomic) engineering may be required to control the effect. At other levels (for example resource security), autonomic approaches may be over-engineering — and a simple layered architecture may suffice. Autonomic solutions are no silver bullet and are rarely the shortest distance between two points.

7.2 Implementation

This section discusses issues that arise during the design, implementation and testing of applications that purposefully *exploit* emergent behavior. Aspects of these systems are compared and contrasted with their counterparts in traditional distributed software systems. Section 7.2.1 discusses issues affecting stability and behavioral scope, and section 7.2.2 focuses on issues to be considered when harnessing and testing emergent behavior. The remaining sections present, in outline, two examples of the purposeful exploitation of emergence. Section 7.2.3 provides a demonstration of how, by exploiting emergence, a distributed systems core service can be realized to be simultaneously more robust, scalable, and lightweight than its traditional, internally deterministic counterparts. Section 7.2.4 provides an example of composition, in which a layered architecture is used as a framework to support several individually autonomous services, yielding a highly self-healing application platform.

7.2.1 Stability and Behavioral Scope

Traditional distributed systems are designed such that each component plays a specific role that contributes directly towards the global goals of the system. Nodes' autonomy is usually restricted within a controlling framework, such as the client-server or three-tier architectures, in which the interaction is fundamentally choreographed at design time. Although it is not known *when* a request will trigger an interaction, the *details* of how the interaction unfolds, once initiated, are built into the distributed application (notwithstanding that AI techniques may be embedded to select one of several possible paths). Such approaches are susceptible to problems arising from the interaction of parts; especially livelock and deadlock. Message loss also is problematic as many systems are built around individually-critical messages. To ensure safety and liveliness, much of the design effort is spent avoiding or resolving these problems. Reliable communication protocols trade time- and communication-complexity to achieve predictable, framed, interaction episodes. Further complexity is introduced by the need to ensure continuity of service despite node failures, with mechanisms ranging from simple primary-backup through to the more-scaleable election algorithm approach. Additionally, acceptable performance must be maintained despite load fluctuations, for example through load-balancing schemes.

In natural emergent systems the issues described above take on different meanings and materialize in different ways. The communication mechanisms are inherently unreliable; there may be no equivalent of an acknowledgement or sequence number, for example, so some messages will inevitably be lost. It does not make sense to try to think in terms of communication episodes or sessions, or of events happening sequentially, concurrently, or following any specific pattern. Emergent systems are truly un-ordered systems with no global precedence order of actions (because of the autonomy and asynchronicity of individuals).

Natural emergent systems lack pre-determined control and synchronization, although these behaviors may emerge. Rather, actors take independent actions based on incomplete knowledge learned from their recent interactions with neighbors. Actors have full local autonomy and work asynchronously with a localized view of the system, towards local goals. This autonomy precludes the occurrence of deadlock, but increases the potential for livelock, as actors are potentially free to choose any available path.

Actors have no understanding of global state or of global goals. For example, an ant that follows a pheromone trail to a food source has no knowledge of alternate food sources, or of their relative value or proximity; it simply "knows" that the trail leads to food. Eventually this condition no longer holds (the supply is exhausted); at this point the ants stop reinforcing the trail, the signal fades, and individuals wander until another trail is picked up. Thus the successful natural systems have tuned the achievement of local goals to yield globally acceptable, but not necessarily globally optimal, behavior.

Actors in emergent systems are loosely-coupled in terms of the influence exerted over their peers. Whereas in a distributed computing application there may be a direct control relationship, such as between a service coordinator and workers, in the emergent system actors can "influence" their peers by leaving pheromone messages or by other stigmergic method.[2] There is no direct feedback to indicate that other actors have noticed or acted on the messages left.

In (at-least-partially) synchronous systems, an event can trigger an update which may propagate through the system by means of one or more "rounds" of communication. At the end of such update, some semblance of known global state can be achieved, for example a routing algorithm that has converged. In emergent systems, there is never true convergence as such because, as the system is totally asynchronous, the "convergence" achieved is more

---

[2] Stigmergic communication is communication that uses the environment as its media. Stigmergy can be direct: for example trophallaxis (the exchange of food or liquid), or chemical contact (recognition of colony members by their odor); or indirect: in which one actor modifies the environment (such as by leaving a pheromone message or scent) and another actor "reads" the "message" at a later time.

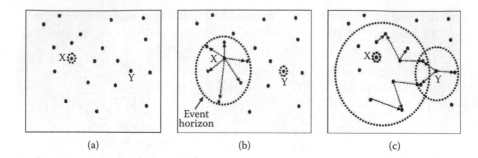

(a)   (b)   (c)

FIGURE 7.1
Three snapshots of information "waves" propagating through an ad-hoc neighborhood by means of neighbor interaction. A: An "event" is observed by actor X. B: Information propagates away from the source X, while another "event" is initiated at Y. C: A new "event" is initiated at X, while an information interference pattern arises where the wavefronts meet.

analogous to ripples on a pond where, momentarily there may be a group of nodes with the same information (on the wavefront). The natural emergent system is constantly striving to adapt to its ever-changing environment. Convergence could imply an interruption of this process. "States" may propagate through an "eco-system" from a local epicenter and there is naturally some lag. However, there can be *many* disturbances (sources of ripples) and thus the information surface of the emergent system is highly dynamic and complex. Figure 7.1 captures the essence of this type of information dissemination.

In reality the "wave" propagates at an uneven rate in various directions (as the actors move) and thus the wavefront is unlikely to be as regular as that shown. The event horizons are non-synchronous because of variable information lag. The "interference" pattern is only of consequence if the events are related — for example if one event indicates that more actors should assume a certain role (such as *gather food*, in the ant colony scenario), and the other event is a contra-indication such as *there is now enough food* — then localized instabilities occur.

The definitions of *stable behavior* and *desirable behavior* are dependent on the specific nature of the system under consideration. Many natural systems depend on randomness to provide an asymmetric aspect to actor's local views of the system. This helps to ensure that actors will make slightly different decisions and/or that there will be variable temporal separation between their actions. The extent of this non-uniformity can be tuned such that systems are self-stabilizing. Identical local state, in contrast, can lead to erratic behavior as all actors have the same reaction to a given external stimulus. This can result in over-compensation and thus oscillation, which destabilizes a system and consumes resources unnecessarily. However, there are some natural systems in which it *is* desirable for all actors to respond to stimuli near-synchronously, such as the synchronization of fire-fly lights. In this case it is emulated synchronous behavior that emerges from an asynchronous system.

7.2.2 Issues in Harnessing and Testing Emergent Behavior

Software developers have historically been trained to think deterministically and this is reflected in the designs of systems that are in common use today. The randomness that occurs in distributed computer systems, evidenced in variation and unpredictability in network delay, message loss, traffic and load patterns, random event timing, etc., is seen as undesirable. The deterministic design paradigm can lead to the elimination of the effects of randomness, at the expense of extra design, test, and run-time costs. For example, several group communication protocols have been developed to ensure partial ordering, e.g. [32], or even total ordering of messages, e.g. [33]. A globally consistent view of message delivery order is ensured at the cost of extra message and time complexity, which impacts on scalability. The additional complexity is justified by the nature of the applications deployed, such as in financial systems in which real-world events are causally related and this relationship must be preserved in the system's transaction ordering. However, it is also important to consider alternative computational models that may be applicable to certain problem domains without such rigid imposition of order (and the implied costs), for example where the causality is weak, or artificial. In particular, this approach could apply to internal behavior in systems and services which still exhibit externally-deterministic behavior, as we show in a later section. When implementing software which targets the exploitation of emergence, a different mindset is required.

Natural emergent systems evolve over very-many generations with fitness pressures acting at each step. There is no "design" as such; not even a specific end-design goal, so it is not possible to predict the outcome. Just surviving in the current environment is enough. Random mutations lead to variants and, over several generations, the more-fit become dominant. It is important to note that the environment in which the natural systems live is itself changing and thus the fitness functions are non-constant. Due to this, it is sometimes acceptable for oscillatory behavior in the design, for example: *grow longer, leaner legs; grow shorter, stronger legs; grow longer, leaner legs...* It is also important to realize that the mutations are random (rather than targeted design inputs) and thus the system is unlikely to ever reach a truly optimum design. The best that can be hoped for is that the system becomes increasingly fit for its purpose, or at least keeps up with the changes in its environment. Sometimes a local minimum attracts the system, requiring a significant evolutionary step to move the system towards the global optimum. This effect has been described as punctuated equilibrium [34], and could be observed in an engineered system as stable, but non-optimized behavior.

The concept of "designing" an emergent system is somewhat a contradiction of terms. However, it is possible to "steer" evolution towards certain outcome characteristics. For example, we can plant a garden to a certain design, making macro-level decisions, *"a tree here, a shrub there ...,"* but we cannot determine exactly how things will develop. The various plants interact, competing for resources such as sunlight and nutrients from the soil, so one plant may prosper at the cost of another. We can re-balance the system towards our intended design, *"a bit of pruning here, some high-grow feed there. . . ."* But no matter how many such iterations we go through, and no matter how attentive we are, we cannot control any aspect at the micro level, and thus we cannot completely control its shape at the macro level. This view of "steered" evolution provides a metaphor for a possible future intelligent development facility that will "evolve" systems; the evolution mediated by macro-level developer influence. *Evolutionary* emergence is somewhat analogous to software design-time issues. In section 7.1.2 we have identified evolutionary emergence as a second order form of emergence.

However, there is another distinct mode of expression of emergence: *emergent behavior* arises from the local autonomy of actors and their random, non-mediated interactions — we refer to this as first order emergence. This could be described as a run-time quality. With current technology it is possible to build a system that exhibits emergent behavior, in which the autonomy of actors leads to complex interaction behavior. For example, by equipping nodes with a discovery capability, they may form associations with other nodes based on pre-programmed criteria. A number of mechanisms can be employed in complementary ways to govern the direction of emergent behavior and the rate of adaptation to a disturbance. Stable, efficient, and *desirable* behavior may emerge if the nodes are supplied with suitably flexible, yet scoped rules of engagement, feedback mechanism(s), and some sources of randomness.

Environmental factors such as random event timing, random interaction pairings of nodes, random message loss, etc., might provide sufficiently diverse local views of the current system state. Sometimes it may be necessary to introduce further randomness to ensure stable behavior. For example, to disrupt accidentally introduced synchronous behavior, such as when several nodes near-simultaneously receive a message[3] which triggers them to act in a particular way; timers with random timeout values can be used to disperse the responses over time. There may also be application-dependent reasons for a short time interval before reacting, for example to ensure that a message has had sufficient time to traverse the system. In this specific case, from the authors' empirical experience, it is often better to have a relatively small random time-component added to a larger fixed element so that nodes' behavior is sufficiently, but not excessively, diverse. This approach is successfully applied in the emergent election algorithm [35][36] which is discussed in a later section.

Randomness can play a role beyond breaking the symmetry of local views. For example, from the combination of random time-delays prior to actions, and the observance of actors in one's neighborhood (which is itself randomly constructed), node-roles can be assigned without the need for explicit

---

[3] This problem can arise in software implementations of emergent behavior because network protocol features such as local broadcast are not a true reflection of the actual communication found in natural systems.

negotiation. This is used in the emergent election algorithm in both of its stages of operation (see later) and is directly mimicking one-way pheromone communication in the sense that an actor decides what to do based on its local determination of what its peers are doing (or not doing).

The loose-coupling of actors implies that feedback will almost always be indirect; this is yet another departure from traditional distributed systems with their acknowledgements, handshakes, and call-backs. Emergent systems use a form of open-loop feedback, in which either the environment or the macro-level behavior of the system provides an indirect link between actors. Actors influence one another anonymously, over time. Consider the indirect interaction of stock traders, mediated only by the markets. Patterns and trends in the market influence the decisions of traders (each interpreting the signals slightly differently, and importantly, *not knowing* how the other traders have interpreted the information). A trader's local view (confidence in her interpretation of market trends, as well as her expectations of the other traders' short-term behavior) influences her next move. The combined effect of thousands of traders moves the market.

The rate of movement of an emergent system towards its goal state can be influenced by adjusting the balance between positive and negative feedback. Depending on implementation, positive feedback is used to encourage continued movement in a certain direction (i.e. *keep doing the same thing*), or to discourage movement away from a desirable state. Negative feedback can be used to cause a change in direction. A system that "adapts" too abruptly is unlikely to be stable (see Figure 7.2-a), some damping is required. It is better to gear the system to continually: move a small step in the (perceived) right direction, settle, and re-assess (see Figure 7.2-b). The balance between the influences of randomness and of the feedback mechanisms is the crux, because desirable outcomes by chance are improbable, as they usually represent a very small proportion of the possible outcomes for a system.

The validation and testing of systems exhibiting or exploiting emergent behavior requires special considerations, due fundamentally to their vast state-space. In many cases the problem is exacerbated by high, and possibly

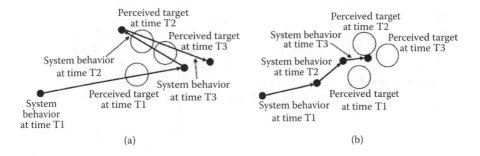

FIGURE 7.2

Two very different possible behaviors of a system. The un-damped behavior (a) is erratic. The damped behavior (b) is more stable and predictable. Both scenarios are shown with the same starting point and the same (moving) perceived target behaviors.

non-linear, sensitivity to environmental conditions. For almost all emergent systems, mapping out a system's possible output behavior in terms of trees of states stretching back to initialization conditions would be an intractable problem in NP. So it is not realistic to validate this by examining a mapping of inputs to outputs in the traditional sense, although it may be possible to do so for very small examples of systems with restricted input vectors to gain an indication of certain behavior "hot-spots." For example we can create a stigmergic communication mechanism in which agents leave "food" for each other, or simple pheromones each having a specific meaning. We can tune this system and see how the behavior unfolds. However, due to the emergent nature of the behaviors it is not possible to observe every possible behavior at the system level, and even if we examine millions of simulations, or empirical traces we cannot guarantee to capture all possible patterns of behavior. So we must look for trends and localized attractor behavior rather than singularities (which leave no scope for the fluctuation that is an intrinsic quality of emergent behavior). Deterministic systems are tested against clearly defined expected outcomes and tests usually have a Boolean result. The emergent system does not lend itself to such scrutiny; its behavior may never be *exactly* the same on any two occasions. Envelopes of acceptable behavior can be described by a set of parameters. We can then devise tests to see how well the system can maintain itself within the permitted envelope, and the extent of the perturbations, along various dimensions, needed to destabilize the system (away from the envelope). On this basis we can establish *confidence factors*, which indicate our level of confidence that the *trends* in behavior will be acceptable.

7.2.3 The Emergent Election Algorithm

The design of the algorithm was inspired by the simple communications mechanisms and strategies of natural systems, especially insect colonies such as harvester ants. A key high-level goal was to achieve similarly-high scalability, robustness, and stability as exhibited by such systems; which are very impressive when compared to traditional distributed software solutions.

The communication mechanism mimics insect colonies' stigmergic communication, using a small pheromone vocabulary. The communication strategy adopted is to use low-value messages, which do not need to be protected by mechanisms such as acknowledgements, sequence numbers, timeouts and retransmissions. In such a system, an "important" message is transmitted at intervals while it remains relevant. These messages are informational but not instructional: i.e. they are restricted to advertising the sender's local knowledge. A receiver locally interprets messages in the context of its own current state. The information is conveyed in the *type* of message, and/or *frequency* of receipt. These messages do not carry *content*, and are thus very short (in this application each message contains only a single character indicating message type). This communication strategy is a major contribution to the efficiency of the algorithm, which in turn contributes to its scalability.

Environmental randomness is embraced and additional randomness has been built in to some of the timing mechanisms. Random message loss and randomness in message delay contribute to the stability of the algorithm by helping to break the symmetry; giving rise to different local system views. Messages are individually expendable; their loss is naturally accommodated as part of the algorithm's normal operation. Nodes' asynchronicity contributes to stability by introducing randomness into the relative timing of events. In some parts of the operation it has been necessary to introduce further randomness in the form of random components added to timeout values to further differentiate the behavior of nodes. For example, several nodes in close-proximity will receive a broadcasted message within a short time of one another. This can introduce accidental near-synchronicity if these nodes then take some immediate action (which can be destabilizing and incur extra corrective work). To avoid this, a timeout period with a random component is used to disperse the actions of the nodes over time. The relative size of the random value range as a ratio of the total timeout value range provides a means of tuning the algorithm in terms of stability versus responsiveness.

Election algorithms are a generic mechanism for reliable or replicated services in which a single coordinator must be maintained, and quickly replaced when it fails. In this sense the algorithm has to have externally-deterministic behavior. However there is no such restriction on the internal behavior. A two-stage operation is used. The first stage incorporates non-deterministic emergent behavior to achieve the non-functional requirements of efficiency, robustness, etc. The second stage ensures that the functional-requirements are met in an externally-deterministic manner. A node can occupy one of four states, which are shown, together with the possible transitions between the states, in Figure 7.3.

The first stage involves all but the master node, separating them into a small, active pool of slaves and a larger passive group (idle). Traditional election algorithms tend to make all nodes "active" in the sense that they monitor the presence of the master and join elections if the master fails. In this algorithm, only the slaves take on these roles. The majority of nodes adopt the "idle" state which means that they do not participate in elections, and they do not transmit messages of any kind. They monitor the slave population, however. The movement of nodes between the slave and idle states is governed by

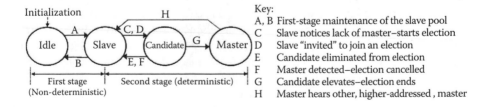

FIGURE 7.3
State-transition diagram, differentiating between the two stages of operation.

carefully balanced positive and negative feedback. Slaves transmit periodic informational messages. By counting these messages, slave and idle nodes can locally estimate the size of the slave pool. To ensure some variation in the local views, and in the reaction-time of nodes, the slave transmissions are counted during a period which has a locally-random time-component. At the end of this time-period, each node compares its estimate against upper and lower thresholds representing the *desired* range of slave-population sizes (the thresholds are separated by a dead-zone to help avoid oscillation). If an idle node locally determines that there are too few slaves it will become a slave. If a slave locally determines that there are too many slaves it will become idle. These are negative feedback examples where the current locally-perceived state differs from the desired state. Idle nodes that determine that the slave pool is sufficiently large, and slave nodes that determine that the slave pool is not too large, take no action. These are examples of positive feedback reinforcing the status quo. The size of the slave-pool is an *emergent feature* and is independent of the size of the system. Stable maintenance of the slave pool (with very low levels of fluctuation) is achieved as a result of the asynchronicity of nodes' actions. Refer to the left hand part of Figure 7.3.

The second stage involves only the small active subset of the nodes. This permits deterministic behavior (accompanied by higher communication and interaction intensity) without negating the overall scalability and robustness achieved in the dominant first stage. The second stage ensures that there is exactly one master. There are two modes of operation. In normal mode (a master exists), the master node periodically transmits a status message. Slaves monitor these messages.

Operation switches to election mode when at least one slave notices the lack of a master. If no messages are received over a time interval (longer than three-times the master transmission period, and containing a local random component) a slave starts an election by elevating to candidate status and sending a candidate message. Other slaves then join the election unless they have received a candidate message containing a higher ID than their own ID. Similarly, candidates demote to slave state if they subsequently receive a higher-addressed candidate message. After sending its candidate message, a candidate waits a short period to allow other nodes to react, and if at the end of this period it has not been eliminated from the election, it elevates to master state and immediately sends its first master-message. Upon receipt of a master-message, all slave and candidate nodes abandon any ongoing election and reset their timers. A master node that receives a master-message from a higher-addressed node demotes immediately, without negotiation. The various mechanisms work together to ensure the stable and reliable convergence of the system to exactly one master node. The second stage of the algorithm is illustrated in the right-hand part of Figure 7.3.

As a result of its design, which purposefully employs emergence and environmental characteristics including randomness, the algorithm has some highly desirable characteristics. The functional goal of an election algorithm — the maintenance of a single master — is achieved under a wide-range of conditions that would defeat deterministic algorithms simply because of the rate of "critical message" loss and/or node failure. The algorithm overcomes quite extreme levels of message loss and node failure because it is designed around the precept that everything is expendable on an individual basis. The design circumvents the typical conflicts between non-functional goals in traditional distributed software, for example robustness is often achieved at the expense of time or message complexity. Thus the typical non-functional requirements for distributed software are met better than in typical deterministic approaches. In particular it is:

1. Scalable. The key contributing factor to the scalability of the algorithm is the way in which the idle-slave interaction (in the first stage of the algorithm) uses pheromone-inspired communication. Low-value, very short, asynchronous, unacknowledged messages are used. The communication consists of one-way messages that inform other nodes of the sender's state (there is no conversation, negotiation, or instruction). The algorithm is also *lightweight* in the sense that no state is stored or exchanged to identify nodes for the purpose of the idle-slave interaction; this further contributes to scalability.

2. Self-stabilizing. The way in which the algorithm balances randomness, and positive and negative feedback, has been tuned such that the algorithm is self-stabilizing over a very wide range of conditions. Only severely abnormal conditions, such as a per-message message-loss-probability of greater than 0.2 (which caused behavioral violation during one part in a thousand of the run time), or per-node mean-time-between-failure (MTBF) values of less than 5 hours (causing behavioral violation during two parts in a thousand of the run time), were able to disturb the system away from its stable envelope.

3. Robust. The self-stabilization characteristic of the algorithm contributes significantly to its high robustness. When operating within realistic node failure probability and message loss probability conditions, the single-master condition is upheld and there is very little fluctuation in the size of the slave-pool. The algorithm functions reliably in large systems as well as in systems containing a single node. The algorithm handles changes in population transparently (nodes can join or depart at any point; the other nodes adjust their states accordingly). There are no individually critical messages, so that the correctness of the algorithm is not impacted by the non-recovered loss of any single message. Nodes are heterogeneous with respect to the algorithm roles so the availability of redundant "spares" increases linearly with the system scale.

4. Satisfies liveness. By our definition, a system must have at least one node (the limit case) and this node will be in one of the states:

idle, slave, candidate, or master. If no master exists then there *must* be at least one node in one of the other states. If this node is in the candidate state an election is underway. A slave that detects the absence of a master (it does not receive transmissions from the master during a limited period containing a random component) either starts an election or joins an ongoing election. In a similar way, an idle node detects the absence of slave nodes and will elevate to slave status. Each of these steps are individually time-limited, so whatever the starting condition, a master node will be elected within a bounded time. The algorithm design is such that additional nodes contribute to the effectiveness of the limit case behavior and do not impede it. For example (due to the random time component), when there are multiple slaves the first slave-timeout is probabilistically earlier.

5. Satisfies safety. Elections have several features to prevent the occurrence of multiple masters (described earlier).

An extended explanation of the algorithm, including performance analysis, is provided in [36]. A summary of the performance results, against three key criteria, is presented here: 1. Sensitivity to message loss. The per-message message-loss probability is found to have a direct impact on the number of state transitions because as the number of lost messages increases, the number of idle nodes erroneously determining that the slave-pool has diminished, rises. The algorithm remains completely stable, with no false elections up to 10% message loss. At message loss levels of 15% and above there is an increasing trend in the occurrence of multiple masters. However, even at extreme message-loss levels of 20% multiple masters only occur for 0.064% of the time in 200-node simulations, and 0.104% of the time (i.e. 1 part in 1000) in 400-node simulations. 2. Sensitivity to node failure. It is found that a reduction in per-node MTBF causes an increase in the number of state transitions and a reduction in the mean slave-pool size. A majority proportion of node failures concern idle nodes, as these are most numerous, having no effect on the number of state-transitions or mean slave-pool size. Failure of a slave temporarily reduces the slave-pool size and generally causes at least one idle-to-slave transition. Failure of a master causes at least one slave-to-candidate transition and one candidate-to-master transition, as well as the resulting idle-to-slave transition to maintain the slave-pool. Over a very wide range of node-failure probabilities, the algorithm maintained the mean size of the slave-pool between the upper and lower threshold values. The number of elections increases as the per-node MTBF is reduced. The algorithm is stable so that elections only occur when the system is initialized and when a master node fails. Each time a master node fails there is a short period, the duration of an election, in which no master exists. With per-node MTBF as low as 300 minutes a master existed 99.88% of the time in a 200-node system, and 99.81% of the time in a 400-node system. 3. The scalability of the algorithm was confirmed by comparing 200-node and 400-node simulations. The behavior is essentially unchanged even under artificially harsh conditions of message loss or node failure.

7.2.4 Layered Support for Higher-Level Composition of Emergence

Using the analogy of self-healing at many layers within biological systems, we envisage layered systems in which emergence is employed to effect self-healing and self-optimization in several service-containing layers which work cooperatively to support composition of higher functionalities, yet also retain autonomy at the service level.

As a vehicle for the investigation of composition of emergent behaviors, we have developed a generic layered architecture in which each layer may contain several services and each node may be a member of, or contribute to, many services. The architecture facilitates two-dimensional autonomous behavior, see Figure 7.4. Vertical autonomy exists between the independently functioning layers (and services within) while horizontal autonomy is achieved between nodes which remain autonomous, even though they interact with other nodes through common service membership (each node having dynamically selected roles in possibly several services in any, or all of the layers). This contributes to very high system robustness and flexibility with respect to dynamically changing node populations. Each service is individually self-managing. Services may interact and cooperate, but are not dependent upon one another. This contributes to very high stability, as, even if a localized instability arises in a particular service, it is not propagated across services. The architecture allows very flexible composition, and thus maps onto a wide variety of problem domains that require simultaneously high scalability, robustness, stability, and efficiency; while exhibiting autonomic or emergent properties.

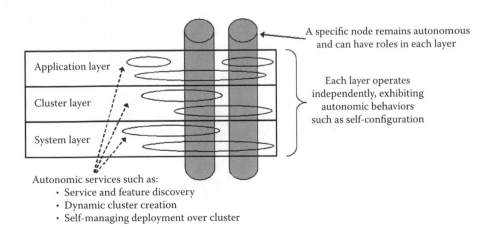

FIGURE 7.4
The generic layered framework permits both horizontal and vertical autonomy.

Each layer of the architecture can contain many autonomous services which individually exhibit emergent self-managing behavior and can be used to compose higher-level systems directed by application-level requirements. The architecture re-uses and extends both our philosophical approach to the engineering of emergent systems and our actual mechanisms, as discussed in the previous section.

Services deployed in the system layer potentially comprise all the nodes in system. The emergent election algorithm, described earlier, is used to maintain a system-wide coordinator. The election algorithm guarantees that this coordinator will be replaced within a few seconds if it fails. This provides a central pillar on which the overall system's reliability is built. Should a partitioning occur, a coordinator will emerge in each partition. The coordinator is only used to initiate other services, so it has a low and infrequent workload and is very unlikely to become a bottleneck. Once other services have been initiated they are self-managing and are not tracked by the system-layer coordinator. Services have different node membership depending on: the nature of the service, system size, workload levels, and the requirements of clients. Some services might be self-healing with respect to depletion of members, by having mechanisms to advertise for new members or to discover available nodes that can operate as workers.

The cluster layer contains and supports services associated with the flexible maintenance, coordination, and operation of generic clusters. A cluster is a subset of the nodes in the system, dynamically assembled to perform or support a specific application-layer task. The cluster layer can be configured in many ways, depending on the services implemented in the system layer, and application requirements. For example: cluster existence may be dynamic or permanent, and they may be isolated or overlapping. The association of nodes in forming a cluster may be based on physical adjacency, logical relationship, or a random selection from the nodes available.

Cluster creation is instigated by services in the application layer, which make requests to the system-layer coordinator, specifying the desired cluster size. The system layer is responsible for initially building the clusters. This comprises determining node availability, deciding cluster topology (such as the extent of overlap), assigning nodes to clusters, and informing nodes of their neighbors (logical or physical) based on common cluster membership.

The system is dynamic in terms of node existence and their membership of clusters and services, so it is not desirable or feasible for the coordinator to keep track of available nodes when building clusters. For clusters that require a specific physical topology, a specific service to map nodes onto cluster membership must be deployed in the system layer. For clusters whose membership is based on some node-local quality (such as low load, or some specific specialized capability — perhaps the ownership of a particular type of sensor), as well as for clusters simply composed of a random collection of available nodes, the *delayed-bid mechanism* [38] is employed (described below). This introduces emergence into the cluster creation process: an unpredictable subset of *suitable*[4] nodes are assigned to the cluster in an efficient and low-latency fashion enhancing stability and scalability by significantly reducing the number of messages required.

On receiving a request to establish a cluster, the system coordinator issues an *InviteMembershipBids(k)* broadcast inviting nodes to join cluster k (membership criteria can be included in the message). The cluster ID (k) is chosen by the system coordinator. When creating relatively small clusters in large systems, it is possible that a much larger number of nodes will send replies to the system coordinator than are needed to form the cluster. To resolve this, a short locally-random time interval is introduced before nodes reply with their unicast *MembershipBid(k)*. This spreads out, in time, the arrival of bids at the system coordinator. The coordinator accepts the required number of bids by sending unicast *AcceptBid(k, role)* messages. The *role* parameter indicates the node's initial role within the newly formed cluster (usually the first node accepted initially coordinates the new cluster). Once the appropriate number of bids have been received, the system coordinator broadcasts a *StopBids(k)* message, which cancels any outstanding unsent (because of the random delay) membership bids.

To illustrate the effectiveness of the mechanism, consider a system of 100 nodes in which 80 are currently available (determined locally) to join a new cluster. Without the delayed bid mechanism, an invitation message to create a cluster of 10 nodes would receive 80 replies. The coordinator would then send 10 accept messages. The total number of messages to create the cluster would be 91. However, with the delayed bid mechanism in place, the *StopBids* transmission will prevent up to 70 of the bid messages, for a cost of one additional broadcast. The random delay used is between 0 and 500ms. Thus the worst-case effect of the mechanism is to add 500ms to the cluster-creation latency. Probabilistically, for large systems, enough messages will arrive at the coordinator to create typical-sized clusters before the majority of messages have been sent, allowing the cancellation message to be highly effective and limiting the effect that the delay mechanism has on the latency of cluster creation. Table 7.1 shows the relationship between system size and the actual numbers of messages required to build clusters of several sizes, based on experimentation with the simulation model. The delayed-bid mechanism has the potential to massively reduce communication complexity, and consequently the time complexity, of cluster creation activities. Further details have been reported in [38].

Once created, clusters operate independently to each other and are self-managing and self-healing; each cluster maintaining its own coordinator (reusing the election algorithm). The requirement for flexibility is that the cluster layer must be able to support as-yet-unknown configurations, requested by

---

[4] Nodes locally determine their suitability to join a particular cluster when they receive the invitation to join.

TABLE 7.1

Message Reduction Achieved by Using the Delayed-Bid Mechanism

		System Size (number of nodes)			
		100	200	300	400
Cluster size (mean number of nodes)	10	78%	87%	89%	90%
	50	37%	57%	68%	75%
	100	—	48%	53%	55%

as-yet-unknown applications. The requirement of robustness and self-healing is that the cluster must maintain its size despite node departures, to meet its service contract with the application layer service it serves, thus it must accept new members appropriately.

The application layer can contain applications and their support services which deploy tasks onto clusters, or use the emergent qualities of the cluster layer to provide fault-tolerance and self-organization. The exact behavior of services in this layer is dependent on the application domain. Some applications (such as replication management) will interact more closely with the cluster, perhaps mapping some application-level functionality such as replication coordination onto the cluster-level coordinator node, while others (such as parallel processing) will submit a task onto the cluster without interest in the current cluster-level roles of nodes within the cluster (e.g. the higher level application is not interested in which node is the current coordinator of the cluster).

The configuration of clusters is application-domain dependent. For example, deployment of parallel tasks over clusters requires dynamic, logical, isolated clusters whose member nodes are selected because they have some spare processing capacity. However, applications that are sensitive to nodes' locations, such as the mapping of sensor-monitoring services; some wireless networking applications including routing; and cell-based problems (such as telephony-network base-station management, vehicle tracking networks, and emergency-services resource-finding networks) need to be deployed onto clusters in which the cluster membership is based on node's "physical" adjacency or location. In this way, the clusters not only provide a computational resource, they can also provide a dynamically assembled communication framework. Location-based clusters need to be static in terms of existence (but not necessarily in terms of membership), and for robustness and communication flexibility clusters may have to be overlapping.

The architecture has been validated through the investigation of composition in a cell-based application domain, Air Traffic Control. The application can be described as a grid of cells, each having fixed (geographically) neighbors. To ensure a very high degree of fault tolerance it is required that sufficient redundant backup capability exists to ensure management coverage of all cells despite potentially high node failure rates. To support this type of application, the cluster layer can be configured as a group of overlapping, independent clusters, each with their own coordinator. The cluster coordination activity is carried out by a single elected master node, per cluster. An emergent election algorithm is used to ensure the continued existence of a master node until cluster exhaustion (all member nodes have failed). The extent of cluster overlap (in this investigation a Moore-neighborhood provided eight neighbors grouped around a central node, each neighbor being a member of the cluster based on the central node; but also each neighbor is the base for its own cluster) has an impact on the extent of self-organization and self-healing exhibited under various conditions of node failures and recoveries. See [11] for further details of the layered architecture and the exemplar application.

Conclusion

The level of interest in emergence has reached critical mass in terms of the concentration of scientists and engineers working to understand its nature and to build systems that exploit emergent behavior. Emergence is now a research discipline in its own right. Interest in the computer science community strongly associates emergence with the concept of autonomic systems and self-* behaviors.

There do remain several issues that must be addressed before emergent engineering becomes a mainstream practice. There is a "deterministic mindset" dominant in the wider software developer community, due to the (several-decades-old) regime of rigid control and fixed logic, which is evidenced in systems architectures, software architectures, development methodologies, and the need for strict adherence to input:output mapped testing. It is not surprising that non-determinism is treated with suspicion is some quarters.

Policy-based computing has longer-standing roots and so has penetrated further into the software development arena than other tools and techniques associated with autonomics. However, there are few, if any, genuinely self-adapting policy-based systems in production software. Almost all policy systems are fundamentally static or involve open-loop adaptation in which inefficiencies of conflicts are identified automatically but the solutions require human mediation.

There is a lack of design methods and implementation tools specifically to aid the modeling and development of emergent and/or autonomic systems. However, these are now starting to appear (see for example, MIT's StarLogo [39], and IBM's Policy Management for Autonomic Computing (PMAC) [40] and it is anticipated, at the current rate of effort, a rich set of tools will be available in the next year or two.

There is a need for a deeper understanding of issues concerning stability, safety, and scoped behaviour, and the extent to which the actions of the system, including its adaptation, need to be, or should be, steered or mediated, by human intervention (a certain paradox arises if we strive to build systems that are equipped with the self-* behaviors, but we then insist on having the last word). The extent to which the behaviors that emerge can be usefully employed — and trusted — is dependent on our ability to design systems that can exhibit adaptive, yet stable behavior. Embracing randomness in the environment, and developing balanced feedback mechanisms, are key precursor requirements.

With respect to the appropriateness of employing emergence, application domains fall into three fundamental categories: Firstly there are certain application domains to which emergent design naturally lends itself, or where emergent behavior is inevitable. This category includes swarming of mobile CE devices (MANETS) built upon now-standard technologies such as Bluetooth communication. Other examples include network management, novel computation, games and entertainment, and modelling to enable better understanding of complex systems such as oceanographic, atmospheric, and weather systems. Some nascent technologies (such as nanotechnology) readily lend themselves to emergent engineering. Even in this "primary" category, there remains the issue as to the extent that emergence is purposefully exploited in a useful, positive way (i.e. designed in) as opposed to merely being an interesting phenomenon.

The second category represents those application domains for which, at least externally, the behavior must be completely predictable. There will always be many applications with deterministic end-goals, such is the nature of society needs. Financial systems, stock management systems, and many others will need to provide absolute non-disputable balances and checks, audits and so on. However, this does not mean to say that every aspect of the behavior of these systems must be deterministic. We have demonstrated (in the design of the emergent election algorithm) how non-deterministic behavior can be usefully employed in "externally" deterministic systems, providing benefits in terms of the achievement of the non-functional goals without disruption to the functional goals. For example, when considering inventory records it may not be necessary to know the exact details of current state in minutia, it may be far more productive to understand the *trends* over time ("can we more or less fill this order — how expensive is the margin for the bit we are uncertain of — here's the resultant price").

The third category represents those application domains for which emergence is not appropriate, or simply unnecessary. This includes those situations where simple, or established solutions are adequate, or where completely deterministic behavior is non-negotiable, for example in the control of a nuclear power station.

In this chapter we have identified that that emergent phenomena fall into first order (behavioral) types and second order (evolutionary or learning) types. Both can be useful in developing systems. In our second implementation example we described a layered architecture, purposefully designed to support composition. Composition is a powerful way in which higher-level emergence arises. This is harnessed in the natural systems and the opportunities that composed emergence offers for software systems are only beginning to be understood. It is intuitively clear that the effects of composition cannot always be predicted, or steered such that improved outcomes can be guaranteed; there is scope for unanticipated negative behavior. Just as evolution rejects regressive mutations, emergent systems need the capability to detect and reject regressive adaptations.

The natural systems provide many examples of emergence, and inspire us with their beauty and perceived simplicity. A favourite question (as yet unanswered) is "how do you measure the intelligence of a shoal of fish?" There is probably no way to actually compare that "intelligence" with the human "intelligence" or the "intelligence" represented by deterministic system design; there is probably no common scale of measurement. However, we are intrigued by how such emergent synchronization and responsiveness (consider the "bait ball" defensive action) could be utilized to build better systems for, perhaps, network traffic management. Similarly, could slime-mold behavior (with its amoeboid movement) lead to the design of an advanced self-organizing data-repository, where different "species" of data congregate together and migrate around the system according to some locality-of-reference requirement?

Present autonomic systems support same-level transitions where the system attempts to tune its own behavior within a scoped behavioral envelope (often this is at least partially externally mediated, for example policy updates). Current autonomic systems do not have governance over their own goals, which are built into the manager components — the adaptation is restricted to behavior. MetaSystem Transitions (MST) describe the emergence of a new control level, through the composition of lower-level subsystems. MSTs were proposed by Turchin [41] to explain a *"quantum of evolution."* MSTs are key evolutionary steps by which new control levels (new levels of sophistication) arise in types of system as diverse as biological, societal, and computer technology. For example, a mechanical calculator has its mathematical programs (functions) "hardwired;" in contrast, Von Neumann's idea of treating the program as data represents an MST: the program controls the operations executed by the hardware [42]. MST could emerge spontaneously (as composition) from *suitably sophisticated* and complex autonomic computing environments (perhaps combining techniques such as self-adaptive policies, dynamic utility functions, and stigmergic communication mechanisms). This will be a turning point from which the evolution of autonomic systems really takes off.

Emergence is now recognized by national and international research funding bodies as a key area for investigation. In many research centers, both academic as well as industrial, there are currently on-going and planned research in autonomic computing, emergence, and related fields such as nanotechnology, which is advancing at a blistering pace. This is a field that directly complements developments in emergence because the miniaturization and dramatic cost reductions per unit facilitate the implementation of composed systems of literally thousands or millions of locally-intelligent artefacts. In the same way that very few hardware artificial neural networks have ever been built because it is far cheaper to use software systems to emulate them, so too most applied "emergence" is derived from software simulations, or systems of hardware nodes working at modest scale. However, nanotechnology offers the possibility of realized hardware systems with millions of independent agents. This opens up a whole new arena of possibilities for the exploitation of emergence.

References

1. Lloyd Morgan, *Emergent Evolution: The Gifford Lectures*, Henry Holt and Company, New York, 1923.
2. Stephen Wolfram, *A New Kind of Science*, Wolfram Media Incorporated, 2002.
3. Richard Anthony, Alun Butler, Mohamed T. Ibrahim, *Layered Autonomic Systems*, 2nd International Conference on Autonomic Computing (ICAC), 2005, IEEE.
4. Steven Strogatz, *Sync — The Emerging Science of Spontaneous Order*, Penguin Books, 2003.
5. T. De Wolf, T. Holvoet, and Y. Berbers, *Emergence as a paradigm to engineer distributed autonomic software*, Department of Computer Science, K.U. Leuven, Report CW 380, Leuven, Belgium, March 2004.
6. P. Corning, *The Re-emergence of "Emergence": A Venerable Concept in Search of a Theory*, Complexity: 7(6): 18–30 (2002).
7. John H. Holland, *Emergence — From Chaos to Order*, Perseus Books 1999.
8. David Fisher and Dennis Smith, *Emergent Issues in Interoperability*, Eye On Integration. Number 3, 2004.
9. Jim Dowling, Raymond Cunningham, Anthony Harrington, Eoin Curran, Vinny Cahill, *Emergent Consensus in Decentralised Systems Using Collaborative Reinforcement Learning*, Self-star Properties in Complex Information Systems, 2005, pp. 63–80.
10. Lee Spector, Jon Klein, Chris Perry, Mark Feinstein, *Emergence of Collective Behavior in Evolving Populations of Flying Agents*, Genetic Programming and Evolvable Machines, 6(1): 111–125 (2005).
11. Richard Anthony, Alun Butler, Mohammad Ibrahim, Composing Emergent Behaviour in a Fault Intolerant System, Workshop on Self-Adaptive and Autonomic Computing Systems (SAACS), DEXA 2005, Copenhagen, Denmark, August 2005, pp.145–149, IEEE.
12. Sergio Camorlinga and Ken Barker. *The Emergent Thinker*. International Workshop on Self-* Properties in Complex Information Systems, Bertinoro, Italy, May 31 – June 2, 2004.
13. SFI, http://www.santafe.edu/index.php
14. Institute for the Study of Complex Systems, http://www.complexsystems.org/
15. Emergent Phenomena Research Group, http://emergent.brynmawr.edu/eprg/?page=EmergentPhenomena
16. Ozalp Babaoglu Home Page, http://www.cs.unibo.it/~babaoglu/
17. Ozalp Babaoglu, *Grassroots Approach to Autonomic Computing*, keynote address at ICAC 2005, available from http://www.caip.rutgers.edu/~parashar/icac2005/presentations/ICAC05%20Babaoglu%20Keynote.pdf
18. Özalp Babaoglu, Hein Meling, and Alberto Montresor, *Anthill: A Framework for the Development of Agent-Based Peer-to-Peer Systems*. ICDCS, 2002, pp. 15–22.

19. Smith, Adam, *An Inquiry into the Nature and Causes of the Wealth of Nations*, 1776, available from http://www.gutenberg.org/etext/3300
20. Jeff Kephart, Software Agents and the Information Economy — Agent and Emergent Phenomena, presentation available from http://www.ima.umn.edu/talks/workshops/11-3-6.2003/kephart/kephart.ppt (downloaded 2005).
21. William E. Walsh, Gerald Tesauro, Jeffrey O. Kephart, Rajarshi Das: Utility Functions in Autonomic Systems. ICAC, 2004, pp. 70–77.
22. Torsten Eymann, Michael Reinicke, Oscar Ardaiz, Pau Artigas, Felix Freitag, and Leandro Navarro, *Self-Organizing Resource Allocation for Autonomic Networks*, DEXA Workshops, 2003, pp. 656–660.
23. Giorgos Cheliotis and Chris Kenyon. *Autonomic Economics*, IEEE International Conference on E-Commerce Technology, p. 120, 2003.
24. D.A. Fisher and H.F. Lipson, *Emergent Algorithms — A New Method for Enhancing Survivability in Unbounded Systems*, Proceedings of 32nd Annual Hawaii International Conference on System Sciences (HICSS-32), Maui, HI, Jan. 5-8, 1999. Los Alamitos, CA: IEEE CS Press, 1999.
25. S. Masoud Sadjadi, Philip K. McKinley, and Betty H. C. Cheng, *Transparent shaping of existing software to support pervasive and autonomic computing*, DEAS '05: Proceedings of the 2005 workshop on design and evolution of autonomic application software, ACM Press, 2005.
26. Jochen Fromm, *The Emergence of Complexity*, Kassel University Press, 2004.
27. Alun Butler, Mohamed T. Ibrahim, Keith Rennolls, Kassel Berma, and Liz Bacon, *On The Persistence of Computer Dreams — An Application Framework for Robust Adaptive Deployment*. DEXA Workshops 2004, pp. 716–720, IEEE.
28. Steven F. Bush and Amit B. Kulkarni, *Engineering Emergent Protocols*, Whitepaper available from http://www.research.ge.com/~bushsf/4D-Emerg Protocol.html
29. *Emergence*, Wikipedia article, available from http://en.wikipedia.org/wiki/Emergence, downloaded September 2005.
30. Moshe Sipper, *Machine Nature*, McGraw-Hill, New York, 2002.
31. William Yurcik, Gregory A. Koenig, Xin Meng, and Joseph Greenseid, *Cluster Security as a Unique Problem with Emergent Properties: Issues and Techniques*, 5th LCI International Conference on Linux Clusters, May 2004.
32. Kojiro Taguchi, Kenichi Watanabe, Tomoya Enokido, Makoto Takizawa, Causally ordered delivery in a hierarchical group of peer processes, *Computer Communications*, 28 (11), Elsevier, 2005.
33. Ge-Ming Chiu, Chih-Ming Hsiao, Wen-Ray Chang, Total ordering group communication protocol based on coordinating sequencers for multiple overlapping groups, *Journal of Parallel and Distributed Computing*, 65 (4), Elsevier, 2005.
34. S. J. Gould and N. Eldredge, Punctuated equilibria: the tempo and mode of evolution reconsidered. *Paleobiology* 3, 1977, pp. 115–151.
35. R. Anthony, Emergence: a Paradigm for Robust and Scalable Distributed Applications, First International Conference on Autonomic Computing (ICAC), New York, USA, May 2004, pp. 132–139, IEEE.
36. R. Anthony, An Autonomic Election Algorithm based on Emergence in Natural Systems, *Journal of Integrated Computer-Aided Engineering*, 12 (2005), IOS Press, Amsterdam, 2005, pp. 1–20.
37. S. Johnson, *Emergence: The Connected Lives of Ants, Brains, Cities and Software*, Penguin Press, London, 2001, pp. 73–100.

38. R. Anthony, Emergent Cluster Configuration, 8th World Multiconference on Systemics, Cybernetics and Informatics (SCI 2004), Vol (II), Orlando, USA, July 18–21, 2004, pp. 116–121, International Institute of Informatics and Systemics (IIIS).

39. http://education.mit.edu/starlogo/

40. R. Lotlikar, R. Vatsavai, M. Mohania, and S. Chakravarthy, Policy Schedule Advisor for Performance Management, *Proceedings of the 2nd International Conference on Autonomic Computing (ICAC)*, IEEE, Seattle, 2005, pp. 183–192.

41. V. Turchin, *The Phenomenon of Science*, Columbia University Press, New York, 1977

42. http://pcp.vub.ac.be/MST.html

8

A Control-Based Approach to Autonomic Performance Management in Computing Systems

Sherif Abdelwahed and Nagarajan Kandasamy

CONTENTS

This chapter describes a model-based control and optimization framework to design autonomic or self-managing computing systems that continually optimize their performance in response to changing workload demands and operating conditions. The performance management problems of interest are posed as one of sequential optimization under uncertainty, and a lookahead control approach is used to optimize the forecast system behavior over a limited prediction horizon. The basic control concepts are then extended to tackle distributed computing systems where multiple controllers must interact with each other to ensure overall performance goals. Two case studies, dealing with processor power management and distributed signal classification, demonstrate the applicability of the control framework.

8.1   Introduction

Distributed computing systems host information technology applications vital to commerce, transportation, industrial process control and military command and control, among others, and typically comprise numerous software and hardware components that must together satisfy stringent quality-of-service (QoS) requirements while operating in highly dynamic environments. For example, the workload to be processed may be time vary-ing, and hardware and software components may fail during system oper-ation. Therefore, to achieve the desired QoS, multiple performance-related parameters must be tuned dynamically and correctly to the prevailing oper-ating conditions. Moreover, as these systems grow in scale and complexity, maintaining the specified QoS via manual tuning will become increasingly hard.

To cope with the complexity expected of future computing systems, it is highly desirable for such systems to manage themselves, given only high-level objectives by administrators. Such *autonomic computing systems* aim to achieve QoS objectives by adaptively tuning key operating parameters with minimal human intervention [17].

Advanced optimization, control, and mathematical programming tech-niques offer a formal framework to design performance management tech-niques for autonomic computing systems. If the system of interest is correctly modeled and the effects of its operating environment accurately estimated, control algorithms can be developed to achieve the desired QoS objectives. The key advantages of using such a framework instead of ad hoc heuristics are: (1) one can systematically pose various performance control problems of in-terest within the same basic framework, and (2) the feasibility of the proposed algorithms with respect to the QoS goals may be verified prior to deployment (here, *feasibility* is defined as the controller's ability to move the system to the desired operating region and maintain it there under dynamic conditions).

Recent research efforts, in both academia and industry, have focused on using control theory as the basis for enabling self-managing behavior in com-puting systems [19, 18]. Control-theoretic concepts have been successfully applied to selected resource management problems such as task scheduling [25, 12], bandwidth allocation and QoS adaptation in web servers [4], load bal-ancing and throughput regulation in email and file servers [29, 24], network flow control [27, 33], and power management [14, 31].

The above methods all use *classical feedback* or *reactive control* to observe the current application state and take corrective action, if necessary, to achieve the specified QoS. In [19], the number of users accessing a Lotus Notes workgroup server during any given time interval is regulated using a feedback controller to achieve a desired response time. Processor load is used as feedback in [4] to guarantee desired response times for HTTP requests arriving at a web server. To avoid overloading the processor, some requests are rejected or their QoS degraded appropriately. Control theory has also been used to sched-ule a dynamic workload on processors under deadline constraints [25, 12]. Maintaining a smooth flow of messages in computer networks without con-gestion can also be formulated in control-theoretic terms. For example, [27] describes mechanisms aimed at preventing senders from overloading both the receivers and the network with messages, while [33] identifies those aspects of congestion avoidance that can be formulated as feedback-control problems.

Recently, the power consumed by high-performance processors has become an important design constraint. To tackle this problem, a number of modern processors allow their operating frequency and supply voltage to be scaled, and control-theoretic methods have been proposed to address the trade-off between performance and power consumption [14, 31, 30].

Feedback control, however, has some inherent limitations. It usually assumes a linear and discrete-time model for system dynamics with an unconstrained state space, and a continuous input and output domain. Recently, we have developed a control and optimization framework to enable self-managing behavior in applications exhibiting hybrid behavior comprising both discrete-event and time-based dynamics [6]. The problems of interest have the following key characteristics.

- *Uncertain operating environments.* The workload to be processed is usually time varying, and system components may fail during system operation.

- *Multivariable optimization in the discrete domain.* Control or tuning options must be chosen from a finite set at any given time. The cost function specifying application QoS requirements can include multiple variables and must typically be optimized under explicit and dynamic operating constraints.

- *Control actions with dead times.* Actions such as dynamically (de)allocating and provisioning computing resources often incur a substantial dead time (the delay between a control input and the corresponding response), requiring *proactive control* where control inputs must be provided in anticipation of future changes in operating conditions.

A performance manager must, therefore, tackle complex combinatorial problems such as selecting operating modes for the application, and allocation and provisioning of hardware and software resources, which may need re-solving with observed environmental events such as time-varying workload patterns and failures of individual components. Also, since the underlying control set is finite, traditional optimal control techniques cannot be applied directly to such systems, and in most cases, a closed expression for a feedback control map cannot be established.

This chapter presents a model-based control framework, wherein the performance management problems of interest are posed as those of sequential and discrete optimization under uncertainty, and solved using a *limited look-ahead control (LLC)* approach. At each time step, the control actions governing system operation are obtained by optimizing its forecast behavior, described by a mathematical model, for the specified QoS criteria over a limited prediction horizon. The LLC framework allows for multiple QoS objectives and system operating constraints to be represented explicitly in the optimization problem and solved for each time step. It can, therefore, be used as a control technique for computing systems with both simple and nonlinear dynamics, as well as those with dead times. It can also accommodate changes to the system model itself, caused by resource failures and/or parameter changes in time-varying systems.

The LLC concept is adopted from model predictive control (MPC) [26], used to solve optimal control problems for which classical feedback solutions are extremely hard or impossible to obtain. The MPC approach is widely used in the process control industry, where a limited time forecast of process behavior is optimized as per given performance criteria. Also related to the framework is the lookahead supervision of discrete event systems [13] where, after each input and output occurrence, the next control action is chosen after exploring a search tree comprising future states over a limited horizon.

As specific applications of the control framework, we present two case studies. First, assuming a processor capable of dynamic voltage scaling and operating under a time-varying workload, we design a controller to address the trade-offs between the corresponding QoS requirements and power consumption. The second case study extends the control approach to distributed applications. We develop a hierarchical controller for distributed signal classification, where incoming digital signals are processed to extract certain features from them in real time. Here, the controller addresses the trade-off between detection accuracy and responsiveness under a time-varying signal arrival rate.

8.2 Online Control Concepts

We now discuss the basic concepts underlying the LLC framework, including the system model, estimation techniques, and the formulation of the control problem.

System Model. We use a *switching hybrid system* model [3] to capture the operating dynamics of an application when only a finite number of control options are available to affect its behavior at any given time instant. The following discrete-time state-space equation describes the dynamics of such systems.

$$x(k + 1) = f(x(k), u(k), \omega(k)), \tag{8.1}$$

where $x(k) \in \mathbb{R}^n$ is the system state at time step k, and $u(k) \in U \subset \mathbb{R}^m$ and $\omega(k) \in \mathbb{R}^r$ denote the set of control inputs and environment parameters at time k, respectively.

The system (or component) model f captures the relationship between the operating parameters, particularly those relevant to QoS, and the control inputs that adjust these parameters. Even if the system is hard to model explicitly, we can still capture its behavior using standard system identification techniques where the parametric form of the model is assumed known, and the corresponding parameters are estimated using the input/output relationships [23]. The model can also be obtained via supervised learning where the system is first simulated for various environmental inputs [10]. The simulation results are then used to train an approximation architecture corresponding to f; for example, a neural network or regression tree.

Estimation. Since the controller explores a set of future states within the lookahead horizon, environmental inputs to the controller within this horizon must be estimated. Though environmental inputs to the computing system are uncontrollable, they can be predicted online. For example, a typical workload in e-commerce and web applications [8, 7] exhibits time-varying cyclical variations that can be predicted using well-known forecasting techniques such as the Box-Jenkins ARIMA modeling approach [11]. A general forecasting model for the environmental input ω has the form $\hat{\omega}(k) = \varphi(\underline{\omega}(k-1, n), a(k))$, where $\underline{\omega}(k-1, n)$ is the set of n previously observed environment vectors $\{\omega(k-1), \ldots, \omega(k-n-1)\}$, $\hat{\omega}(k)$ denotes the estimated value, and $a(k) \in \mathbb{R}^p$ denotes the parameters relevant to the estimation method. The estimator may update $a(k)$ at each time step to minimize the forecasting error $\|\hat{\omega}(k) - \omega(k)\|$.

Since $\omega(k)$ cannot be measured until the next sampling instant, the subsequent system state cannot be computed precisely but only estimated as

$$\hat{x}(k+1) = f(x(k), u(k), \hat{\omega}(k)) = f(x(k), u(k), \varphi(\underline{\omega}(k-1), a(k))). \quad (8.2)$$

Performance Specifications. Computing systems must achieve specific QoS objectives while satisfying certain operating constraints. A basic control action in such systems is *set-point regulation* where key operating parameters must be maintained at a specified level or follow a certain trajectory; for example, a desired HTTP-request response time in web servers. The controller, therefore, aims to drive the system to within a close neighborhood of the desired operating state $x^* \in X$ in finite time and maintain the system there.

The computing system must also operate within strict constraints on both the system variables and control inputs. These constraints can be expressed as a feasible domain for the composite space of a set of system variables, possibly including control inputs themselves. We consider a general form of operating constraints described by $H(x) \leq 0$ and $U(x) \subseteq U$, where $U(x)$ denotes the permissible input set when in state x, and the inequality $H(x) \leq 0$ defines the system state space X.

It is also possible to consider transient costs as part of the system operating requirements, indicating that certain trajectories toward the desired state are preferable to others in terms of their cost or utility to the system. Then, the overall performance measure can be given by a function $J(x(k), u(k), \Delta u(k))$, where $x(k)$ is the current state, and $u(k)$ and $\Delta u(k) = u(k) - u(k-1)$ denote the control inputs and the corresponding change in these inputs, respectively. We use the following norm-based function to define the overall cost:

$$J(x(k), u(k), \Delta u(k)) = \alpha_1 \cdot \|x(k) - x^*\| + \alpha_2 \cdot \|u(k)\| + \alpha_3 \cdot \|\Delta u(k)\|, \quad (8.3)$$

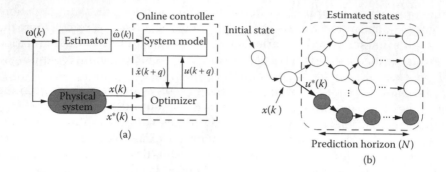

FIGURE 8.1
(a) The online controller architecture and (b) the LLC approach where the shaded states lie along the optimal trajectory within the prediction horizon.

where α_1, α_2, and α_3 are user-defined weights denoting the relative importance of the variables in the cost function.

Control Problem. Figure 8.1(a) shows the overall controller architecture. The LLC approach constructs a set of future states from the current state $x(k)$ up to a prediction horizon N, as shown in Figure 8.1(b). A trajectory $\{u^*(q)|q \in [k+1, k+N]\}$ that minimizes the cumulative cost while satisfying both state and input constraints is selected within this horizon. The first input leading to this trajectory is chosen as the next control action. The process is repeated at time $k+1$ when the new system state $x(k+1)$ is available. The problem formulation in Figure 8.2 accounts for future changes in the desired system trajectory and measurable disturbances (e.g., environment inputs) and includes them as an intrinsic part of the overall control design.

Computational Issues. In a switching hybrid system where control inputs must be chosen from a set of discrete values, the optimization problem in Figure 8.2 will show an exponential increase in worst-case complexity with an increasing number of control options and longer prediction horizons. Since the execution time available for the controller is often limited by hard bounds, it is necessary to consider the possibility that we may have to deal with suboptimal solutions. For control purposes, however, it is not critical

$$\text{Compute: } \min_{U} \sum_{q=k}^{k+N} J(x(q), u(q))$$
$$\text{Subject to: } \hat{\omega}(q) = \phi(\omega(q-1), a(q)),$$
$$H(f(x(q), u(q), \hat{\omega}(q))) \leq 0,$$
$$u(q) \in U(x(q))$$

FIGURE 8.2
Formulation of the limited lookahead optimization problem.

to find the global optimum to ensure system stability—a feasible suboptimal solution will suffice. We would still like to use the available time exploring the most promising solutions leading to optimality.

The specific methods used to solve the lookahead optimization problem are application dependent. For example, in a computing system with few control options, the controller can simply generate a tree of all possible future states from the current state $x(k)$ up to a specified depth N (also assumed to be small) and then select the trajectory minimizing the cost function [21]. Other advanced search methods have been proposed for hybrid systems having a richer control-input set [28]: (1) The search space can be pruned using branch and bound strategies; (2) We can approximate the control-input domain as continuous and use linear or quadratic programming to quickly find the best solution, which is then mapped to the closest discrete value; (3) Taking advantage of the fact that system dynamics do not change drastically over a short period of time, we can obtain (suboptimal) solutions using local search methods, where, given the current state $x(k)$, the controller searches a bounded neighborhood of $x(k)$ for a solution.

Performance Characterization. As noted earlier, adaptive behavior in an application, if enforced by a formal control and optimization framework, can be characterized and analyzed in terms of its optimality, stability, and feasibility. In this regard, the LLC problem presents a number of interesting challenges. The proposed method belongs to a wider class of online algorithms [1, 15] where not all problem parameters are known *a priori*, and in general, the optimality of such methods under dynamic operating conditions cannot be shown. Given the uncertain operating environment, control decisions cannot be shown to be optimal, since the controller does not have perfect knowledge of future environmental parameters, and control decisions are made after exploring only a limited prediction horizon. Similarly, there is no general technique to analyze the convergence (or stability) properties of a constrained optimization problem such as the one in Figure 8.2, since the presence of constraints typically leads to nonlinear controller behavior [26]. Therefore, in practice, only nominal stability is shown (e.g., by ignoring all constraints, assuming perfect future knowledge, etc.), for such controllers and their real-world performance is tuned by experimentation.

In a practical sense, one must, therefore, focus on the feasibility of LLC algorithms in terms of their ability to achieve system objectives, perhaps in suboptimal fashion. A *feasible* operating region is defined as a bounded region around the set-point that the controller can drive the system into, from any initial state in a finite number of steps, and maintain it there. In [34], the feasibility of a special case of LLC—a one-step lookahead policy—is formulated as a joint containability and attraction problem. A procedure based on nonlinear programming is used to compute a *containable* region, a bounded neighborhood around the set-point x^* such that any trajectory inside it cannot move out under the given control policy. Then, we decide if this region is reachable in a finite number of steps from an external operating state.

Determining the feasibility of a general N-step lookahead controller is a topic of ongoing research by the authors.

8.3   Case Study: Processor Power Management

This section uses the concepts introduced in Section 8.2 to develop an online controller managing the power consumed by a processor operating under a time-varying web workload. Assuming a processor with multiple operating frequencies, an online controller is developed to achieve a specified response time for these requests while minimizing the operating frequency. (Note that power consumption relates quadratically to the supply voltage, which can be reduced at lower frequencies.) Furthermore, many processors, such as the StrongArm SA-1100 [16], Pentium M [20], and AMD-K6-2+ [5], support a limited number of operating frequencies—eleven, six, and seven, respectively. This case study is summarized from our previous work in [21].

Processor Dynamics. We use a queuing model to capture the dynamics of a processor P where $\lambda(k)$ denotes the arrival rate of incoming requests and $q(k)$ the queue size at time k. Each P operates within a limited set of frequencies U. Therefore, if the time required to process a request while operating at the maximum frequency u_{max} is c, then the corresponding processing time while operating at some frequency $u(k) \in U$ is $c/\phi(k)$ where $\phi(k) = u(k)/u_{max}$ is the scaling factor. The average response time achieved by P during time step k is denoted by $r(k)$; this includes both the waiting time in the queue and the processing time on P. We use the model proposed in [32] to estimate the average power consumed by P while operating at $u(k)$ as $\psi(k) = \phi^2(k)$. The following equations describe the dynamics of a processor.

$$\hat{q}(k+1) = q(k) + \left(\hat{\lambda}(k) - \frac{\phi(k)}{\hat{c}(k)} \right) \cdot T \tag{8.4}$$

$$\hat{r}(k+1) = (1 + \hat{q}(k)) \cdot \frac{\hat{c}(k)}{\phi(k)} \tag{8.5}$$

$$\hat{\psi}(k+1) = \phi^2(k). \tag{8.6}$$

Given the observed queue length $q(k)$, the estimated length $\hat{q}(k+1)$ for time $k+1$ is obtained using the predicted workload arrival and processing times. The average response times of requests arriving during the time interval $[k, k+1]$ is estimated as $\hat{r}(k+1)$, and the corresponding power consumption estimate is $\hat{\psi}(k+1)$. Note that T denotes the sampling duration of the controller.

We use an ARIMA model [11] implemented via a Kalman filter to provide the environmental estimate $\hat{\omega}(k) = (\hat{\lambda}(k), \hat{c}(k))$ to the controller. The average processing time per request is estimated using an exponentially weighted moving-average filter as $\hat{c}(k) = \beta \cdot c(k) + (1 - \beta) \cdot \hat{c}(k - 1)$, where β is a smoothing constant.

Control Problem. The cost $J(x(k), u(k))$ corresponding to a system state $x(k)$ is determined by the response time $r(k)$ and the corresponding power consumption $\psi(k)$. Soft constraints are added to the cost function using *slack variables*, defined such that they are nonzero only if the corresponding constraints are violated. Their nonzero values may be heavily penalized in the cost function. Therefore, the controller has a strong incentive to keep them at zero if possible. If r^* is the desired average response time, we define $\epsilon(k)$ as

$$\epsilon(k) = \begin{cases} 0 & : \quad r(k) \leq r^* \\ (r(k) - r^*)/r^* & : \quad \text{otherwise.} \end{cases}$$

The cost function is therefore given by

$$J(x(k), u(k)) = \alpha_1 \cdot \|\epsilon(k)\| + \alpha_2 \cdot \|\psi(k)\| + \alpha_3 \cdot \|\Delta u(k)\|$$

and the power management problem is posed as

$$\min_{U} \sum_{j=k+1}^{k+N} J(x(j), u(j)) \tag{8.7}$$

$$\text{subject to} \quad \hat{x}(j+1) = f(x(j), u(j), \hat{\omega}(j)). \tag{8.8}$$

The controller exhaustively evaluates all possible operating states within the prediction horizon N to determine the best input u^* to apply at time k. The number of explored states is given by $\sum_{i=1}^{N} |U|^i$, where $|U|$ denotes the number of available operating frequencies. However, when both N and $|U|$ are small, as is the case with the processor of interest, the computational overhead is negligible.

Performance Evaluation. The performance of the LLC scheme is now evaluated using a representative e-commerce workload. We simulated a busy server processing the workload shown in Figure 8.3(a), comprising HTTP requests made to a computer at an Internet Service Provider over a week [8]. A processor with possible operating frequencies of 532, 665, 798, 1197, and 1529 MHz is assumed.

The distribution of individual requests within the arrival sequence was determined using two important characteristics of most web workloads: popularity and temporal locality. We generated a virtual store comprising 10,000 objects, and the time needed to process an object request was randomly chosen from a uniform distribution between $(10, 25)$ milliseconds. Simulated requests to the store had the following characteristics:

- *Popularity*: It has been widely observed that some files are more popular than others and that the popularity distribution commonly follows Zipf's law [8]. (A few files are extremely popular, while many others are very rarely requested.) Therefore, we partitioned the virtual store in two—a "popular" set with 1000 objects receiving 90% of all requests, and a "rare" set containing the remaining objects in the store receiving only 10% of requests.

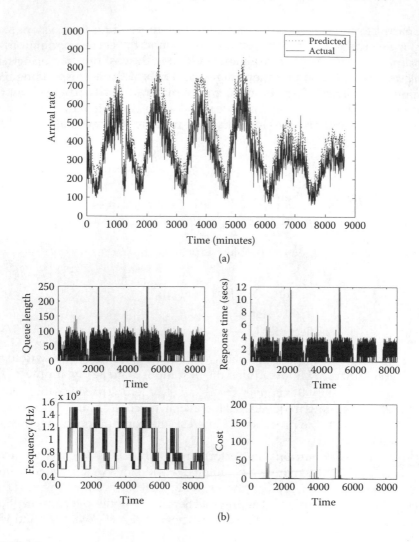

FIGURE 8.3
(a) Workload arrivals and the corresponding prediction, and (b) key operating parameters of the controller for a prediction horizon of $N = 2$.

- *Temporal locality*: This is the likelihood that once an object is requested, it will be requested again in the near future. In many web
  workloads, temporal locality follows a lognormal distribution [9].

Figure 8.3(a) shows both the actual and estimated arrival rates for a prediction horizon of $N = 2$. The arrival-rate predictions are obtained using a
constant-velocity Kalman filter. Parameters of this filter such as the process,
noise, and covariance matrices are first tuned using an initial portion of the

workload, and then used to forecast the remainder of the load during controller execution. When the workload is noisy, the observed dynamic error between predicted and actual values is used to obtain conservative estimates for $\hat{\lambda}$ by providing some "spare capacity" to the original predictions. Predictions covering 99% of the noise in the original workload are shown in Figure 8.3(a).

Request processing times are estimated using an exponentially weighted moving-average filter with a smoothing constant of $\beta = 0.10$. The performance of the controller is evaluated over the entire portion of the workload in Figure 8.3(a), where the requests received by the computer during a week are plotted in one-minute intervals. This interval is sufficient to smooth the variabilities in arrival rates and adequately track these using the prediction model. Therefore, the sampling period of the controller was set to $T = 30$ seconds. Also, the overhead due to controller execution as well as the system dead time (the delay between changing the operating frequency and its completion) is negligible and therefore ignored in the experiments. The response time to be achieved by the controller was set to $r^* = 4$ sec. The weights were set to $\alpha_1 = 100$ and $\alpha_2 = 1$ in the cost function to heavily penalize the controller if a chosen frequency did not satisfy the above response time. Also, we set $\Delta u(k) = 0$ to ignore the negligible penalty incurred when switching between different frequencies.

Figure 8.3(b) summarizes the performance of the controller for a prediction horizon $N = 2$. The figure shows the queue length at various time instants and how the controller changes the processor frequency to accommodate the time-varying workload. Note that the controller does not achieve the desired response time during very brief time periods, since it cannot predict sudden and short-term spikes in the arrival rate. However, the overall performance of the controller is very good, and a short prediction horizon of depth two appears appropriate for this case study.

8.4 Hierarchical Control

We now consider the problem of managing the QoS requirements of a distributed computing system comprising several processing components, each specifying a desired operating region. A feasible control scheme must, therefore, effectively coordinate interactions between the various components to ensure system-wide QoS. To tractably solve performance management problems in such a distributed setting, the overall optimization problem can be functionally decomposed and mapped to a hierarchical control structure where high-level controllers manage interactions between lower-level ones to ensure QoS goals. Each low-level controller is responsible for optimizing the performance of a specific component (or subsystem) it controls under the operating constraints imposed by the higher levels. Such hierarchical structures occur naturally in many key distributed computing systems; for example, end-to-end QoS management in multi-tier server clusters for e-commerce applications and power-aware load balancing in computing clusters.

To more clearly illustrate the key concepts, we will restrict this discussion to a simple two-level hierarchical structure where controller C_g manages interactions between n lower-level controllers using forecast operating and environmental parameters to achieve the desired QoS goals for the overall system. Each local controller $LC_i, i \in [1, n]$ is responsible for managing the component i under its control within the operating constraints imposed by C_g while contributing toward this QoS goal. The discrete-time system dynamics equation for C_g is given as

$$y(k_g + 1) = g(y(k_g), v(k_g), \mu(k_g)), \qquad (8.9)$$

where the global state vector $y_g(k_g)$ is given as $\Psi(\underline{x}_1(k_l, m), \ldots, \underline{x}_n(k_l, m))$, where Ψ is the abstraction map, $k_g = m \cdot k_l$, and $\underline{x}_i(k_l, m) = \{x_i(k_l - m + 1), \ldots, x_i(k_l)\}$ is the set of local states for the ith component. Similarly, we can define the global environment inputs, $\mu(k_g)$, as seen by the global controller at time k_g as an aggregation of the local environment inputs $\omega_j(k_l)$, over the global time frame, namely, $\mu(k_g) = [\underline{\omega}_1(k_l, m), \ldots, \underline{\omega}_n(k_l, m)]$. The global input $v(k_g) \in V$ and V is a set of discrete inputs representing the available control settings for each LC_i.

The *abstraction map* Ψ approximates the composite behavior of the local components under C_g's control in response to the environment inputs $\mu(k_g)$ and operational settings $v(k_g)$. Since an analytical model capturing the complex dynamics of a component with its controller in the loop is typically hard to obtain, Ψ is initially learned in offline fashion by simulating the local controller using various values from the input set V and a quantized approximation of the domain of μ. The simulation results are then used to obtain an approximation architecture for g which can be adjusted online using observed system behavior.

The objective of C_g is to minimize a global cost function $J_g(y, v)$, posed as a set-point specification. We can, then, directly apply the control concepts discussed in 8.2 to optimize $J_g(y, v)$. Also, since C_g makes decisions using the aggregate behavior of local controllers, it typically operates on a longer time frame compared with the local ones. Based on the assumption that global specifications are of higher priority than local ones, the decisions made by C_g are communicated to the local controllers, which then aim to optimize the performance of the components under their control while ensuring that conditions imposed by C_g are not violated. The feasibility of the hierarchical control structure can be checked as the conjunction of the feasibilities of the global and local controllers.

We now apply the abstract concepts discussed in this section to enforce self-managing behavior in a distributed signal detection application, a case study summarized from [2].

8.5   Case Study: Signal Detection System

Figure 8.4 shows the distributed signal detection application comprising multiple components processing digital signals to extract features such as human voice and speech from them. Incoming signals are stored in a global buffer and distributed to individual detectors where they are locally queued. Each detector then examines a chunk of these signals to identify various features of interest, including the signal type (e.g., PSK, FSK, MSK), baud rate, carrier frequency, and the number of phase-shift keying (PSK) levels. Since detection accuracy improves with chunk size at the cost of increased computational complexity, the control scheme must address the trade-off between accuracy and responsiveness to optimize overall system performance.

We assume that the local controllers on individual detectors can operate in the following qualitative modes: (1) *low-load*, used when the arrival rate is low and large signal chunks may therefore be processed; (2) *medium-load*, which applies to medium arrival rates; and (3) *high-load*, used when signals

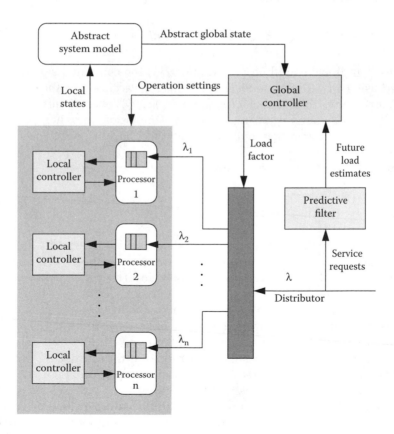

FIGURE 8.4
Architecture of the signal detection system.

arrive at a high rate and processing must be done in small chunks. Based on the overall signal arrival rate, the global controller C_g selects the appropriate operating mode for each local controller and the fraction of incoming signals to distribute to a local detector to satisfy the global QoS goals.

Global Controller Design. The global controller in Figure 8.4 receives forecasts for the signal arrival rate, and the average queue size and detection accuracy from each component over the past sampling time period. It then uses this information to distribute a fraction of the new arrivals to each detector and set its operating mode. New signals are distributed to the ith component using $\underline{v} = \{ v_i \in \Omega \mid i \in [1, n]\}$, where Ω is a finite set of positive reals in $[0, 1]$ and $\sum_i v_i = 1$. The next operating mode is given by $\underline{m} = \{m_i \in M \mid i \in [1, n]\}$, where M is the set of possible modes. For the ith component, its average processing rate and accuracy under the current operating mode depend on the incoming signal arrival rate and are specified by the functions $p_i(m_i, \lambda_i)$ and $y_i(m_i, \lambda_i)$, respectively.

The global controller decides on an appropriate control action using the average estimated queue size and detection accuracy of the system components. The estimated queue size during time $k_g + 1$ is given by

$$\hat{q}(k_g + 1) = q(k_g) + \left(\hat{\lambda}(k_g) - \sum_{i=1}^{n} p_i(m_i(k_g), v_i(k_g) \cdot \hat{\lambda}(k_g)) \right) \cdot T_g$$

where T_g is global sampling period and $\hat{\lambda}(k_g)$ the estimated signal rate obtained using an ARIMA model [11]. The average accuracy of the overall system $\hat{\bar{y}}(k_g)$ is simply the sum of $\bar{y}_i(m_i(k_g), v_i(k_g) \cdot \hat{\lambda}(k_g))$ over all the components. Initially, the functions p_i, y_i are learned via simulation, and their parameters are updated during system operation using feedback from the local components.

The controller C_g aims to maximize the cost function given by

$$\sum_{i=k_g+1}^{k_g+N_g} \alpha_1 \, [\hat{\bar{q}}(i)]^2 + \alpha_2 \, [\hat{\bar{y}}(i)]^2, \tag{8.10}$$

where N_g is the lookahead depth and the weights α_1 and α_2 specify the relative importance of system responsiveness and accuracy, respectively.

Local Controller Design. At each time step, local detectors remove a chunk of enqueued signals, store it in a temporary buffer, and process this buffer to extract features. If this chunk is sufficient to correctly classify the signal, then those signals are removed from the queue. Otherwise, more signals from the queue are added to existing ones in the temporary buffer and detection is reattempted. A typical end result of this process is the estimated symbol rate, the computation time required, and a confidence measure. Figure 8.5 shows the structure of a signal detector and its controller.

The controller shown in Figure 8.5 adjusts the feature level, determined by the chunk size, in a closed loop to maximize the utility of the local detector. It operates at a sampling rate T_l where $T_g = m \cdot T_l$ and $m > 1$ is a positive integer. The ith detector in Figure 8.4 receives signals from the distributor

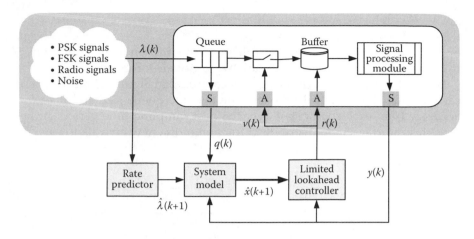

FIGURE 8.5
The structure of a local detection component.

at a rate $\lambda_i(k_l)$ (estimated several local time steps ahead using an ARIMA model) and stores them in a queue where the current size is denoted by $q_i(k_l)$. Depending on the switch $v_i(k_l)$ set by the controller, new signals are retrieved from the queue and stored in a temporary buffer, and a chunk $r_i(k_l)$ is dispatched to the signal detector from this buffer; alternatively, a fraction of only the temporarily buffered signals may be sent to the detector without removing any from the queue. Once detection is performed on the signal chunk delivered by the buffer, the signal processing module provides $y_i(k_l)$ as the confidence measure.

The utility of a detector depends both on the quality of the results as well as the detection latency in terms of response time or queue size. The local queue size at $k_l + 1$ is estimated using its current size and predicted signal arrival rate as

$$\hat{q}_i(k_l + 1) = q_i(k_l) + (\hat{\lambda}_i(k_l) \cdot c_i(r_i) - v_i(k_l)) \cdot T_l, \qquad (8.11)$$

where $c_i(r_i)$ is the estimated processing time for a given chunk of signals r_i. The corresponding instantaneous confidence measure $y_i(k_l)$ depends on the size of the chunk $r_i(k)$ and whether or not the signal is new $v_i(k_l)$. The objective of the local controller is to maximize the following utility function:

$$\hat{J}_i(k_l) = \sum_{i=k_l+1}^{k_l+N_l} b_1 \, [\hat{q}_i(i)]^2 + b_2 \, [\hat{y}_i(i)]^2, \qquad (8.12)$$

where N_l is the lookahead depth. The tuple $(b_1, b_2) \in M$ defines the current mode of the system as assigned by the global controller at each global time step.

Performance Analysis. We evaluated the performance of the multilevel controller using a synthetic workload generated by a proprietary application simulating multiple signal sources. A distributed architecture comprising two signal detectors is assumed.

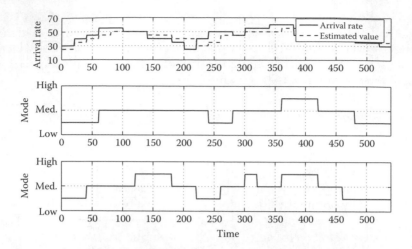

FIGURE 8.6
Decisions made by the global controller in response to the signal arrival rate, affecting the operating modes of the two local detectors.

Figure 8.6 shows the mode-change decisions made by the global controller for the two local detectors in response to the time-varying signal arrival rate. The prediction horizon was set to $N_g = 2$ and the controller sampling duration was $T_g = 90$ time units. Figure 8.7 shows the performance of a local controller

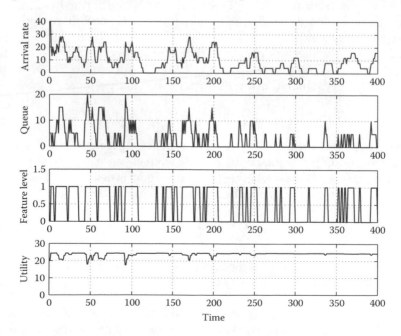

FIGURE 8.7
Decisions made by a local controller to maximize the utility of the underlying signal detector.

when operating in the low-load mode, i.e., when the signal arrival rate is low. Under this scenario, larger chunks of signal data are processed, as evidenced by the achieved feature level. (Note that the feature level is an inverse function of data size.) The prediction horizon for the local controller was $N_l = 3$ and its sampling duration was $T_l = 5$ time units.

8.6 Conclusions

We have presented a generic control and optimization framework to design autonomic computing systems where actions governing system operation are obtained by optimizing its behavior, as forecast by a mathematical model, over a lookahead horizon. As a specific application, we developed an online controller to efficiently manage power consumption in processors under a time-varying workload. We then described a hierarchical control structure to tackle performance management problems in distributed computing systems and applied it to a signal classification application. The results reported in these case studies are very encouraging and show the feasibility of applying the LLC framework to optimize the performance of applications exhibiting hybrid behavior.

The proposed framework can be used to enable self-managing behavior in a variety of business and scientific applications. Utility or on-demand computing is one such example, where a service provider makes computing resources and infrastructure available to customers as needed. This model is becoming increasingly common in enterprise computing systems, and realizing this model requires making dynamic, and sometimes risky, resource provisioning and allocation decisions in a dynamic and uncertain operating environment to maximize revenue while reducing operating cost. In the scientific domain, self-managing data-streaming services can be enabled using the LLC scheme, as part of Grid-based simulation workflows. Such workflows comprise long-running simulations, executing on remote supercomputing sites and generating several terabytes of data, which must then be streamed over a wide-area network for live analysis and visualization [22]. In this setting, a self-managing data-streaming service must minimize the data-streaming overhead on the simulation, adapt to dynamic network bandwidth, and prevent data loss.

References

1. R. El-Yaniv and A. Borodin. *Online Computation and Competitive Analysis*. Cambridge University Press, 1998.
2. S. Abdelwahed, N. Kandasamy, and S. Neema. Online control for self-management in computing systems. In *IEEE Real-Time and Embedded Technology and Applications Symposium*, 2004.

3. S. Abdelwahed, G. Karsai, and G. Biswas. Online safety control of a class of hybrid systems. In *41st IEEE Conference on Decision and Control*, 1988–1990, 2002.

4. T. F. Abdelzaher, K. G. Shin, and N. Bhatti. Performance guarantees for web server end-systems: A control theoretic approach. *IEEE Transactions on Parallel and Distributed Systems*, 13(1):80–96, Jan. 2002.

5. Advanced Micro Devices Corp. *Mobile AMD-K6-2+ Processor Data Sheet*, publication 23446 edition, June 2000.

6. P. Antsaklis and A. Nerode, editors. *Special Issue on Hybrid Control Systems*, volume 43 of *IEEE Transactions Automatic Control*, April 1998.

7. M. Arlitt and T. Jin. Workload characterization of the 1998 World Cup web site. Technical Report HPL-99-35R1, Hewlett-Packard Labs, September 1999.

8. M. F. Arlitt and C. L. Williamson. Web server workload characterization: The search for invariants. In *Proceedings ACM SIGMETRICS Conference*, 126–137, 1996.

9. P. Barford and M. Crovella. Generating representative web workloads for network and server performance evaluation. In *Proceedings ACM SIGMETRICS Conference*, 151–160, 1998.

10. D. P. Bertsekas. *Nonlinear Programming*. Athena Scientific, Nashua, NH, 1995.

11. G. P. Box, G. M. Jenkins, and G. C. Reinsel. *Time Series Analysis: Forecasting and Control*. Prentice-Hall, Upper Saddle River, NJ, 3rd edition, 1994.

12. A. Cervin, J. Eker, B. Bernhardsson, and K. Arzen. Feedback-feedforward scheduling of control tasks. *J. Real-Time Syst.*, 23(1–2), 2002.

13. S. L. Chung, S. Lafortune, and F. Lin. Limited lookahead policies in supervisory control of discrete event systems. *IEEE Trans. Autom. Control*, 37(12):1921–1935, December 1992.

14. Z. Lu et al. Control-theoretic dynamic frequency and voltage scaling for multimedia workloads. In *International Conference Compilers, Architectures, & Synthesis Embedded System (CASES)*, 156–163, 2002.

15. A. Fiat and G. J. Woeginger, editors. *Online Algorithms: The State of the Art*, vol. 1442 of *Lecture Notes in Computer Science*. Springer, 1998.

16. J. Flinn, K. I. Farkas, and J. Anderson. Power and energy characterization of the itsy pocket computer (version 1.5). Technical Report TN-56, Western Research Lab, 2000.

17. A. G. Ganek and T. A. Corbi. The dawn of the autonomic computing era. *IBM Systems Journal*, 42(1):5–18, 2003.

18. J. L. Hellerstein, Y. Diao, S. Parekh, and D. M. Tilbury. *Feedback Control of Computing Systems*. Wiley-IEEE Press, Hoboken, NJ, 2004.

19. J. L. Hellerstein, Y. Diao, and S. S. Parekh. Applying control theory to computing systems. Technical Report RC23459 (W0412-008), IBM Research, December 2004.

20. Intel Corp. *Enhanced Intel SpeedStep Tecnology for the Intel Pentium M Processor*, 2004.

21. N. Kandasamy, S. Abdelwahed, and J. P. Hayes. Self-optimization in computer systems via online control: Application to power management. In *Proceedings IEEE International Conference Autonomic Computing*, 54–62, 2004.

22. Z. Lin, T. S. Hahm, W. W. Lee, W. M. Tang, and R. B. White. Turbulent transport reduction by zonal flows: Massively parallel simulations. *Science*, 1835–1837, 1998.

23. L. Ljung. *System Identification: Theory for the User*. Prentice Hall, Englewood Cliffs, NJ, 2nd edition, 1998.

24. C. Lu, G. A. Alvarez, and J. Wilkes. Aqueduct: Online data migration with performance guarantees. In *Proceedings USENIX Conference File Storage Tech.*, 219–230, 2002.

25. C. Lu, J. Stankovic, G. Tao, and S. Son. Feedback control real-time scheduling: Framework, modeling and algorithms. *Journal of Real-Time Systems*, 23(1/2):85–126, 2002.

26. J. M. Maciejowski. *Predictive Control with Constraints*. Prentice Hall, London, 2002.

27. S. Mascolo. Classical control theory for congestion avoidance in high-speed internet. In *Conference Decision & Control*, 2709–2714, 1999.

28. D. Mignone, A. Bemporad, and M. Morari. A framework for control, fault detection, state estimation, and verification of hybrid systems. In *Proceedings American Control Conference*, 1999.

29. S. Parekh, N. Gandhi, J. Hellerstein, D. Tilbury, T. Jayram, and J. Bigus. Using control theory to achieve service level objectives in performance management. In *Proc. IFIP/IEEE International Symposium on Integrated Network Management*, 2001.

30. V. Sharma, A. Thomas, T. Abdelzaher, K. Skadron, and Z. Lu. Power-aware qos management in web servers. In *Real-Time Systems Symposium*, 63–72, 2003.

31. T. Simunic and S. Boyd. Managing power consumption in networks on chips. In *Proceedings on Design, Automation, & Test Europe (DATE)*, 110–116, 2002.

32. A. Sinha and A. P. Chandrakasan. Energy efficient real-time scheduling. In *Proceedings International Conference Computer Aided Design (ICCAD)*, 458–463, 2001.

33. R. Srikant. Control of communication networks. *Perspectives in Control Engineering: Technologies, Applications, New Directions*, 462–488, 2000.

34. R. Su, S. Abdelwahed, and S. Neema. Computing finitely reachable containable region for switching system. *IEE Proceedings—Control Theory and Applications*, 152(4):477–486, July 2005.

9

Transparent Autonomization in Composite Systems

S. Masoud Sadjadi and Philip K. McKinley

CONTENTS

9.1 Introduction

The ever-increasing complexity of computing systems has been accompanied by an increase in the complexity of their management. Contributing factors include the increasing size of individual networks and the dramatic growth of the Internet, the increasing heterogeneity of software and hardware components, the deployment of new networking technologies, the need for mobile access to enterprise data, and the emergence of pervasive computing. In this chapter, we focus on the management complexity resulting from integrating existing, heterogeneous systems to support corporate-wide, as well as Internet-wide, connectivity of users, employees, and applications.

Autonomic computing [1] promises to solve the management problem by embedding the management of complex systems inside the systems themselves. Instead of requiring low-level instructions from system administrators in an interactive and tightly coupled fashion, such self-managing systems require only high-level human guidance—defined by goals and policies—to work as expected. However, if the code for self-management and application integration is entangled with the code for the business logic of the original systems, then the complexity of managing the integrated system may actually increase, contradicting the purpose of autonomic computing.

To integrate heterogeneous applications, possibly developed in different programming languages and targeted to run on different platforms, requires conversion of data and commands between the applications. The advent of middleware—which hides differences among programming languages, computing platforms, and network protocols [2–4]—in the 1990s mitigated the difficulty of application integration. Indeed, the maturity of middleware technologies has produced several successful approaches to *corporate-wide* application integration [5,6], where applications developed and managed by the same corporation are able to interoperate with one another.

Ironically, the difficulty of application integration, once alleviated by middleware, has reappeared with the proliferation of *heterogeneous* middleware technologies. As a result, there is a need for a "middleware for middleware" to enable Internet-wide and business-to-business application integration [7]. Successful middleware technologies such as Java Remote Method Invocaton (RMI), Common Object Request Broker Architecture (CORBA), and DCOM/.NET Remoting have been used to integrate corporate-wide applications. However, such middleware technologies are often unable to integrate applications managed by different corporations connected through the Internet. The reasons are twofold: (1) different corporations select different middleware technologies, which are more appropriate to integrate their own applications; and (2) middleware packets often cannot pass through Internet firewalls.

Web services [8] offer one approach to addressing these problems. A *web service* is a program delivered over the Internet that provides a service described in the Web Service Description Language (WSDL) [9] and communicates with other programs using messages in Simple Object Access Protocol (SOAP) [10]. WSDL and SOAP are both independent of specific platforms, programming languages, and middleware technologies. Moreover, SOAP leverages the optional use of the HTTP protocol, which can bypass firewalls, thereby enabling Internet-wide application integration.

Although Web services have been successfully used to integrate heterogeneous applications, by themselves they do not provide a *transparent* solution. A challenging problem is to enable integration of existing applications without entangling the integration and self-management concerns with the business logic of the original applications. In this chapter, we describe a technique to enable self-managing behavior to be added to composite systems transparently, that is, without requiring manual modifications to the existing code. The technique uses *transparent shaping* [11], developed previously to enable dynamic adaptation in existing programs, to weave self-managing behavior into existing applications. We demonstrate that combining transparent shaping with Web services provides an effective solution to the transparent application integration problem.

The remainder of this chapter is organized as follows. Section 9.2 provides background on Web services. Section 9.3 describes two approaches to transparent application integration through Web services. Section 9.4 overviews two instances of transparent shaping that can be used in application integration. Section 9.5 presents a case study, where we use these transparent shaping techniques to integrate two existing applications, one developed in Microsoft .NET and the other in CORBA, in order to construct a fault-tolerant surveillance application. Section 9.6 discusses related work, and Section 9.7 summarizes the chapter.

9.2 Web Services Background

A service-oriented architecture [8], as depicted in Figure 9.1, is composed of at least one *provider program*, which is capable of performing the actions associated with a service defined in a service description, and at least one *requester program*, which is capable of using the service provided by a service provider.[1] In this model, we assume that a program is executed inside a process, with a boundary distinguishing local and remote interactions, and is composed of a number of software components, which are units of software composition hosted inside a program process.[2] A component implementing a service is called a *provider component*, and a component requesting a service is called a *requester component*.

Figure 9.1 also shows that the application-to-application (A2A) interaction is accomplished through the use of a middleware technology over a network. The network can be the Internet, an intranet, or simply an interprocess communication (IPC) facility. In the case of Web services, the middleware is composed of two layers: a SOAP messaging layer governed by a WSDL layer (described below). *Web services* are software programs accessible over the Internet by other programs using the SOAP messaging protocol and service descriptors defined in WSDL.

SOAP. SOAP [10] is an XML-based messaging protocol designed to be independent of specific platforms, programming languages, middleware

---

[1] We use the terms "provider program" and "requester program" instead of the terms "provider agent" and "requester agent," as used in [8], to avoid confusion with agents in agent-based systems.

[2] The example programs provided in this chapter are all developed in object-oriented languages. For simplicity, the terms "component" and "object" have been used interchangeably. However, this does not imply that a service-oriented system must either be implemented using object-oriented languages or be designed using an object-oriented paradigm.

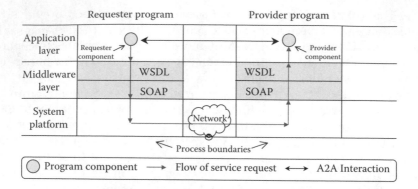

FIGURE 9.1
A simplified Web service architecture.

technologies, and transport protocols. SOAP messages are used for interactions among Web service providers and requesters. Unlike object-oriented middleware such as CORBA, which requires an object-oriented model of interaction, SOAP provides a simple message exchange among interacting parties. As a result, SOAP can be used as a layer of abstraction on top of other middleware technologies (effectively providing a middleware for middleware).

A SOAP message is an XML document with one element, called an envelope, and two children elements, called the header and body. The contents of the header and body elements are arbitrary XML. Figure 9.2 shows the structure of a SOAP message. The header is an optional element, whereas the body is not optional; there must be exactly one body defined in each SOAP message. To provide the developers with the convenience of a procedure-call abstraction, a pair of related SOAP messages can be used to realize a request and its corresponding response. SOAP messaging is *asynchronous*, that is, after sending a request message, the service requester will not be blocked waiting for the response message to arrive. For more information about details of SOAP messages, please refer to [10,12].

```
1    <?xml version="1.0" encoding="UTF-8" ?>
2    <soap:Envelope xmlns:soap=
3      "http://schemas.xmlsoap.org/soap/envelope/ ...>
4      <soap:Header>
5        <!- Header contents in defined in arbitrary XML. ->
6      </soap:Header>
7      <soap:Body>
8        <!- Body contents in defined in arbitrary XML. ->
9      </soap:Body>
10   </soap:Envelope>
```

FIGURE 9.2
SOAP message structure.

WSDL. WSDL [9] is an XML-based language for describing valid message exchanges among service requesters and providers. The SOAP messaging protocol provides only basic communication and does not describe what pattern of message exchanges are required to be followed by specific service requesters and providers. WSDL addresses this issue by describing an interface to a Web service and providing the convenience of remote procedure calls (or more complicated protocols). For more information about details of WSDL, please refer to [9,12].

9.3   Transparent Application Integration

Several different approaches have been described in the literature to integrate applications. Regardless of the specific technique, integration of two heterogeneous applications requires translating the syntax and semantics of the two applications, typically during execution. Providing direct translations for N heterogeneous middleware technologies requires $(N^2 - N)/2$ translators to cover all possible pairwise interactions. Using a common language reduces the number of translators to N, assuming that one side of the interaction (either requester or provider program) always uses the common language. Web services provide one such language. Depending on *where* the translation is performed (e.g., inside or outside the requester and provider programs), we distinguish two approaches to transparent application integration: the bridge approach and the transparent shaping approach.

Bridge Approach. An intuitive approach to transparent application integration is to use *bridge programs*, which sit between requesters and providers, intercepting the interactions and translating them from application-specific middleware protocols to Web services protocols, and vice versa. The architecture for this approach is illustrated in Figure 9.3. A bridge program hosts one or more *translator components*, which encapsulate the logic for translation. A translator component plays the role of a provider component for the requester component, as well as the role of a requester component for the provider component.

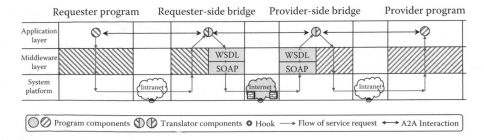

FIGURE 9.3
Transparent application integration using bridge programs.

Using this architecture is beneficial for the following reasons. First, hosting translator components inside a separate process (the bridge program) does not require modifications to the requester and provider programs. Second, a bridge program can host several translator components, where each translator component may provide translation to one or more requester and provider programs. Third, the localization of translator components in one location (the bridge program) simplifies the maintenance of application integration. For example, security policies can be applied to the bridge program once, and will be effective to all the translator components hosted by the bridge.

The main disadvantage of this architecture is the overhead imposed on the interactions due to process-to-process redirection (in case the bridge programs are located on the same machine as the requester and/or provider) or machine-to-machine redirection (in case the bridge programs are located on separate machines). Other disadvantages include a potential single point of failure and a communication/processing bottleneck at the bridge. Furthermore, this approach may not even be possible in some situations—for example, if the provider address is hard-coded in the requester program.

Transparent Shaping Approach. To avoid these problems, the translator components can instead be hosted *inside* the requester and provider programs, as illustrated in Figure 9.4. However, providing intercepting and redirecting interactions transparently to the application is challenging. *Transparent shaping* provides a solution to this problem by generating adaptable programs from existing applications. We call a program *adaptable* if its behavior can be changed with respect to the changes in its environment or its requirements. An adaptable program can be thought of as a managed element, as described in [1].

We developed transparent shaping originally to enable reuse of existing applications in environments whose characteristics were not necessarily anticipated during the design and development [13,14]. An example is porting applications to wireless networks, which often exhibit relatively high packet loss rates and frequent disconnections. In this chapter, however, we show how transparent shaping can be used to enable transparent application integration. The integration is performed in two steps.

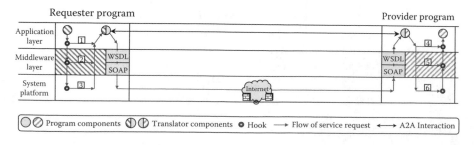

FIGURE 9.4
Transparent application integration using transparent shaping.

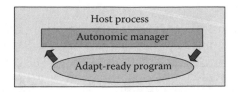

FIGURE 9.5
Transparent adaptation of an adapt-ready program at runtime.

In the first step, an *adapt-ready* program is produced at compile, startup, or load time using static transformation techniques. An adapt-ready program is a managed element whose behavior is initially equivalent to the original program but which can be adapted at runtime by inserting or removing adaptive code at certain points in the execution path of the program, called *sensitive joinpoints*. For application integration, we are interested only in those joinpoints related to remote interactions. To support such operations, the first step of transparent shaping weaves interceptors, referred to as *hooks*, at the remote interaction joinpoints. As illustrated in Figure 9.4, hooks may reside inside the program code itself (arrows 1 and 4), inside its supporting middleware (arrows 2 and 5), or inside the system platform (arrows 3 and 6). Example techniques for implementing hooks include weaving aspects into the application (compile time) [14], inserting portable interceptors into a CORBA program [15] (startup time), and byte-code rewriting in a virtual machine [16] (load time).

In the second step, executed at runtime, the hooks in the adapt-ready program are used by an autonomic manager to redirect the interactions to adaptive code, which in this case implements the translator. As illustrated in Figure 9.5, the autonomic manager may introduce new behavior to the adapt-ready program according to the high-level user policies by inserting or removing adaptive code using the generic hooks.

9.4 Transparent Shaping Mechanisms

In this section, we briefly describe two concrete instances of transparent shaping. The first one is a language-based approach that uses a combination of aspect weaving and meta-object protocols to introduce dynamic adaptation to the application code directly. The second one is a middleware-based approach that uses middleware interceptors as hooks. Both instances adhere to the general model described above.

Language-Based Transparent Shaping. Transparent Reflective Aspect Programming (TRAP) [14] is an instance of transparent shaping that supports dynamic adaptation in existing programs developed in class-based, object-oriented programming languages. TRAP uses generative techniques to create

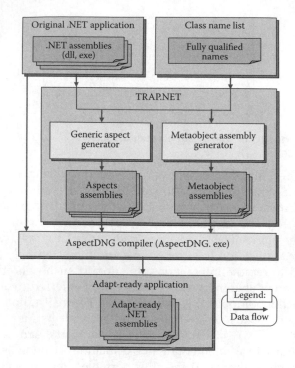

FIGURE 9.6
TRAP.NET compile-time process.

an adapt-ready application, without requiring any direct modifications to the existing programs. To validate the TRAP model, previously we developed TRAP/J [14], which supports dynamic adaptation in existing Java programs. To support existing programs developed in C++, members of our group have implemented TRAP/C++ using compile-time meta-object protocols supported by Open C++ [17].

For the case study described in Section 9.5, we developed TRAP.NET, which supports dynamic adaptation in .NET applications developed in any .NET programming language (e.g., C#, J#, VB, and VC++). As illustrated in Figure 9.6, the developer selects at compile time a subset of classes in the existing .NET assemblies[3] that are to be reflective at runtime. A class is *reflective* at run time if its behavior (i.e., the implementation of its methods) can be inspected and modified dynamically [18]. Since .NET does not support such functionality inherently, TRAP.NET uses generative techniques to produce adapt-ready assemblies with hooks that provide the reflective facilities for the selected classes. Next, we use *AspectDNG* version 0.6.2, a recently released .NET aspect weaver [19], to weave generated aspect and meta-object assemblies into the original application. Finally, we execute the adapt-ready

---

[3] A .NET assembly is simply a .NET executable file (i.e., a .EXE file) or a .NET library file (i.e., a .DLL file).

application together with an autonomic manager using a host application (explained in Section 9.5.3).

As the adapt-ready assemblies execute, the autonomic manager may introduce new behavior to the adapt-ready assemblies according to the high-level user policies by insertion and removal of adaptive code via interfaces to the reflective classes. Basically, the hooks inside a reflective class wrap the original methods, intercept all incoming calls, and can forward the calls to new implementations of the methods, as needed. The new implementations can be inserted dynamically by the autonomic manager.

Middleware-Based Transparent Shaping. For those applications written atop a middleware platform, transparent shaping can be implemented in middleware instead of this application itself. The *Adaptive CORBA Template (ACT)* [13,20] is an instance of transparent shaping that enables dynamic adaptation in existing CORBA programs. ACT enhances CORBA Object Request Brokers (ORBs) to support dynamic reconfiguration of middleware services transparently not only to the application code, but also to the middleware code itself. Although ACT itself is specific to CORBA, the concepts can be applied to many other middleware platforms. To evaluate the performance and functionality of ACT, we constructed a prototype of ACT in Java, called *ACT/J* [13,20]. To support CORBA programs developed using C++ ORBs, we plan to develop ACT/C++. In addition, we are planning to develop similar frameworks for Java/RMI and Microsoft's .NET.

Figure 9.7 shows the flow of a request/reply sequence in a simple CORBA application using ACT/J. For clarity, details such as stubs and skeletons are not shown. ACT comprises two main components: a generic interceptor and an ACT core. A *generic interceptor* is a specialized request interceptor that is registered with the ORB of a CORBA application at startup time. The *client* generic interceptor intercepts all outgoing requests and incoming replies (or exceptions) and forwards them to its ACT core. Similarly, the *server* generic

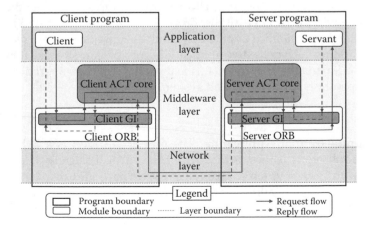

FIGURE 9.7
ACT configuration in the context of a simple CORBA application.

interceptor intercepts all the incoming requests and outgoing replies (or exceptions) and forwards them to its ACT core. The CORBA application is *adapt-ready* if a generic interceptor is registered with all its ORBs at startup time.

Implementing transparent shaping in middleware, as in ACT, can produce greater transparency than a language-based approach such as TRAP, which requires recompilation of the application. To produce an adapt-ready version of an existing CORBA program, we do not need to transform the original program as we do in TRAP. By introducing generic hooks inside the ORB of a CORBA application at startup time, we can intercept all CORBA remote interactions. Specifically in ACT/J, a generic host program (this program is part of the ACT/J framework) is first instructed to follow the settings in a configuration file.[4] The configuration file instructs the host program to incorporate generic CORBA portable interceptors [15] into the ORB of the host program. Next, the host program loads both the original application together with an autonomic manager.

Later at runtime, these hooks can be used by an autonomic manager to insert and remove adaptive code with respect to the adapt-ready program, which in turn can adapt the requests, replies, and exceptions passing through the ORBs. In this manner, ACT enables runtime improvements to the program in response to unanticipated changes in its execution environment.

9.5   Case Study: Fault-Tolerant Surveillance

To demonstrate how transparent shaping and Web services can be used to integrate existing applications, while introducing new autonomic behavior, we conducted a case study. Specifically, we created a fault-tolerant surveillance application by integrating existing .NET and CORBA image retrieval applications and adding a self-management component to switch among the two image sources in response to failures. The integration and self-management code is transparent to the original applications. The resultant integrated system is a *fault-tolerant* surveillance application. In the remainder of this section, we briefly introduce each of the two applications, review our strategy for integration and self-management, and describe the details of the integration process and the operation of the integrated system.

9.5.1   Existing Applications

The Sample Grabber Application. The first application is a .NET sample grabber application (called `SampleGrabberNET`). Basically, it captures a video stream from a video source (e.g., a WebCam) and displays it on the screen. In addition, it allows the user to grab a live frame and save it to a bitmap file. This application is a .NET standalone application written in

---

[4] The corresponding command line instruction used at startup time is the following: java Host.class Host.class.config OriginalApplication.class AutonomicManger.class

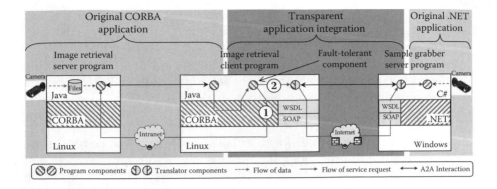

FIGURE 9.8
The configuration of a fault-tolerant surveillance application.

C# and freely available at the Code Project Web site (`http://www.code-project.com`). It is part of the `DirectShow.NET` framework developed by NETMaster.[5]

The Image Retrieval Application. The second application is a CORBA image retrieval application written in Java. It has two parts, a *client program* (called `SlideClient`) that requests and displays images, and a *server program* (called `SlideService`) that stores images and replies to the client program requests. The client program continuously sends requests to the server program asking for images and displays each returned image. In our implementation, the images maintained by the server program are live shots taken periodically using a WebCam and stored in a file. This application was developed by BBN Technologies and is distributed with the QuO framework [21]. The image retrieval application by itself can benefit from the QuO framework, which supports several types of adaptive behavior. In this study we disabled the QuO framework and used the application only as a CORBA application.

9.5.2 Integration and Self-Management Strategy

Our goal is to construct a surveillance application that can use either application as the source of images. For example, the application should enable the CORBA client application to retrieve live images from the .NET frame grabber application, when the CORBA server application is not available (e.g., due to server crash, network disconnection, etc.). Figure 9.8 shows the configuration of the integrated application. The .NET frame grabber application plays the role of a provider program and must be exposed as a frame grabber Web service. On the other side, the client program of the image retrieval

---

[5] NETMaster is the name for a group of developers who are actively contributing to the Code Project. NETMaster uses the interfaces provided in the `DirectShow.NET` framework to interoperate with DirectShow. The Code Project (`http://www.codeproject.com`) is a repository for a large number of free C++, C#, and .NET articles, code snippets, discussions, and news on the Internet.

application plays the role of a requester program and must be shaped to use both the frame grabber Web service as well as the CORBA server program.

In the rest of this section, we describe how transparent shaping is used to expose the frame grabber application as a Web service, followed by how the image retrieval client program is shaped to use this Web service.

9.5.3   Exposing the Frame Grabber Application

.NET Remoting is an enabler for distributed applications to communicate with one another, thus allowing interprocess communication. We note that the server side of a distributed application that uses .NET remoting can be exposed as a Web service without modification if none of the types to be exposed by the .NET server application is a .NET specific type [22,23]. However, our frame grabber application is a .NET *standalone* application (as opposed to a .NET remoting application). Therefore, no .NET remoting service is exposed by the frame grabber application itself. Hence, we first need to shape the .NET frame grabber application to become a .NET remoting application, and then use it as a Web service.

Following the first step of TRAP.NET (see Figure 9.6), we generate an adapt-ready version of the frame grabber application. To transform this application into a .NET remoting server, we need to weave hooks inside the main class of the application (MainForm). Therefore, we list only the name of the main class to be passed to TRAP.NET. The code for .NET remoting and the self-management functionality are located in a separate assembly (AutonomicManager.exe). At startup time these two programs are loaded inside another program (Host.exe), which is listed in Figure 9.9 (lines 1

```
1  // The host application defined in  Host.cs
2  public class Host  {
3    static private string configFilename, managedElement, autonomicManager;
4    public static void Main(string [] args)  {
5    if (args.Length != 3) return;
6    configFilename = args[0]; managedElement = args[1]; autonomicManager = args[2];
7    try  {
8      RemotingConfiguration.Configure(configFilename);
9      AppDomain ad = AppDomain.CurrentDomain;
10     ad.ExecuteAssembly(autonomicManager);
11     ad.ExecuteAssembly(managedElement);
12     } catch(Exception e)  { }
13   String keyState = ''''; keyState = Console.ReadLine();
14  }
15
16  // The configuration file defined in Host.exe.config
17  <configuration> <system.runtime.remoting> <application name="Server">
18    <service>
19      <wellknown mode="Singleton" type="SampleGrabberWebService.
20        SampleGrabberObject, SampleGrabberObject" objectUri="SampleGrabberObject" />
21    </service>
22    <channels> <channel port="9000" ref="http" /> </channels>
23  </application> </system.runtime.remoting> </configuration>
```

FIGURE 9.9
Excerpted code for the Host program hosting the adapt-ready application and its autonomic manager.

to 14).[6] The modified .NET frame grabber program is inside the `Sample-GrabberNET.exe` assembly, the .NET remoting code is inside the `AutonomicManager.exe` assembly, and the configuration is inside the `Host.exe.config` file. The excerpted code for the `Host.exe.config` configuration file is listed in Figure 9.9 (lines 16 to 23).

As shown in Figure 9.9, first, the configuration file is parsed and the instructions are followed (line 8), which provides flexibility to configure the `Host` program at startup time (e.g., the port address at which the Web service can be reached can be defined in this configuration file). Next, the `autonomicManager` and the `managedElement` are executed using the .NET reflection facilities (lines 10 and 11).

Now that the provider program is ready to run, we need to generate the Web service description of our provider program (to be used in the shaping of the CORBA client program). We used the .NET framework `SOAPsuds.exe` utility with the `-sdl` option, which generates a WSDL schema file. The excerpted WSDL description is listed in Figure 9.10.

The abstract description part (lines 3 to 16) describes the interface to the Web service using `message` elements (lines 3 to 8), which define what type of messages can be sent to and received from the Web service, and a `portType` element (lines 9 to 16), which defines all the operations that are supported by the Web service. The `GrabFrame` operation (lines 10 to 15) defines the valid message exchange pattern supported by the Web service.

The concrete part of the WSDL description (lines 18 to 33) complements the abstract part using a `binding` element (lines 18 to 26), which describes *how* a given interaction is performed over *what* specific transport protocol, and a `service` element (lines 28 to 33), which describes *where* to access the service. The *how* part describes how marshalling and unmarshalling is performed using the `operation` element inside the `binding` element (lines 21 to 25). The *what* part is described in line 20 using the `transport` attribute. The *where* part is described using the `port` element (lines 29 to 32).

9.5.4 Shaping the Image Retrieval Client

We follow the architecture illustrated in Figure 9.8 to shape the CORBA client program to interoperate with the .NET frame grabber program. We use the ACT/J framework to host a Web service translator in the adapt-ready client application.

Figure 9.11 lists the Interactive Data Language (IDL) description used in the original CORBA image retrieval application. The `SlideShow` interface defines six methods (lines 4 to 9). As listed in Figure 9.12 (lines 21 to 26), all the `read*()` methods defined in the IDL file are mapped to the `GrabFrame()` method of the Web service exposed by the provider program. The `getNumberOfGifs()` method simply returns -1 (line 27) to indicate that

---

[6] The corresponding command line instruction used at startup time is the following: `Host.exe Host.exe.config AdaptReadyApp lication.exe AutonomicManger.exe`

```
1    <?xml version='1.0' encoding='UTF-8'?>
2    <definitions name='SampleGrabberObject' ...> <types> ...</types>
3      <message name='SampleGrabberObject.GrabFrameInput'>
4        <part name='nQuality' type='xsd:int'/>
5      </message>
6      <message name='SampleGrabberObject.GrabFrameOutput'>
7        <part name='return' type='ns2:ArrayOfShort'/>
8      </message>
9      <portType name='SampleGrabberObjectPortType'>
10       <operation name='GrabFrame' parameterOrder='nQuality'>
11         <input name='GrabFrameRequest'
12           message='tns:SampleGrabberObject.GrabFrameInput'/>
13         <output name='GrabFrameResponse'
14           message='tns:SampleGrabberObject.GrabFrameOutput'/>
15       </operation>
16     </portType>
17
18     <binding name='SampleGrabberObjectBinding'
19       type='tns:SampleGrabberObjectPortType'>
20       <soap:binding style='rpc' transport='http://schemas.xmlsoap.org/soap/http'/> ...
21       <operation name='GrabFrame'>
22         <soap:operation soapAction='...'/> ...
23         <input name='GrabFrameRequest'> <soap:body .../> </input>
24         <output name='GrabFrameResponse'> <soap:body .../> </output>
25       </operation>
26     </binding>
27
28     <service name='SampleGrabberObjectService'>
29       <port name='SampleGrabberObjectPort' binding='tns:SampleGrabberObjectBinding'>
30         <soap:address location=
31           'http://haydn.cse.msu.edu:9000/Server/SampleGrabberObject'/>
32       </port>
33     </service>
34   </definitions>
```

FIGURE 9.10
The excerpted WSDL description of the sample grabber service.

the images being retrieved are live images (as opposed to being retrieved from a number of stored images at the server side).

Using the ACT/J framework, the calls to the original CORBA server application are intercepted and redirected to the translator component. The translator component is defined as SlideService_ClientLocalProxy class

```
1    // The slide show interface defined in SlideShow.idl
2    module com { module bbn { module quo { module examples { module bette {
3      interface SlideShow {
4        void readSmall ( in long gifNumber, out string size, out octetArray buf );
5        void readSmallProcessed ( in long gifNumber, out string size, out octetArray buf );
6        void readBig ( in long gifNumber, out string size, out octetArray buf );
7        void readBigProcessed ( in long gifNumber, out string size, out octetArray buf );
8        void read ( in long gifNumber, out string size, out octetArray buf );
9        long getNumberOfGifs ( );
10     };
11   }; }; }; }; };
```

FIGURE 9.11
The slide show IDL file used in the original CORBA application.

```
1    // The proxy defined in SlideService_ClientLocalProxy.java
2    public class SlideService_ClientLocalProxy extends SlideShowPOA
3      implements Serializable, SlideShowOperations {
4      private SampleGrabberObjectPortType sampleGrabberObject = null;
5      public SlideService_ClientLocalProxy(ORB orb) { ...
6        string endpoint = "http://haydn.cse.msu.edu:9000/Server/SampleGrabberObject";
7        try {
8          Stub stub = (Stub)(new SampleGrabberObjectService_Impl().
9            getSampleGrabberObjectPort());
10         stub._setProperty(javax.xml.rpc.Stub.ENDPOINT_ADDRESS_PROPERTY, endpoint);
11         sampleGrabberObject = (SampleGrabberObjectPortType)stub;
12       } catch (Exception ex) {...}
13     }
14     private byte[] grabFrame( int nQuality ) {
15       byte [] frameByteArray = null; short[] frameShortArray = null;
16       try { frameShortArray = sampleGrabberObject.GrabFrame( nQuality ); }
17       catch(Exception e) {...}
18       frameByteArray = convertShortArray2ByteArray( frameShortArray );
19       return frameByteArray;
20     } ...
21     public void readBig(int gifNum, StringHolder sizeHolder, octetArrayHolder pixHolder) {
22       pixHolder.value = grabFrame( 75 ); sizeHolder.value = "big";
23     } ...
24     public void readSmall(int gifNum, StringHolder sizeHolder, octetArrayHolder pixHolder) {
25       pixHolder.value = grabFrame( 25 ); sizeHolder.value = "small";
26     } ...
27     public int getNumberOfGifs() { return -1; }
28   }
```

FIGURE 9.12
Excerpted code for the Web service translator component defined as a proxy object in the
ACT/J framework.

listed in Figure 9.12. First, a reference to the `SampleGrabberObject` Web
service is obtained (lines 4 to 13). We used the Java WSDP framework to generate the stub class corresponding to the Web service using the WSDL file,
generated in the previous part and listed in Figure 9.10. Next, all calls to the
original CORBA object are forwarded to the Web service (lines 14 to 27).

9.5.5 Self-Managed Operation

The configuration of the resulting fault-tolerance surveillance application is
illustrated in Figure 9.8. The target setting is to have two (or more) cameras
monitoring the same general area (a parking lot, building entrance, etc.).
The cameras and supporting software might have been purchased for other
purposes and now are being combined to create a fault-tolerant, heterogenous
application. Although this application is relatively simple, it demonstrates
that it is possible to integrate application in a transparent manner without
using bridging.

In our experiments, we first start the adapt-ready .NET frame-grabber application (shown in the right side of Figure 9.8), next the original CORBA
slide-server (shown in the left side), and finally the adapt-ready CORBA
slide-client application (shown in the middle). We have also implemented
a simple user interface that enables a user to enter policies to be followed

by the application. By default, the initial policy regarding the data source is to retrieve images from the CORBA server, since our experiments show it is more responsive than the .NET application. If the CORBA server is not available or does not respond for a certain interval (1 second by default), then the client should try to retrieve images from the .NET server. While the images are being retrieved from the .NET server, the client continues to probe the CORBA server to see if it is available. If the CORBA server returns, the client should stop retrieving images from the .NET server and switch to the CORBA server. All these operations are carried out completely transparently to the original CORBA and .NET applications.

Some limited functionality has been provided to enable the user to change the high-level policy via the Graphical User Interface (GUI). For example, the user can change the initial policy and configure the system so that every other image is retrieved from the .NET server. In addition, the user can change the timeout parameter to be other than 1 second. The user can also set the frequency of images being retrieved and ask the system to maintain this frequency. For example, if the user asks for 2 frames per second, then the system automatically monitors the round-trip delay of retrieving images and correspondingly inserts interval delays (if the current frequency is too high) or requests smaller images with lower resolution (if the current frequency is too low). This feature has been tested over a wireless ad hoc network, where a user with the client application is walking about a wireless cell and experiencing different packet loss rates in the wireless network. Although the round-trip time was constantly changing in this experiment, the system was able to maintain the frequency of images being retrieved and displayed.

Finally, we note that once the adapt-ready applications are executed, we can dynamically modify even the self-management functionality itself—for example, introducing a more sophisticated GUI, without the need to stop, modify, compile, and restart the system.

9.6   Related Work

In this section, we review several research projects, standard specifications, and commercial products that support application integration. Based on the transparency and flexibility of the adaptation mechanisms used to support application integration, we identify three categories of related work.

In the first category, we consider approaches that provide transparency with respect to either an existing provider program or an existing requester program, but not both. Therefore, the programs hosting translator components are required to be either developed from scratch or modified directly by a developer. Examples of research projects in this category include the Automated Interface Code Generator (AIAG) [24], the Cal-Aggie Wrap-O-Matic project (CAWOM) [25], and the World Wide Web Factory (W4F) [26]. AIAG [24] supports application integration by providing an interface wrapper model, which enables developers to treat distributed objects as local objects. AIAG is an automatic wrapper generator built on top of JavaSpaces that can be used to generate the required glue code to be used in client programs. CAWOM [25] provides a tool that generates wrappers, enabling command-line systems to be accessed by client programs developed in CORBA. This approach provides transparency for existing command-line systems. Examples of the use of CAWOM include wrapping the JDB debugger, which enables distributed debugging, and wrapping the Appache Web server, which enables remote administration. Finally, W4F [26] is a Java toolkit that generates wrapper for Web resources. This toolkit provides a mapping mechanism for Java and XML.

In the second category, we consider approaches that provide transparency with respect to both the provider and requester programs using a bridge program hosting the translator components. Although such approaches provide transparency, they suffer from extra overhead imposed by another level of process-to-process or machine-to-machine redirection, as discussed in Section 9.3. An example of a research project in this category is on-the-fly wrapping of Web services [27]. In this project, Web services are wrapped to be used by Java programs developed in Jini [28], a service-based framework originally developed to support integration of devices as services. The wrapping process is facilitated by the `WSDL2Java` and `WSDL2Jini` generator tools, which generate the glue code part of the bridge program and the translator component. A developer is required to complete the code for the bridge and to ensure that the semantics of translations are correct. Using the Jini lookup service, the bridge publishes the wrapped Web service as a Jini service, which can be used transparently by Jini client programs.

We consider transparent shaping in a third category. Similar to the approaches in the second category, transparent shaping provides transparency to both provider and requester programs, and in addition provides flexibility with respect to *where* the translator components are hosted. To our knowledge, transparent shaping is the only application integration technique exhibiting both features. That said, transparent shaping is intended to complement, rather than compete with, the approaches in the second category. Specifically, we plan to employ the automatic translation techniques provided by those approaches in our future work.

9.7 Summary

In this chapter, we have demonstrated how transparent shaping can be used to facilitate transparent application integration in the construction of autonomic systems from existing applications. Transparent shaping enables integration of existing applications—developed in heterogeneous programming languages, middleware frameworks, and platforms—through Web services while the integration and self-management concerns are transparent to the original applications. A case study was described, in which we constructed a fault-tolerant surveillance application by integrating existing applications and adding self-management functionality. We used transparent shaping to enable two existing image retrieval applications, one of them developed in .NET and the other in CORBA, to interact as required, without modifying either application directly.

We note that several challenges remain in the domain of transparent application integration, including automatic translation of the semantics of heterogeneous applications and automatic discovery of appropriate Web services. The increasing maturity of business standards, which have supported automated interactions in business-to-business application integration over the past 20 years, addresses these issues to some extent [6]. Examples of electronic businesses based on Web services include ebXML, RosettaNet, UCCNet, and XMethods. Also, the automatic location of services, which is one of the goals of Web services, has been specified in the Universal Description, Discovery, and Integration (UDDI) specification. UDDI is a Web service for registering other Web service descriptions. Together with tools and techniques for transparently integrating existing applications, as described in this chapter, these developments promise to significantly increase the degree to which autonomic computing is used in the Internet.

Further Information. A number of related papers and technical reports of the Software Engineering and Network Systems Laboratory can be found at `http://www.cse.msu.edu/sens`. Papers and other results related to the RAPIDware project, including a download of the TRAP/J, TRAP.NET, ACT/J toolkits, and their corresponding source code, are available at `http://www.cse.msu.edu/rapidware`.

Acknowledgments

This work was supported in part by the U.S. Department of the Navy, Office of Naval Research under Grant No. N00014-01-1-0744, and in part by National Science Foundation grants CCR-9912407, EIA-0000433, EIA-0130724, and ITR-0313142.

References

1. J. O. Kephart and D. M. Chess, "The vision of autonomic computing," *IEEE Computer*, vol. 36, no. 1, 41–50, 2003.
2. D. E. Bakken, *Middleware*. Kluwer Academic Press, 2001.

3. W. Emmerich, "Software engineering and middleware: a roadmap," in *Proceedings of the Conference on The Future of Software Engineering*, 117–129, 2000.

4. A. T. Campbell, G. Coulson, and M. E. Kounavis, "Managing complexity: Middleware explained," *IT Professional, IEEE Computer Society*, 22–28, September/October 1999.

5. A. Gokhale, B. Kumar, and A. Sahuguet, "Reinventing the wheel? CORBA vs. Web services," in *Proceedings of International World Wide Web Conference*, (Honolulu, Hawaii), 2002.

6. S. Vinoski, "Where is middleware?" *IEEE Internet Computing*, 83–85, March–April 2002.

7. S. Vinoski, "Integration with Web services," *IEEE Internet Computing*, 75–77, November–December 2003.

8. D. Booth, H. Haas, F. McCabe, E. Newcomer, M. Champion, C. Ferris, and D. Orchard, *Web Services Architecture*. W3C, 2004.

9. R. Chinnici, M. Gudgin, J.-J. Moreau, J. Schlimmer, and S. Weerawarana, *Web Services Description Language (WSDL) Version 2.0*. W3C, 2.0 ed., March 2004.

10. M. Gudgin, M. Hadley, N. Mendelsohn, J.-J. Moreau, and H. F. Nielsen, *SOAP Version 1.2*. W3C, 1.2 ed., 2003.

11. S. M. Sadjadi, P. K. McKinley, and B. H. Cheng, "Transparent shaping of existing software to support pervasive and autonomic computing," in *Proceedings of the First Workshop on the Design and Evolution of Autonomic Application Software 2005 (DEAS'05), in conjunction with ICSE 2005*, (St. Louis, Missouri), May 2005. To appear.

12. F. Curbera, M. Duftler, R. Khalaf, W. Nagy, N. Mukhi, and S. Weerawarana, "Unraveling the Web services web: An introduction to SOAP, WSDL, and UDDI," *IEEE Internet Computing*, vol. 6, no. 2, 86–93, 2002.

13. S. M. Sadjadi and P. K. McKinley, "Transparent self-optimization in existing CORBA applications," in *Proceedings of the International Conference on Autonomic Computing (ICAC-04)*, (New York), 88–95, May 2004.

14. S. M. Sadjadi, P. K. McKinley, B. H. Cheng, and R. K. Stirewalt, "TRAP/J: Transparent generation of adaptable java programs," in *Proceedings of the International Symposium on Distributed Objects and Applications (DOA'04)*, (Agia Napa, Cyprus), October 2004.

15. Object Management Group, Framingham, Massachusetts, *The Common Object Request Broker: Architecture and Specification Version 3.0*, July 2003.

16. G. A. Cohen, J. S. Chase, and D. Kaminsky, "Automatic program transformation with JOIE," in *1998 Usenix Technical Conference*, June 1998.

17. S. Chiba and T. Masuda, "Designing an extensible distributed language with a meta-level architecture," *Lecture Notes in Computer Science*, vol. 707, 1993.

18. J. Ferber, "Computational reflection in class based object-oriented languages," in *Conference Proceedings on Object-Oriented Programming Systems, Languages and Applications*, 317–326, ACM Press, 1989.

19. AspectDNG Project Summary. Available at `http://sourceforge.net/projects/aspectdng`

20. S. M. Sadjadi and P. K. McKinley, "ACT: An adaptive CORBA template to support unanticipated adaptation," in *Proceedings of the 24th IEEE International Conference on Distributed Computing Systems (ICDCS'04)*, (Tokyo, Japan), March 2004.

21. J. A. Zinky, D. E. Bakken, and R. E. Schantz, "Architectural support for quality of service for CORBA objects," *Theory and Practice of Object Systems*, vol. 3, no. 1, 1997.

22. T. Thangarathinam, ".NET remoting versus Web services," Online article, 2003. Available at `http://www.developer.com/net/net/article.php/11087_2201701_1`

23. P. Dhawan and T. Ewald, "Building distributed applications with microsoft .net (ASP.NET Web services or .NET remoting: How to choose)," Online article, September 2002.

24. N. Cheng, V. Berzins, Luqi, and S. Bhattacharya, "Interoperability with distributed objects through java wrapper," in *Proceedings of the 24th Annual International Computer Software and Applications Conference* (Taipei, Taiwan), October 2000.

25. E. Wohlstadter, S. Jackson, and P. Devanbu, "Generating wrappers for command line programs: the Cal-Aggie Wrap-O-Matic project," in *Proceedings of the 23rd International Conference on Software Engineering*, 243–252, IEEE Computer Society, 2001.

26. A. Sahuguet and F. Azavant, "Looking at the Web through XML glasses," in *Proceedings of the Fourth IFCIS International Conference on Cooperative Information Systems*, 148–159, September 1999.

27. G. C. Gannod, H. Zhu, and S. V. Mudiam, "On-the-fly wrapping of web services to support dynamic integration," in *Proceedings of the 10th IEEE Working Conference on Reverse Engineering*, November 2003.

28. W. K. Edwards, *Core Jini*. Prentice-Hall, Upper Saddle River, NJ 1999.

10

Recipe-Based Service Configuration and Adaptation

Peter Steenkiste and An-Cheng Huang

CONTENTS

In the last few years, several grid infrastructures [12, 13] have been deployed that support the dynamic discovery, monitoring, and allocation of distributed computing resources. These systems have simplified the development of a variety of distributed systems such as scientific visualization [28], distributed simulation [27], and video conferencing [20]. Increasingly, these services are developed by composing service components that implement functions such as multicast and transcoding. Optimizing the execution of such services remains, however, a significant challenge. One reason is that the quality of the delivered service depends strongly on how effectively the available resources are used. However, since resource availability can differ significantly from invocation to invocation and can even change during execution, careful service configuration and runtime adaptation are needed to achieve high quality of service. Another reason is that the optimization of a service is highly service specific, so it is not possible to provide canned solutions.

Recent research efforts to support automatic service configuration can be divided into two categories. *Service-specific* architectures are designed for particular classes of services [7,27,17,32,28]. Examples include resource selection for resource-intensive applications and resource allocation for services consisting of a set of multifidelity applications. Since the service-specific knowledge is hard-wired into the service composition module by the developer, service-specific approaches do not support the reuse of functionality in other services. In contrast, *generic* architectures can compose a broader class of services, typically through type-based service composition [16, 35, 33, 14, 23]. In these systems, components have well-defined input and output interfaces, and the service is created as a sequence of components with matching input and output types. While generic architectures can be reused by a broad range of services, they cannot exploit service-specific knowledge, often resulting in higher overhead and suboptimal configurations.

We can similarly identify several approaches to runtime adaptation. Most systems today adopt an "internalized" approach in which developers integrate their service-specific *adaptation strategies* into the target system. Examples include general "parameter-level" adaptation techniques (e.g., [11, 4, 32, 3, 10]), communication adaptation in a client-server system (e.g., [25, 29, 31, 8, 2]), and adaptive resource allocation (e.g., [9, 36, 26]). This approach is flexible and can be very effective, but it forces a developer to embed the adaptation rules into the system, increasing design complexity and development cost. An alternative is to use generic solutions, e.g., process migration. These solutions are general, but they are limited in scope. In a third approach, adaptive behavior is specified as a set of "externalized" adaptation strategies that define under what condition specific adaptation operations should be executed [15, 36]. By separating the specification of the adaptive behavior from the normal service functionality, it is easier to understand and analyze the dynamics of the system. It also becomes possible to reuse some of the runtime adaptation infrastructure across services.

In this chapter we present an architecture that supports automatic service optimization for a broad class of service, while still allowing service-specific optimizations. The key idea is to carefully separate the service-specific knowledge from generic functionality. The service-specific knowledge is represented as a *service recipe* that is used by the service infrastructure to perform customized configuration and adaptation. Using this approach, services can share a significant amount of functionality, yet get the benefit of service-specific optimizations.

This chapter is organized as follows. In Section 10.1 we use an example to illustrate the automatic service composition and adaptation problem. Section 10.2 describes how our architecture uses service-specific knowledge to configure and adapt service instances. We elaborate on how recipes specify service-specific knowledge in Section 10.3 and briefly describe our prototype implementation in Section 10.4. We evaluate our system in Section 10.5 and summarize in Section 10.6.

10.1 Example of Automatic Service Optimization

Assume we have a self-optimizing videoconferencing service that was built by a service developer with expertise in videoconferencing. A group of five users submits a request to hold a video conference: P1 and P2 use the MBone conferencing applications vic/SDR (VIC), P3 and P4 use NetMeeting (NM), and P5 uses a receive-only handheld device (HH). The service developer's service-specific knowledge is used to construct the following customized configuration (Figure 10.1). A videoconferencing gateway (VGW) is needed for protocol translation and video forwarding between VIC and NM. A handheld proxy (HHP) is needed for P5 to join the conference. The service uses an End System Multicast (ESM) [6] overlay consisting of three ESM proxies (ESMPs) to enable wide-area multicast among P1, P2, VGW, and HHP. The criterion for selecting all the components is to minimize the shown objective function to reduce the network resource usage. The weights in the function reflect the bandwidth usage, e.g., NetMeeting receives only one video stream.

At runtime, the service configuration needs to be adapted to accommodate environmental changes. For example, the developer may decide that

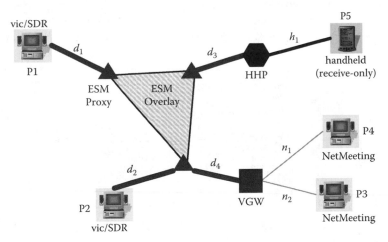

Objective: minimize $(5(d_1 + d_2 + d_3 + d_4) + (n_1 + n_2) + 2h_1)$
d_i, n_i, h_i: network latencies

FIGURE 10.1
A videoconferencing service.

runtime problems such as "VGW overload" and "VGW congestion" should be handled by replacing the VGW with a "high-capacity VGW" and "high-bandwidth VGW," respectively. These strategies are service specific because other developers could use different adaptation strategies, e.g., reducing the codec quality and bit rate, to handle overload and congestion. Another important aspect of the adaptation knowledge is how the strategies should be "coordinated." For example, suppose that a VGW becomes simultaneously overloaded and congested. Clearly only one of the two adaptation strategies should be allowed to execute.

This example shows that service-specific knowledge is important for automatic service optimization. More examples can be found elsewhere [21, 22, 19]. We now describe how service-specific knowledge is specified in our service framework.

10.2   Recipe-Based Service Composition and Adaptation

Our architecture for recipe-based service composition and adaptation is shown in Figure 10.2. A *service recipe* is written by a service developer and contains an operational description of the service-specific configuration and adaptation knowledge. At runtime, users can send requests to the synthesizer. Upon receiving the request, the synthesizer extracts the user requirements (e.g., the IP addresses and conferencing applications of the five participants) and invokes the interpreter module which executes the recipe to perform service composition. Similar to previous work [16, 35, 33, 14, 23, 17], composition involves two steps, *abstract mapping* and *physical mapping*, each of which benefits from service-specific knowledge. During the *abstract mapping* step, the synthesizer generates an *abstract configuration* specifying what types

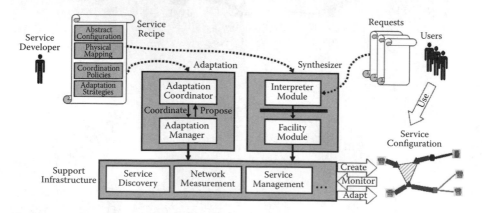

FIGURE 10.2
An architecture for recipe-based service composition and adaptation.

of components (e.g., VGW, HHP, ESMP) are needed to satisfy a request. Next, the *physical mapping* step maps each *abstract component* in the abstract configuration to a *physical component* (e.g., VGW$\Rightarrow$192.168.1.1, HHP$\Rightarrow$192.168.2.2, etc.). The outcome is customized service instance. The synthesizer is service independent and also includes a *facility module* that implements common, reusable service composition functions, e.g., component selection algorithms. The *facility interface* exports the facility module functionality in the forms of an application program interface (API) and libraries that can be invoked from recipes by developers (see next section). In our prototype system, recipes are written in Java, allowing a very flexible representation of service-specific knowledge.

The recipe also specifies service-specific adaptation knowledge in the form of *adaptation strategies* and *coordination policies*. This knowledge is used at runtime by the self-adaptation framework, which consists of an *adaptation manager* (AM) and an *adaptation coordinator* (AC). The AM realizes a developer's adaptation strategies by monitoring the configuration to detect runtime problems. When a problem occurs, e.g., a component becomes overloaded, and one of the developer's strategies is designed to handle the problem, the AM invokes the strategy to adapt the configuration, e.g., replace the overloaded component. When a strategy is executed, it submits a *proposal* specifying how it wants to change the configuration to the AC. If the proposal does not conflict with other proposals, the AC directs the AM to change the configuration accordingly. Otherwise, the AC rejects the proposal.

The supporting infrastructures provide common functions required for runtime configuration and adaptation, e.g., a network measurement infrastructure for measuring critical network performance metrics, a service discovery infrastructure for finding new components, a component management and deployment infrastructure for controlling/deploying the components, etc.

10.3   Recipe Specification

In this section, we describe how a service recipe can be used to represent both self-configuration (abstract and physical mapping) and adaptation (adaptation strategies and coordination policies) information.

10.3.1   Abstract Mapping Knowledge

An abstract configuration is a graph consisting of nodes representing the abstract components and of links representing the connections between the components. While the synthesizer needs to use the knowledge in the recipe to decide what components and connections to use, the data structures for the graph are generic and can be reused across services. Therefore, we provide a developer with generic data structures and functions (Figure 10.3)

Abstract Configuration API	
Data structure	
`AbsConf`	Represent abstract configurations; contains components and connections.
`AbsComp`	Represent abstract components; contains component properties.
`PhyComp`	Represent fixed components (e.g., users); contains component properties.
Function	
`addComp(spec)`	Add an abstract component with specification `spec` to the abstract configuration.
`addConn(c1,c2)`	Add a connection between components c1 and c2 to the abstract configuration.
`addSubComp(n,spec)`	Add *n* identical sub-components (with `spec`) to a component.
`getProperty(prop)`	Get the value of the property named `prop` of a component.

Objective Function API	
Data structure	
`LatencyM`	Represent the network latency between two components.
`BandwidthM`	Represent the available bandwidth between two components.
`Function`	Represent an objective function; contains a `Term`.
`Term`	Contain a metric or a floating point number; also provides member functions listed below for appending other instances of `Term` to "this" instance.
Function (member functions of `Term`)	
`add(t)`	Add t (an instance of `Term`) to "this" instance.
`subtract(t)`	Subtract t from this instance.
`multiplyBy(t)`	Multiply this instance by t.
`divideBy(t)`	Divide this instance by t.
`pow(t)`	Raise this instance to the t-th power.

Adaptation API	
Data structure	
`RelationOp`	Represent relation operators such as "==", ">=", etc.
`BooleanOp`	Represent boolean operators such as "AND", "OR", etc.
`Condition`	Represent a combination of "`Term RelationOp Term`".
`Constraint`	Represent a combination of "`Condition BooleanOp Condition`".
`Tactic`	Represent the base class for a tactic.
`Strategy`	Represent the base class for a strategy.
Function	
`replaceComponent(c)`	Represent an adaptation action that replaces an existing component c in the current configuration.
`changeParameter(pn,pv)`	Represent an adaptation action that changes the value of the parameter pn of a component to pv.
`connect(c1,c2)`	Represent an adaptation action that connects components c1 and c2.
`setTacticObjective(obj)`	Set the component selection objective for a tactic to obj, which will be used for, e.g., the `replaceComponent` actions.
`setConstraint(C)`	Associate the constraint C with an adaptation strategy.
`invokeTactic(T)`	Invoke the tactic T.
`addProblemConflict(t1,t2)`	Register a conflict between two tactics.

FIGURE 10.3
API used by recipes.

that can be used to construct abstract configurations in the recipe. This functionality is provided by the facility module.

Using the video conferencing service as an example, Figure 10.4 shows how the API can be used to create a customized abstract configuration. The right-hand side is the recipe, written in Java. Here we focus on the unshaded lines, which specify the abstract mapping; the physical mapping (shaded lines) are explained in the next section. The recipe has three segments, dealing with three different service features, which are illustrated by the figures on the left. The first segment inserts a VGW into the abstract configuration and connects it with every participant who uses NetMeeting. The second segment

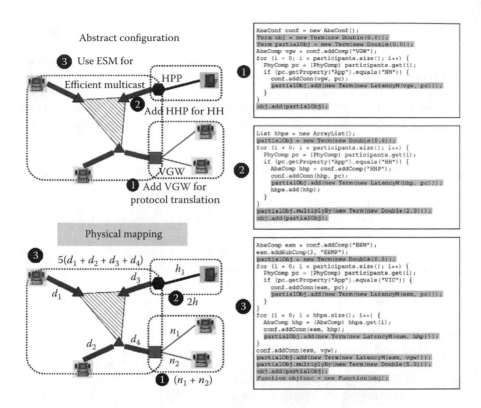

FIGURE 10.4
A videoconferencing recipe.

adds an HHP for each handheld participant and connects them. The third segment inserts an ESM component with three ESMP subcomponents and connects ESM with the multicast endpoints. Note that `participants` is the list of participants (specifically, a `List` of `PhyComp` objects) extracted from a user request by the synthesizer; for simplicity, we use symbolic names (e.g., "VGW") to represent complete specifications of component types and attributes (e.g., "(serviceType = VideoGateway) (protocols = H323,SIP)...").

When a user request is received, the synthesizer simply executes the recipe to construct the abstract configuration `conf`.

10.3.2 Physical Mapping Knowledge

Given an abstract configuration, the physical mapping problem involves mapping each abstract component onto a physical component so that an *objective* function is optimized. An important observation is that while the optimization objective is service-specific, the actual optimization is a generic operation that can be reused by different services. Therefore, our approach is to let a service developer specify a service-specific *objective function* in the recipe using the facility interface. The facility module then formulates an optimization problem accordingly and solves it using built-in algorithms.

An objective function is a function of *metric terms*, e.g., "5*latency(server, transcoder) + latency(transcoder,user)." Metrics represent properties of components or properties of connections between components. To use a metric in objective functions, the synthesizer must be able to obtain the represented property from the support infrastructure. For example, our prototype support infrastructure provides network latency information, so the synthesizer supports "latency" as a metric in objective functions. Other metrics such as speed of the central processing unit (CPU), memory size, bandwidth, and cost can be added by extending the support infrastructure to provide information on these properties.

The Java data structures and functions for constructing objective functions are shown in Figure 10.3. As an example, the videoconferencing developer's objective function (Figure 10.1) is a weighted sum of several latencies in the abstract configuration. The shaded lines in the videoconferencing recipe in Figure 10.4 show how this objective function is constructed using the above API. On the left-hand side of the figure, we illustrate which part of the objective function each segment of the recipe constructs. These partial objectives are added together in `obj`, which is finally used to construct the function `objfunc`. Note that developers can specify different objective functions, for example allowing them to make different tradeoffs between the cost of the optimization and the optimality of the resulting service instance [21].

After constructing the abstract configuration `conf` and the objective function `objfunc`, the recipe execution is complete. The synthesizer can then solve the physical mapping optimization problem and finally instantiate the service instance. Before we describe these runtime components in Section 10.4, we first elaborate on the adaptation and coordination information in the recipe.

10.3.3   Adaptation Strategies

Our framework uses an externalized approach to adaptation [15, 36] in which adaptation strategies and mechanisms are separated from the application or system. This allows us to build a general adaptation framework that can be reused by different systems. Adaptation strategies are specified as event-action rules that let a developer specify what "actions" should be taken when a particular "event" occurs. For example, when the event "component X becomes overloaded" occurs, the appropriate action is "replace X with a higher capacity one." This approach has been used for systems involving heterogeneous components [15, 10] because it is more flexible than solutions based on utility functions [36]. An adaptation strategy has three parts: a constraint, a problem determination module, and a tactic. We now elaborate on these three components and explain how they are specified in the recipe (see Figure 10.3). Note that our focus is on localized adaptation involving one component.

A strategy is invoked when its *constraint* is "violated." A constraint is a condition on certain properties of the configuration, e.g., "load(X) < C" where X is a component in the configuration. The properties can be connection performance metrics, e.g., bandwidth and latency, or component properties. At runtime, the AM monitors the constraints of the strategies and invokes a strategy when its condition becomes false. Constraints can be specified using the objective function API, extended with support for Boolean and relational operators.

A constraint violation may be caused by multiple *triggering problems*, so we need a *problem determination* step to determine the cause of the violation. This often involves collecting runtime and service configuration information. For example, in the videoconferencing example, a strategy triggered by "low HH video quality" needs to determine whether the actual problem is HHP failure, low-quality codec used by the HHP, or congestion at the HHP. Since our framework is based on Java, developers can naturally use Java to implement problem determination.

Finally, a *tactic* addresses a triggering problem by executing a set of *actions*. Actions range from changing a runtime parameter of a component to changing the configuration by inserting/removing components. In the example above, "HHP failure" can be addressed by a tactic that replaces the failed HHP with a new one; "low-quality codec" can be addressed by "increase the codec quality"; and so on. Tactics are specified using an extension of the abstract configuration API, which provides support for replacing components, changing parameters, etc. When a tactic requires a new component, it uses an objective function for physical mapping, similar to the initial configuration.

Finally, we provide a data structure to represent a strategy. The constraint of a strategy can be assigned using `setConstraint`, and a strategy can invoke a particular tactic using `invokeTactic`.

We believe our self-adaptation framework can also be applied to other component-based service frameworks such as Ninja [16], SWORD [33], and ACE [1]. The major requirement is that such a framework provide (1) a representation of the service configuration allowing our framework to reference the components in the configuration and (2) an interface allowing our framework to make changes to the configuration according to developers' knowledge.

10.3.4 Adaptation Coordination

As seen from earlier examples, different adaptation strategies may attempt to change the service configuration in conflicting ways. Therefore, adaptation coordination mechanisms are needed to detect and resolve such conflicts. Similar to the specification of configuration and adaptation, our goal is to require a service developer to specify only the service-specific coordination knowledge without worrying about the underlying mechanisms. In this section we focus on the detection of conflicts between proposals. We describe runtime conflict resolution in Section 10.4.

We categorize conflicts into two types: *action-level* and *problem-level*. An action-level conflict occurs when two proposals want to make configuration changes involving the same target component. For example, if one proposal wants to replace server A with B, and another proposal wants to replace A with C, obviously only one should be accepted. The AC can automatically detect such conflicts by looking at the actions in different proposals.

A problem-level conflict occurs when the intentions of two proposals conflict with each other, i.e., they are addressing two problems that should not be addressed simultaneously. For example, assume there are three ESMPs (A, B, and C) in the videoconferencing configuration. User U wants to join the videoconference and at the same time C fails, so two strategies, S1 and S2, are invoked. S1 proposes to connect U to B, since it is the closest to U. At the same time, S2 proposes to replace C with D. Since D may be closer to U than B is, the developer may want to delay S1 until after C has been replaced with D. Of course, another developer may prefer that S1 and S2 are executed together so that the new user join will not be delayed. This illustrates that problem-level conflicts are service specific: They cannot be detected automatically but must be specified explicitly by developers. We provide the "addProblemConflict" method for specifying conflicts between tactics.

10.4   Implementation

We briefly describe a prototype implementation of the recipe-based service optimization framework. We also elaborate on the implementation of the physical mapping optimization and the conflict resolution module.

10.4.1   Service Framework

We have implemented the synthesizer, including the interpreter and facility modules, the AM and AC, in Java. Service recipes are also written in Java and they use the classes and interface provided by the facility module. When it receives a request, the interpreter module dynamically loads the corresponding compiled recipe and invokes an entry function. The facility module accesses the support infrastructure to carry out actual operations. The AM and AC handle the strategies and coordination at runtime, respectively. We also implemented the self-optimizing videoconferencing system depicted in Figure 10.1.

We use existing solutions for the support infrastructure (see Section 10.4). We use the Global Network Positioning (GNP) [30] system to estimate the network latencies between nodes, supporting the LatencyM metric. We also designed and implemented the Network-Sensitive Service Discovery (NSSD) [20] infrastructure that provides, in addition to traditional service discovery, the capability to return the best m candidates given a local optimization criterion for a component, which is needed for some of optimization algorithms (described below).

10.4.2 Physical Mapping Optimization

The physical mapping problem involves two steps: For each abstract component (e.g., VGW), the synthesizer uses the support infrastructure to find candidates (e.g., machines running the gateway software) and it then selects the best mapping such that the objective function is optimized. The complexity of the mapping problem depends on the format of the objective function. In general, the placements of the components are mutually dependent. For example, if the objective function includes a single metric $M(c_1, c_2)$ (c_1 and c_2 are abstract components with n candidates each), the selection of c_1 *depends* on that of c_2, i.e., the synthesizer needs to select one of n^2 possible combinations to find the optimal selection. Furthermore, this dependency is *transitive*, e.g., if the objective is $M_1(c_1, c_2) + M_2(c_2, c_3)$, then all three are mutually dependent. In addition, some physical mapping problems involve *semidependent* components, e.g., in $M_1(c_1, c_2) + M_2(c_2)$, c_2 is semidependent (independent in the second term but dependent on c_1 in the first term).

If some or all components are mutually dependent, physical mapping is a *global optimization* problem with a worst-case problem size n^m, where m is the number of abstract components. If every abstract component is independent, e.g., if objective is $M_1(c_1) + M_2(c_2) + \ldots + M_m(c_m)$, physical mapping becomes a series of *local optimization* problems with a a complexity of mn. Researchers have also developed many specialized techniques for particular forms of the resource selection problem, e.g., [14, 5, 17, 27, 34].

In our architecture, developers can specify a broad range of objective functions, so the synthesizer must be able to solve a variety of mapping problems. Our approach is to implement a variety of optimization functions and to have the synthesizer pick an appropriate function for a given objective function. The most challenging problems are of course global optimizations. In general, one needs to implement heuristics that offer a tradeoff between the computational cost and quality of the resulting solution. We give the developers control over this tradeoff through a "quality" metric. Currently, this is a qualitative metric, e.g., high/low. The definition of a more precise quantitative metric is future work.

We have implemented the following optimization algorithms in our prototype:

- *Exhaustive search*: This algorithm always yields the actual optimal configuration, but the optimization cost grows rapidly with the problem size.

- *Sim-anneal(R)*: The simulated annealing heuristic is based on the physical process of "annealing" [18]. We use the temperature reduction ratio R as a parameter to control the cost/optimality tradeoff.

- *Hybrid(m)*: This heuristic is for problems with semidependent components, e.g., HHP and VGW in videoconferencing. The synthesizer reduces the problem size by choosing m "local candidates" for each semidependent component according to its "independent metric," and global optimization is performed on the reduced search space. Preliminary evaluation results for this heuristic are presented in [20].

- *HybridSA(m)*: This is the same as Hybrid(m) except that the final global optimization is performed using simulated annealing. The temperature reduction ratio increases with m to achieve better optimality (at higher costs).

10.4.3 Conflict Resolution

The AC must identify and resolve conflicts between proposed adaptations. It does this using a *First-Come, First-Serve (FCFS)* procedure, i.e., it accepts/rejects proposals as they are received. If no other proposals are being executed, the proposal is accepted. If one or more other proposals are being executed, the AC performs conflict detection between the new and running proposals, checking both for action-level and problem-level conflicts. If a conflict is detected, the new proposal is rejected.

An alternative to the FCFS resolution procedure is to use *epochs*, discrete time intervals. With this approach, the AC collects proposals during each epoch and performs conflict detection only at the end of the epoch. If proposals conflict, the one with the highest priority is accepted and all the others are rejected. Priorities can be assigned by the developer. The epoch-based approach supports more flexible conflict resolution, but the cost is reduced "agility" [31] since proposed actions always have to wait until the end of the epoch before they are considered for execution. Epoch-based conflict resolution has, for example, been used to coordinate update rules in active database systems [24]. We decided to use the FCFS approach because agility is very important in our application domain.

10.5   Evaluation

Features of the proposed framework that must be evaluated include its expressiveness for diverse services, the quality of the generated service instances, and its overhead. Here, we briefly look at the cost/optimality tradeoff supported by the physical mapping and the expressiveness of the adaptation strategies. More evaluation results can be found elsewhere [19]. In order to be able to evaluate large deployments, evaluation is based on a simulator that runs our framework over a simulated network.

10.5.1   Physical Mapping: Cost/Optimality Trade-Offs

We simulate the videoconferencing scenario in Figure 10.1, using the different optimization algorithms listed in the previous section. The simulated network consists of 869 Internet nodes whose GNP coordinates correspond to those of real nodes [30], so the latencies between the nodes are realistic. The problem size n in our simulations ranges from 5 to 200 (each of VGW, HHP, and ESMP has n candidates). We generate 200 requests (5 participants each) in 10 different candidate distributions. For each request, the synthesizer executes the recipe (Figure 10.4) and uses the implemented algorithms to find the optimal configuration. Our evaluation metrics are the optimization cost, expressed as the average *optimization time per request*, and quality of the generated service instance, expressed as the optimality relative to the best-performing algorithm. Given that the optimization time is measured using an unoptimized Java prototype running on a low-end desktop machine (Pentium III 933MHz CPU, Red Hat Linux 7.1, and J2SE 1.4.2), the measurements should be used only for comparison purposes. Production environments would achieve much better performance using optimized implementations.

For low values of n, Exhaustive search and the Hybrid algorithm give the best results, but their costs increase rapidly with n ($O(n^5)$ for Exhaustive; $O(m^2n^3)$ for Hybrid). Here we focus on larger values of n. Figure 10.5(a) and (b) show the optimality and optimization cost for the different algorithms. HybridSA generally performs better than SA, and the costs of both increase slowly with n (with SA growing faster due to the larger search spaces). Note: the fact that the curves become flat for n between 100 and 200 means only that all algorithms deteriorate at roughly the same rate, since we normalize to the best-performing algorithm.

These results can be used as guidelines for the synthesizer to select the appropriate algorithm for each instance of this mapping problem based on both the specified cost/optimality tradeoff and the properties of the problem. For example, if the problem size is large ($n = 100$), the cost constraint is 0.75, and the optimality constraint is 1.3, then the synthesizer will select HybridSA(1).

10.5.2 Adaptation: Conflict Resolution

To evaluate adaptation coordination, we applied our framework to a simulated massively multiplayer online gaming service, depicted in Figure 10.6. The simulation models both users and servers. Users move around randomly in the virtual game space, and each server handles the users in a partition of the space. We generate traces of user arrivals and departures where the interarrival time has an exponential distribution, and the stay duration has a bounded Pareto distribution. Each simulation has a duration of 30K seconds and there are on average about 142 users.

We defined five adaptation strategies for the gaming service: *join* connects a new user to the corresponding server, *leave* disconnects a user from its server, *cross* moves a user's state from one server to another, *split* adds a new server when an existing one becomes overloaded, and *merge* removes a server by merging two underutilized servers. Since each strategy invokes only one tactic, the names also refer to the tactics. We present simulation results for two scenarios, assuming different sets of conflicts between tactics.

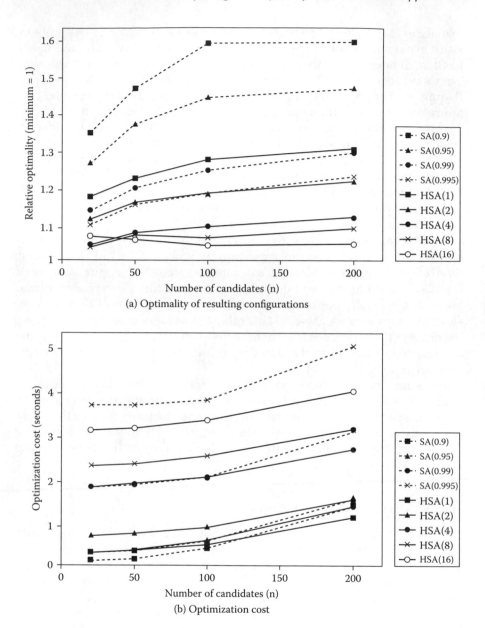

FIGURE 10.5
Optimization for larger-scale problems.

First, let us assume that the gaming service developer uses the "addProb-lemConflict" call to specify the following conflicts between tactics: *join-split*, *join-merge*, *cross-split*, *cross-merge*, *leave-split*, and *leave-merge*. We assume also that when a proposal is rejected, the proposing tactic will repropose as soon as possible, i.e., a rejected adaptation is delayed. Table 10.1 shows the number

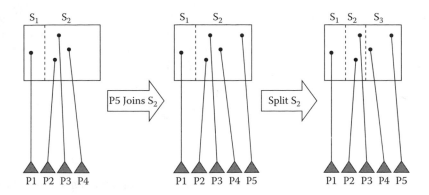

FIGURE 10.6
A massively multiplayer online gaming service.

and percentage of each type of adaptation that is delayed. The percentage of delayed adaptations is much higher for split/merge than that for join/leave/cross. This is because there are many more join/leave/cross operations than split/merge. However, overall, the impact of the delays is small because (1) few join/leave/crosses are delayed and (2) although most split/merges are delayed, they are much less frequent and are expensive (requiring 520 ms without delays), so the delay (maximum about 100 ms) is not significant.

To see how easily a developer can apply a different set of coordination policies, consider the following scenario. Suppose that the above developer improves the server implementation such that it can handle a leaving user in parallel with other adaptations. To take advantage of this new capability, the developer can simply remove the lines specifying conflicts that involve "leave," leaving the following four conflicts: *join-split, join-merge, cross-split*, and *cross-merge*. Table 10.1 shows the simulation results for this reduced set of conflicts. As expected, no leave adaptations are delayed with the new policies. Furthermore, eliminating "leave conflicts" results in fewer split/merge adaptations. This is because having fewer leave conflicts stabilizes the configuration such that fewer split/merge adaptation are necessary. In turn, this allows more join/cross adaptations to be executed without delay. Finally,

TABLE 10.1

Number and Percentage of Delayed Adaptations

	With Leave Conflicts			Without Leave Conflicts		
	Num. Adapt.	Adapt. Delayed	Percent Delayed	Num. Adapt.	Adapt. Delayed	Percent Delayed
Join	50149	258	0.51	50149	184	0.37
Leave	49992	384	0.77	49992	0	0
Cross	6194675	38146	0.62	6375261	28528	0.45
Split	789	715	90.6	623	573	92.0
Merge	783	746	95.3	613	587	95.8

because the users now spend less time "waiting" (i.e., delayed), they have more time to move around in the game space, resulting in more cross adaptations, as seen in Table 10.1. This example demonstrates that our approach allows developers to easily implement different service-specific coordination policies.

10.6   Conclusion

We presented a general architecture for building self-optimizing services. Our framework supports both the dynamic composition of service configurations that are customized for particular user requirements and system characteristics and the runtime adaptation of the configuration in response to changes in the system or requirements. We designed a recipe representation that can be used by developers to capture their service-specific knowledge. This recipe is used by a generic runtime infrastructure to realize both the initial service configuration and adaptation. The runtime infrastructure includes a synthesizer, which constructs an abstract service configuration and maps it onto physical nodes, an adaptation manager that monitors the service and applies adaptation strategies when needed, and an adaptation coordinator that resolves conflicts between adaptation strategies. We used a videoconferencing example to illustrate the capabilities and performance of our framework.

Acknowledgments

This research was sponsored in part by the National Science Foundation under award number CCR-0205266, and in part by the Defense Advanced Research Project Agency and monitored by AFRL/IFGA, Rome, NY 13441-4505, under contract F30602-99-1-0518.

References

1. M. Agarwal and M. Parashar. Enabling Autonomic Compositions in Grid Environments. In *Proceedings of the Fourth International Workshop on Grid Computing (Grid 2003)*, 34–41, Nov. 2003.
2. R. K. Balan, M. Satyanarayanan, S. Park, and T. Okoshi. Tactics-Based Remote Execution for Mobile Computing. In *Proceedings of the First International Conference on Mobile Systems, Applications, and Services (MobiSys '03)*, 273–286, May 2003.

3. J. W. Cangussu, K. Cooper, and C. Li. A Control Theory Based Framework for Dynamic Adaptable Systems. In *Proceedings of the 2004 ACM Symposium on Applied Computing (SAC '04)*, 1546–1553, Mar. 2004.

4. F. Chang and V. Karamcheti. Automatic Configuration and Run-time Adaptation of Distributed Applications. In *Proceedings of the Ninth IEEE International Symposium on High Performance Distributed Computing (HPDC-9)*, 11–20, Aug. 2000.

5. S. Choi, J. Turner, and T. Wolf. Configuring Sessions in Programmable Networks. In *Proceedings of IEEE INFOCOM 2001*, Apr. 2001.

6. Y. Chu, S. G. Rao, and H. Zhang. A Case for End System Multicast. In *Proceedings of the ACM International Conference on Measurement and Modeling of Computer Systems (SIGMETRICS 2000)*, June 2000.

7. K. Czajkowski, I. Foster, N. Karonis, C. Kesselman, S. Martin, W. Smith, and S. Tuecke. A Resource Management Architecture for Metacomputing Systems. In *Proceedings of the IPPS/SPDP '98 Workshop on Job Scheduling Strategies for Parallel Processing*, 62–82, Mar. 1998.

8. E. de Lara, D. S. Wallach, and W. Zwaenepoel. Puppeteer: Component-Based Adaptation for Mobile Computing. In *USENIX Symposium on Internet Technologies and Systems (USITS 2001)*, Mar. 2001.

9. R. Doyle, J. Chase, O. Asad, W. Jin, and A. Vahdat. Model-Based Resource Provisioning in a Web Service Utility. In *Proceedings of the 4th USENIX Symposium on Internet Technologies and Systems (USITS '03)*, Mar. 2003.

10. C. Efstratiou, A. Friday, N. Davies, and K. Cheverst. Utilising the Event Calculus for Policy Driven Adaptation on Mobile Systems. In *Proceedings of the Third International Workshop on Policies for Distributed Systems and Networks (POLICY 2002)*, June 2002.

11. B. Ensink and V. Adve. Coordinating Adaptations in Distributed Systems. In *Proceedings of the 24th International Conference on Distributed Computing Systems (ICDCS 2004)*, Mar. 2004.

12. I. Foster and C. Kesselman. Globus: A Metacomputing Infrastructure Toolkit. *The International Journal of Supercomputer Applications and High Performance Computing*, 11(2):115–128, 1997.

13. I. Foster, C. Kesselman, J. M. Nick, and S. Tuecke. Grid Services for Distributed System Integration. *IEEE Computer*, 35(6), June 2002.

14. X. Fu, W. Shi, A. Akkerman, and V. Karamcheti. CANS: Composable, Adaptive Network Services Infrastructure. In *Proceedings of the Third USENIX Symposium on Internet Technologies and Systems (USITS '01)*, Mar. 2001.

15. D. Garlan, S.-W. Cheng, A.-C. Huang, B. Schmerl, and P. Steenkiste. Rainbow: Architecture-Based Self-Adaptation with Reusable Infrastructure. *IEEE Computer*, 37(10), Oct. 2004.

16. S. D. Gribble, M. Welsh, R. von Behren, E. A. Brewer, D. Culler, N. Borisov, S. Czerwinski, R. Gummadi, J. Hill, A. Joseph, R. Katz, Z. Mao, S. Ross, and B. Zhao. The Ninja Architecture for Robust Internet-Scale Systems and Services. *IEEE Computer Networks, Special Issue on Pervasive Computing*, 35(4), Mar. 2001.

17. X. Gu and K. Nahrstedt. A Scalable QoS-Aware Service Aggregation Model for Peer-to-Peer Computing Grids. In *Proceedings of the 11th IEEE International Symposium on High Performance Distributed Computing (HPDC-11)*, July 2002.

18. J. Hromkovič. *Algorithmics for Hard Problems: Introduction to Combinatorial Optimization, Randomization, Approximation, and Heuristics*. Springer, Berlin, 2001.

19. A.-C. Huang. Building Self-configuring Services Using Service-specific Knowledge, Dec. 2004. Phd Thesis, Department of Computer Science, Carnegie Mellon University.

20. A.-C. Huang and P. Steenkiste. Network-Sensitive Service Discovery. In *The Fourth USENIX Symposium on Internet Technologies and Systems (USITS '03)*, Mar. 2003.

21. A.-C. Huang and P. Steenkiste. Building Self-configuring Services Using Service-specific Knowledge. In *Proceedings of the Thirteenth IEEE International Symposium on High-Performance Distributed Computing (HPDC-13)*, 45–54, June 2004.

22. A.-C. Huang and P. Steenkiste. Building Self-Adapting Services Using Service-specific Knowledge. In *Proceedings of the Fourteenth IEEE International Symposium on High-Performance Distributed Computing (HPDC-14)*, 34–43, July 2005.

23. A.-A. Ivan, J. Harman, M. Allen, and V. Karamcheti. Partitionable Services: A Framework for Seamlessly Adapting Distributed Applications to Heterogeneous Environments. In *Proceedings of the 11th IEEE International Symposium on High Performance Distributed Computing (HPDC-11)*, July 2002.

24. H. V. Jagadish, A. O. Mendelzon, and I. S. Mumick. Managing Conflicts between Rules. In *Proceedings of the 15th ACM SIGACT/SIGMOD Symposium on Principles of Database Systems (PODS '96)*, 192–201, June 1996.

25. A. D. Joseph, A. F. deLespinasse, J. A. Tauber, D. K. Gifford, and M. F. Kaashoek. Rover: A Toolkit for Mobile Information Access. In *Proceedings of the Fifteenth Symposium on Operating Systems Principles*, 156–171, Dec. 1995.

26. C. Lee, J. Lehoczky, D. Siewiorek, R. Rajkumar, and J. Hansen. A Scalable Solution to the Multi-Resource QoS Problem. Technical Report CMU-CS-99-144, Carnegie Mellon University, May 1999.

27. C. Liu, L. Yang, I. Foster, and D. Angulo. Design and Evaluation of a Resource Selection Framework for Grid Applications. In *Proceedings of the 11th IEEE International Symposium on High Performance Distributed Computing (HPDC-11)*, July 2002.

28. J. López and D. O'Hallaron. Evaluation of a resource selection mechanism for complex network services. In *Proceedings of the Tenth IEEE International Symposium on High Performance Distributed Computing*, Aug. 2001.

29. J. P. Loyall, R. E. Schantz, J. A. Zinky, and D. E. Bakken. Specifying and Measuring Quality of Service in Distributed Object Systems. In *Proceedings of the First IEEE International Symposium on Object-Oriented Real-Time Distributed Computing (ISORC '98)*, Apr. 1998.

30. T. S. E. Ng and H. Zhang. Predicting Internet Network Distance with Coordinates-Based Approaches. In *Proceedings of IEEE INFOCOM 2002*, June 2002.

31. B. D. Noble, M. Satyanarayanan, D. Narayanan, J. E. Tilton, J. Flinn, and K. R. Walker. Agile Application-Aware Adaptation for Mobility. In *Proceedings of the Sixteenth ACM Symposium on Operating Systems Principles*, Oct. 1997.

32. V. Poladian, J. P. Sousa, D. Garlan, and M. Shaw. Dynamic Configuration of Resource-Aware Services. In *Proceedings of the 26th International Conference on Software Engineering (ICSE '04)*, 604–613, May 2004.

33. S. R. Ponnekanti and A. Fox. SWORD: A Developer Toolkit for Web Service Composition. In *Proceedings of the Eleventh World Wide Web Conference (WWW 2002), Web Engineering Track*, May 2002.

34. R. Raman, M. Livny, and M. Solomon. Policy Driven Heterogeneous Resource Co-Allocation with Gangmatching. In *Proceedings of the Twelfth IEEE International Symposium on High-Performance Distributed Computing*, June 2003.

35. P. Reiher, R. Guy, M. Yarvis, and A. Rudenko. Automated Planning for Open Architectures. In *Proceedings of the Third IEEE Conference on Open Architectures and Network Programming (OPENARCH 2000)—Short Paper Session*, 17–20, Mar. 2000.

36. W. E. Walsh, G. Tesauro, J. O. Kephart, and R. Das. Utility Functions in Autonomic Systems. In *Proceedings of the International Conference on Autonomic Computing (ICAC'04)*, 70–77, May 2004.

Part III

Achieving *Self-** Properties — Enabling Systems, Technologies, and Services

11

A Programming System for Autonomic Self-Managing Applications

Hua Liu and Manish Parashar

CONTENTS

The emergence of pervasive wide-area distributed computing environments, such as pervasive information systems and computational Grids, has enabled new generations of applications that are based on seamless access, aggregation, and interaction. However, the inherent complexity, heterogeneity, and dynamism of these systems require a change in how the applications are developed and managed. In this chapter we describe Accord, a programming system that extends existing programming models/systems to support the development of autonomic self-managing applications. Accord enables the development of autonomic elements and the formulation of autonomic applications as the dynamic composition of autonomic elements. The design and evaluations of prototype implementations of Accord are presented.

11.1 Introduction

Pervasive wide-area distributed computing environments, such as pervasive information systems and computational Grids, are emerging dominant platforms for a new generation of applications that combine intellectual and physical resources spanning many disciplines and organizations and provide vastly more effective solutions to important scientific, engineering, business, and government problems. For example, it is possible to conceive of a new generation of scientific and engineering simulations of complex physical phenomena that symbiotically and opportunistically combine computations, experiments, observations, and real-time data and can provide important insights into complex systems such as interacting black holes and neutron stars, formations of galaxies, and subsurface flows in oil reservoirs and aquifers, etc. Other examples include pervasive applications that leverage the pervasive information Grid to continuously manage, adapt, and optimize our living context, crisis management applications that use pervasive conventional and unconventional information for crisis prevention and response, medical applications that use in-vivo and in-vitro sensors and actuators for patient management, and business applications that use anytime-anywhere information access to optimize profits.

However, these emerging Grid computing environments, which are based on seamless access to and aggregation of geographically distributed resources, services, data, instruments, and expertise, present significant challenges at all levels. The key characteristics of Grid environments and applications include: (1) Heterogeneity: Both Grid environments and applications aggregate multiple independent, diverse, and geographically distributed elements and resources; (2) Dynamism: Grid environments are continuously changing during the lifetime of an application. Applications similarly have dynamic runtime behaviors, including the organization and interactions of its elements; (3) Uncertainty: Uncertainty in Grid environments is caused by multiple factors, including dynamism that introduces unpredictable and changing behaviors, failures that have an increasing probability of occurrences as system/application scales increase, and incomplete knowledge of global state, which is intrinsic to large distributed environments; (4) Security: A key attribute of Grids is secure resource sharing across organization boundaries, which makes security a critical challenge.

The characteristics listed above impose key requirements on programming models and systems for Grid applications. Specifically, Grid applications must be able to detect and dynamically respond during execution to changes in both the state of execution environment and the state and requirements of the application. This suggests that (1) Grid applications should be composed from discrete, self-managing elements (components/services), which incorporate separate specifications for functional, nonfunctional, and interaction/coordination behaviors; (2) the specifications of computational (functional) behaviors, interaction and coordination behaviors, and nonfunctional behaviors (e.g., performance, fault detection and recovery) should be separated so that their combinations are composeable; and (3) policy should be separated from mechanisms and used to orchestrate a repertoire of mechanisms to achieve context-aware adaptive runtime behaviors. Given these features, a Grid application requiring a given set of computational behaviors may be integrated with different interaction and coordination models or languages (and vice versa) and different specifications for nonfunctional behaviors such as fault recovery and quality of service (QoS) to address the dynamism and heterogeneity of application state and the execution environment.

While there exist several programming models and frameworks on which current Grid programming systems have been built, these existing systems either make very strong assumptions about the behavior of the entities, their interactions, and the underlying system, or have limited support for runtime adaptation, which prevents them from meeting the requirements outlined above. These limitations of current programming systems have led researchers to investigate alternate approaches where applications and application elements are capable of managing themselves based on high-level guidance, with minimal human intervention.

This chapter describes the Accord programming system that extends existing programming systems to enable the development of self-managing Grid applications by relaxing the requirement for static (at the time of instantiation) definition of application and system behaviors and allowing them to be dynamically specified at runtime. Further, it enables the behaviors, organizations, and interactions of elements and applications to be sensitive to the dynamic state of the system and the changing requirements of the application, and to adapt to these changes at runtime. The Accord programming system (1) enables the definition of autonomic elements that encapsulates functional and nonfunctional specifications as well as policies (in the form of high-level rules) and mechanisms for self-management, (2) enables the formulation of self-managing applications as dynamic compositions of autonomic elements, and (3) provides a runtime infrastructure for consistently and efficiently enforcing policies to enable autonomic self-managing functional, interaction, and composition behaviors based on current requirements, state, and execution context. Accord supports two levels of adaptations: (1) at the element level to monitor and control behaviors of individual elements according to their internal state and execution context, (2) at the application level to change application topologies, communication paradigms, and coordination models used among elements to respond to changes in the environments and user requirements. The design and evaluations of prototype implementations of Accord based on a component and service programming models is presented. Note that Accord is part of Project AutoMate [21], which provides required middleware services.

The rest of this chapter is organized as follows. Section 11.2 provides a brief overview of existing programming systems and discusses their capabilities and limitations with respect to the programming requirements of Grid environments. The section also discusses existing adaptation technologies and related efforts. Section 11.3 describes the Accord programming system, including the definition of autonomic elements and rules, and the runtime infrastructure that executes rules to enable adaptation behaviors. This section also presents the design and evaluation of two prototypes of Accord, Accord-CCA and Accord-WS. Section 11.4 concludes the chapter with a summary of the research and current status.

11.2 Background and Related Work

11.2.1 Programming Models and Systems

There has been a significant body of research on programming frameworks for parallel and distributed computing over the last few decades, which has been adapted/extended to support Grid environments. These can be broadly classified as communication frameworks, distributed object systems, component-based systems, and service-based systems.

Current **communication frameworks for distributed and parallel computing** (i.e., message passing models and shared memory models) supplement existing programming systems to support interactions between distributed entities. These systems typically make very strong assumptions about the behavior of the entities, their interactions, and the underlying system, especially about their static nature and reliable behaviors, which limit their applicability to highly dynamic and uncertain computing environments. In **distributed object systems** such as CORBA [9], objects and interactions are tightly coupled, and the model assumes a priori knowledge of the syntax and semantics of interfaces and interactions. However, the model can potentially support adaptation by providing portable request interceptors. Adaptation in this case is achieved by manipulating and redirecting messages, but the direct adaptation of individual objects is not supported. Current **component-based programming systems**, such as CORBA Component Model (CCM) [9], JavaBeans [27], and the Ccaffeine Common Component Architecture (CCA) framework [3], also do not directly support adaptation. However, they do provide some core mechanisms, such as interceptors in CORBA, the container mechanism in

JavaBeans, and the BuilderService in CCA, which can be extended to support adaptation. Note that communication patterns between components and their coordination are typically statically defined. **Service-based systems** such as the Web service and Grid service models [10] allow applications to be constructed in a more flexible and extensible way, but the runtime behaviors of services and applications are still rigid and they implicitly assume that context does not change during the lifetime of the service instance—services are customized when they are instantiated. Current orchestration and choreography mechanisms for Web and Grid services are also static and must be defined a priori.

11.2.2   Adaptation Technologies

Adaptation technologies can be integrated with existing programming systems to realize self-managing applications. This approach involves three key steps: (1) **Specifying adaptation behaviors:** Adaptation behaviors can be either statically specified using templates [5], classes [6] or scripts [4], or dynamically specified in the form of code/scripts/rules [26, 31]. A key drawback of static specification is that all the possible adaptation must be known a priori and coded into the applications, and the application has to be modified and possibly recompiled if new adaptation behaviors are required or application requirements change. (2) **Enforcing adaptation behaviors:** Adaptation enforcement mechanisms in case of legacy applications, where application source code is not accessible, view application units as black box and infer and adapt application behaviors based on low level observations as in [12], and/or use wrapping [30], filters [24], or proxies [25] to introduce adaptation behaviors. However, adaptability using this approach is very limited. If source code is accessible, adaptation behaviors can be integrated within the application execution using techniques such as instrumentation [22] and superimposition [7]. (3) **Conflict detection and resolution:** Conflicts can occur between multiple adaptation behaviors, and between adaptation behaviors and application execution. Conflicts may be detected and resolved statically [5, 6]; however, this approach may not detect conflicts that depend on runtime state. Runtime conflict resolution has been achieved using microeconomic techniques or legal reasoning, as in [1].

11.3   The Accord Programming System

The Accord programming system [18, 19] addresses Grid programming challenges by extending existing programming systems to enable autonomic Grid applications. Accord realizes three fundamental separations: (1) a separation of computations from coordination and interactions; (2) a separation of nonfunctional aspects (e.g., resource requirements, performance) from functional behaviors, and (3) a separation of policy and mechanism—policies in the form of rules are used to orchestrate a repertoire of mechanisms to achieve context-aware adaptive computational behaviors and coordination and interaction relationships based on functional, performance, and QoS requirements at runtime. The Accord conceptual model, the design and evaluation of two prototype implementations, and illustrative applications are presented in this section.

11.3.1 Conceptual Model

Accord extends existing distributed programming models, i.e., object-, component-, and service-based models, to support autonomic self-management capabilities. Specifically it extends the entities and composition strategies defined by the underlying programming model to enable computational and composition/interaction behaviors to be defined at runtime using high-level rules. The resulting autonomic elements and their autonomic composition are described below. Note that other aspects of the programming model, i.e., operations, model of computation, and rules for composition, are inherited and maintained by Accord.

Autonomic Element. An autonomic element extends a programming element (i.e., object, component, service) to define a self-contained modular software unit with specified interfaces and explicit context dependencies. Additionally, an autonomic element encapsulates rules, constraints, and mechanisms for self-management and can dynamically interact with other elements and the system. An autonomic element is illustrated in Figure 11.1 and is defined by three ports. (1) The *functional port* defines a set of functional behaviors that are provided and used by the element. (2) The *control port* is a set of tuples composed of the sensors and actuators exported by the element, and constraints that control access to the sensors/actuators. Sensors are interfaces that provide information about the element while actuators are interfaces for modifying the state of the element. Constraints are based on state, context, and/or high-level access policies. (3) The *operational port* defines the interfaces for formulating, dynamically injecting, and managing rules that are used for controlling the runtime behaviors of the elements, the interactions between elements, between elements and their environments, and the coordination within an application.

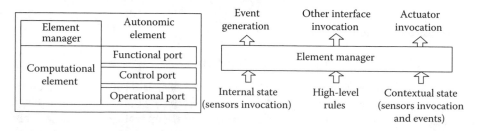

FIGURE 11.1
An autonomic element in Accord.

The control and operational ports enhance element interfaces to export information about their behaviors and adaptability to system and application dynamics. Each autonomic element is associated with an (possibly embedded) element manager that is delegated to manage its execution. The element manager monitors the state of the element and its context, and controls the execution of rules. Note that element managers may cooperate with other element managers to fulfill application objectives.

Rules in Accord. Rules incorporate high-level guidance and practical human knowledge in the form of if-then expressions, i.e., IF *condition* THEN *actions*, similar to production rule, case-based reasoning and expert systems. *Condition* is a logical combination of element (and environment) sensors, function interfaces, and events. *Actions* consist of a sequence of invocations of element and/or system sensors/actuators, and other interfaces. A rule fires when its condition expression evaluates to be true and causes the corresponding actions to be executed. Rule execution and conflict resolution mechanism are defined for different application types, and are discussed in Sections 11.3.2 and 11.3.3.

Two classes of rules are defined in Accord: (1) *behavioral rules* that control the runtime functional behaviors of an autonomic element (e.g., the dynamic selection of algorithms, data representation, input/output format used by the element); and (2) *interaction rules* that control the interactions between elements, between elements and their environment, and the coordination within an autonomic application (e.g., communication mechanism, composition and coordination of the elements). Note that behaviors and interactions expressed by these rules are defined by the model of computation and the rules for composition of the underlying programming model. Behavioral rules are executed by an element manager embedded within a single element without affecting other elements. Interaction rules define interactions among elements. For each interaction pattern, a set of interaction rules are defined and dynamically injected into the interacting elements. The coordinated execution of these rules results in the desired interaction and coordination behaviors between the elements.

Dynamic Composition in Accord. Dynamic composition enables relationships between elements to be established and modified at runtime. Operationally, dynamic composition consists of a composition plan or workflow generation and execution. Plans may be created at runtime, possibly based on dynamically defined objectives, policies and applications, and system context and content. Plan execution involves discovering elements, configuring them, and defining interaction relationships and mechanisms. This may result in elements being added, replaced, or removed or the interaction relationships between elements being changed.

In Accord, composition plans may be generated externally, possibly using the Accord Composition Engine (ACE) [2], and element discovery uses services provided by the AutoMate content-based middleware. A composition relationship between two elements is defined by the control structure (e.g., loop, branch) and/or the communication mechanism (e.g., remote procedure call (RPC), sharedspace) used. Interaction rules are generated from application workflows to describe the composition relationship among elements and are executed by peer element managers to establish control and communication relationships among these elements in a decentralized manner. Rules can be similarly used to add or delete elements. Note that the interaction rules must be based on the core primitives provided by the system. Accord defines a library of rule-sets for common control and communications relationships between elements. The interaction rule generation procedure will guarantee that the local behaviors of individual elements will coordinate to achieve the application's objectives.

Accord Runtime Infrastructure. The Accord runtime infrastructure consists of a portal, peer element and application composition/coordination managers, the autonomic elements, and a decentralized rule enforcement engine. An application composition manager decomposes incoming application workflows (defined by the user or a workflow engine) into interaction rules for individual elements and forwards these rules to corresponding element managers. Element managers execute these rules to establish interaction relationships among elements by negotiating communication protocols and mechanisms and dynamically constructing coordination relationships in a distributed and decentralized manner. Application managers also forward incoming adaptation rules to appropriate element managers. Element managers execute these rules to adapt the functional behaviors of the managed elements. These adaptations are realized by invoking appropriate control (sensors, actuators) and functional interfaces.

Accord Implementation Issues. Accord assumes the existence of common knowledge in the form of an ontology and taxonomy that defines the semantics for specifying and describing application name-spaces, and element interfaces, sensors and actuators, and system/application context and content. This common semantics is used for formulating rules for autonomic management of elements and dynamic composition and interactions between the elements. Further, it assumes the existence of an execution environment that provides (1) support for element communication and coordination, and (2) service for element discovery. Three prototypes of Accord have been implemented as follows:

- An object-based prototype of Accord, named DIOS++ [17], implements autonomic elements as autonomic objects by associating objects with sensors, actuators, and rule agents, and providing a run-time hierarchical infrastructure consisting of rule agents and rule engines for the rule-based autonomic monitoring and control of parallel and distributed applications.

- A component-based prototype of Accord, named Accord-CCA [16], based on the DoE CCA and the Ccaffeine framework in the context of component-based high-performance scientific applications. This prototype extends CCA components to autonomic components by associating them with control and operation ports and component managers, and provides a runtime infrastructure of component managers and composition managers for rule-based component adaptation and dynamic replacement of components.

- A service-based prototype of Accord, named Accord-WS [14], based on the WS-Resource specifications, the Web service specifications, and the Axis framework. Autonomic elements are implemented as autonomic services by extending traditional WS-Resources with service managers for rule-based management of runtime behaviors and interactions with other autonomic services, and coordination agents for programmable communications. A distributed runtime infrastructure is investigated to enable decentralized and dynamic compositions of autonomic services.

Accord-CCA and Accord-WS are described in more detail in this section. A description of DIOS++ can be found in [13, 17]. Accord is currently being used to enable autonomic simulations in subsurface modeling, combustion, and other areas [16, 14]. Further, the prototype implementations interface with advanced feature-based visualization techniques to enable both interactive [8] as well as rule-based automated [15] visualization and feature tracking.

11.3.2 Accord-CCA: A Programming System for Autonomic Component-Based Applications

Accord-CCA is the component-based prototype of Accord. It builds on the DoE CCA and the Ccaffeine framework [3] and supports the development of self-managing component-based high-performance parallel/distributed scientific applications. Further, using the TAU (transport and up) framework [28], Accord-CCA enables performance-driven component and application self-management.

11.3.2.1 Defining Managed Components

In order to monitor and control the behaviors and performance of CCA components, the components must implement and export appropriate "sensor" and "actuator" interfaces. Note that the sensor and actuator interfaces are similar to those used in monitoring/steering systems. However, these systems focus on interactive management where users manually invoke sensors and actuators, while this research focuses on automatic management based on user-defined rules. Adding sensors requires modification/instrumentation of the component source code. In case of third-party and legacy components, where such a modification may not be possible or feasible, proxy components [25] are used to collect relevant component information. A proxy provides the same interfaces as the actual component and is interposed between the caller and callee components to monitor, for example, all the method invocations for the callee component. Actuators can similarly be either implemented as new methods that modify internal parameters and behaviors of a component or defined in terms of existing methods if the component

```
class RulePort: public virtual Port {
public:
    RulePort(): Port() { }
    virtual ~RulePort() { }
    virtual void loadRules(const char* fileName) throw(Exception) = 0;
    virtual void addSensor(Sensor *snr) throw(Exception) = 0;
    virtual void addActuator(Actuator *atr) throw(Exception) = 0;
    virtual void setFrequency() throw(Exception) = 0;
    virtual void fire() throw(Exception) = 0;
};
```

FIGURE 11.2
The Accord-CCA *RulePort* specification.

cannot be modified. The adaptability of the components may be limited in the latter case. In the CCA-based implementation, both sensors and actuators are exposed by invoking the 'addSensor' or 'addActuator' methods defined by a specialized *RulePort*, which is shown in Figure 11.2. Management and adaptation behaviors can be dynamically specified by developers in the form of rules. Behavior rules are implemented as component rules, and interaction rules are implemented as composition rules.

11.3.2.2   The Accord-CCA Runtime Infrastructure

Accord-CCA defines two specialized component types to enable runtime self-management (see Figure 11.3): (1) a component manager that monitors and manages the behaviors of individual components, e.g., selecting the optimal algorithms or modifying internal states, and (2) a composition manager that manages, adapts, and optimizes the execution of an application at runtime. Both component and composition managers are peers of user components and other system components, providing and/or using ports that are connected to other ports by the Ccaffeine framework. The two managers are not part of the Ccaffeine framework, which prevents the framework from being 'overweight' and thus avoids the resulting performance and maintenance implications. Further, programmers can flexibly integrate them into their applications only as needed.

Component managers provide the *RulePort* shown in Figure 11.2. They are instantiated only after the other application components are composed together. Their instantiation consists of two steps: first, instances of managed components expose their sensors and actuators to their respective component manager instances by invoking the 'addSensor' and 'addActuator' methods, and second, component rules are then loaded into component manager instances, possibly from disk files, by invoking the 'loadRules' method. This initialization of component manager instances is a onetime operation.

Management operations are performed during application *quiet intervals*. The managed components (or their proxies) invoke the 'fire' method of the *RulePort* to inform the component managers that they have entered into a quiet interval. This behavior must be explicitly programmed, possibly at the beginning/end of a computation phase or once every few phases, to establish

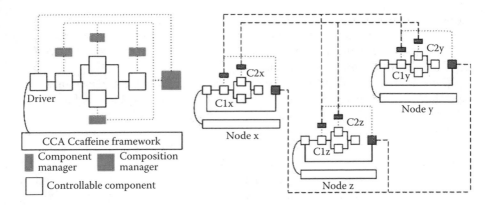

FIGURE 11.3

(a) A self-managing application composed of five components. The solid lines denote computational port connections between components, and the dotted lines are port connections constructing the management framework. (b) Distributed self-managing application shown in (a) executed on three nodes. The solid lines across nodes denote the interactions among manager instances. The dotted lines are port connections constructing the management framework within one node.

the self-management frequency. Adaptations made during a quiet interval will be applied during the next computation phase.

The **composition manager** also provides the *RulePort* (shown in Figure 11.2). Composition manager instances are initialized by the CCA driver component to load in composition rules (possibly from a disk file) using the 'loadRules' method. These rules are then decomposed into subrules and delegated to corresponding component managers. The driver component notifies composition manager instances of quiet intervals by invoking the 'fire' method. During execution of the composition rules, composition manager instances collect results of subrule execution from component manager instances, evaluate the combined rule, and notify component managers of actions to be performed. Possible actions include adding, deleting, or replacing components. When replacing a managed component, the new component does not have to provide and use the exact same ports as the old one. However, the new component must at least provide all the active ports (those used by other components in the application) that are provided by the old component.

11.3.2.3 Rule Execution Model

A three-phase rule execution model is used to ensure correct and efficient parallel rule execution. This model provides mechanisms to dynamically detect and handle rule conflicts for both behavior and interaction rules. Rule execution proceeds as follows. After the evaluation, a precondition is constructed. Rule conflicts are detected at runtime when rule execution changes the precondition (a sensor-actuator conflict), or the same actuator will be invoked multiple times with different values (an actuator-actuator conflict).

Sensor-actuator conflicts are resolved by disabling the rules that change the precondition. Actuator-actuator conflicts are resolved by relaxing the precondition according to user-defined strategies until no actuator is invoked multiple times with different values.

The framework also provides mechanisms for reconciliation among manager instances, which is required to ensure consistent adaptations. For example, in parallel SCMD (Single Component Multiple Data) applications, since each processing node may independently propose different and possibly conflicting adaptation behaviors based on its local state and execution context. Rules are statically assigned one of two priorities. A high priority means that the execution of the rule is necessary, for example, to avoid an application crash. A low priority means that the execution of the rule is optional. During reconciliation, actions associated with the rule with high priority are propagated to all the managers. If there are multiple conflicting high-priority rules, a runtime error is generated and reported to the user. If only low-priority rules are involved, reconciliation uses cost functions to select the most appropriate action at all involved managers. Details of the design and operation of the Accord rule engine can be found in [13].

11.3.2.4 Supporting Performance-Driven Self-Management

The TAU [28] framework provides support for monitoring the performance of components and applications and is used to enable performance-driven self-management. TAU can record inclusive and exclusive wall-clock time; process virtual time, hardware performance metrics such as data cache misses and floating point instructions executed, as well as a combination of multiple performance metrics; and help track application and runtime system level atomic events. Further, TAU is integrated with external libraries such as Precision Approach Path Indicator (PAPI) [20] or Performance Counter Library (PCL) [23] to access low-level processor-specific hardware performance metrics and low latency timers.

In Accord-CCA, TAU application program interfaces (APIs) are directly instrumented into the computational components or into proxies in case of third-party and legacy computational components, and performance data are exported as sensors to component managers. Optimizations are used to reduce the overheads of performance monitoring. For example, as the cache-hit rate will not change unless a different algorithm is used or the component is migrated to another system with a different cache size and/or cache policies, monitoring of cache-hit rate can be deactivated after the first a few iterations, reactivating only when an algorithm is switched or the component is migrated. Similarly, interprocessor communication time is measured per message by default, but this can be modified using the 'setFrequency' method in the *RulePort* to reduce overheads. Another possibility is to restrict monitoring to only those components that significantly contribute to the application performance. Composition managers can identify these components at runtime using mechanisms similar to those proposed in [29] and enable or disable

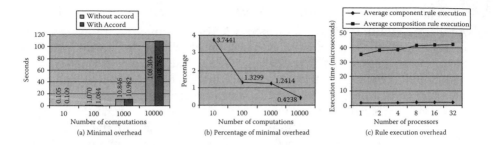

FIGURE 11.4

(a) and (b) The runtime overhead introduced in the minimal rule mode. (c) The overhead introduced by executing component and composition rules.

monitoring as required. Finally, in case of homogeneous execution environments, only a subset of nodes may be monitored.

11.3.2.5 An Experimental Evaluation

The overheads of adding adaptation behaviors is presented in this section on a 64 node beowulf cluster. The cluster contains 64 Linux-based computers connected by 100 Mbps full-duplex switches. Each node has an Intel(R) Pentium-4 1.70GHz central processing unit (CPU) with 512MB RAM and is running Linux 2.4.20-8 (kernel version). The overheads associated with the initialization and runtime rule execution were evaluated.

Experiment 1 (Figure 11.4 (a) and (b)): This experiment measures the runtime overhead introduced in a minimal rule execution mode, i.e., rules are loaded but the execution is disabled. The application execution times with and without the framework are plotted in Figure 11.4 (a), and the percentage overhead is plotted in Figure 11.4 (b). The major overhead in this case is due to the loading and parsing of rules.

Experiment 2 (Figure 11.4 (c)): This experiment evaluates the average execution time of component and composition rules. The component rules are used to dynamically switch algorithms within a component, and the composition rules are used to dynamically replace a component. As the number of processors increases, the average execution times of both rules increase slightly. This is reasonable, since nodes must communicate with each other during reconciliation. The figure also shows that the average execution time of the composition rules is much larger than that of the component rules. This is because, in execution of composition rules, a new component will be instantiated, connected to other components, and loaded with new rules. However, the execution of component rules involves invoking only component actuators.

Note that while the framework does introduce overheads, the benefits of adaptation would outweigh these overheads. Further, the overheads are not significant when compared with the typical execution time of scientific applications, which can be in hours, days, and even weeks.

11.3.2.6   *Illustrative Application Scenarios*

A Self-Managing Hydrodynamics Shock Simulation: This application simulates the interaction of a hydrodynamic shock with a density-stratified interface. The Runge-Kutta time integrator (**RK2**) with an **InviscidFlux** component supplies the right-hand side of the equation on a patch-by-patch basis. This component uses a **ConstructLRStates** component to set up a Riemann problem at each cell interface, which is then passed to **GodunovFlux** for the Riemann solution. A **ConicalInterfaceIC** component sets up the problem—a shock tube with Air and Freon (density ratio 3) separated by an oblique interface that is ruptured by a Mach 10.0 shock. The shock tube has reflecting boundary conditions above and below and outflow on the right. The **AM-RMesh** and **GodunovFlux** are the significant components in this simulation from the performance point of view, and are used to illustrate self-managing behaviors in the discussion below.

Scenario 1: Component Replacement: An Equilibrium Flux Method (EFM) algorithm may be used instead of the Godunov method within **RK2**. **GodunovFlux** and **EFMFlux** demonstrate different performance behaviors and mean execution times as the size of the input array size increases. The appropriate choice of algorithm (Godunov or EFM) depends on simulation parameters, its runtime behaviors, and the cache performance of the execution environment, and is not known a priori. In this scenario we use information about cache misses for **GodunovFlux** obtained using TAU/PCL/PAPI to trigger self-optimization, so that when cache misses increase above a certain threshold, the corresponding instance of **GodunovFlux** is replaced with an instance of **EFMFlux**. To enable the component replacement, one component manager is connected to **GodunovFlux** through the *RulePort* to collect performance data, evaluate rules, and perform runtime replacement. The component manager (1) locates and instantiates **EFMFlux** from the component repository, (2) detects all the provider and user ports of **GodunovFlux**, as well as all the components connected to it, (3) disconnects **GodunovFlux** and deletes all the rules related to **GodunovFlux**, (4) connects **EFMFlux** to related components and loads in new rules, and finally (5) destroys **GodunovFlux**. The replacement is performed at a *quiet interval*. From the next calculation step, **EFMFlux** is used instead of **GodunovFlux**. However, other components in the application do not have to be aware of the replacement, since the abstract interfaces (ports) remain the same. After replacement, the cache behavior improves, as seen in Figure 11.5 (a).

Scenario 2: Component Adaptation: The **AMRMesh** component supports structured adaptive mesh refinement and provides two communication mechanisms. The first exchanges messages on a patch-by-patch basis and results in a large number of relatively small messages. The second packs messages from multiple patches to the same processor and sends them as a single message, resulting in a small number of much larger messages. Depending on the current latency and available bandwidth, the component can be dynamically adapted to switch the communication mechanism used.

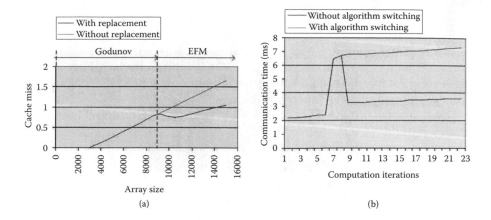

FIGURE 11.5
(a) Replacement of **GodunovFlux** with **EFMFlux** to decrease cache misses. (b) Dynamically switch algorithms in **AMRMesh**.

In this scenario, we use the current system communication performance to adapt the communication mechanism used. As PAPI [20], PCL [23], and TAU [28] do not directly measure network latency and bandwidth, this is indirectly computed using communication times and message sizes. **AMRMesh** exposes communication time and message size as sensors, which are used by the component manager to get the current bandwidth as follows:

$$bandwidth = \frac{commTime_1 - commTime_2}{msgSize_1 - msgSize_2} \tag{11.1}$$

Here, '$commTime_1$' and '$commTime_2$' represent the communication times for messages with sizes '$msgSize_1$' and '$msgSize_2$', respectively. When the bandwidth falls below a threshold, the communication mechanism switches to patch-by-patch messaging (i.e., algorithm 1). This is illustrated in Figure 11.5 (b). The algorithm switching happens at iteration 9 when channel congestion is detected, and results in comparatively smaller communication times in the following iterations.

A Self-Managing Adaptive CH_4 Ignition Simulation: This experiment focuses on the overall performance improvement of the CH_4 ignition simulation. The ignition process is represented by a set of chemical reactions, which appear and disappear when the fuel and oxidizer react and give rise to the various intermediate chemical species. In the simulation application, the chemical reactions are modeled as repeatedly solving the ChemicalRates equation (G) with different initial conditions and parameters using one of a set of algorithms (called backward difference formula, or BDF). The algorithms are numbered from 1 to 5, indicating the order of accuracy of the algorithm. BDF_5 is the highest-order method and is the most accurate and robust. It may,

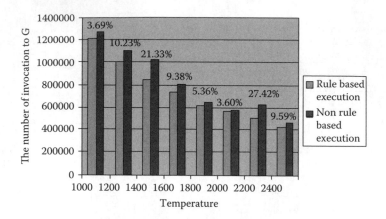

FIGURE 11.6
Comparison of rule-based-and non-rule-based execution of CH_4 ignition.

however, not always be the quickest. As a result, the algorithm used for solving the equation G has to be selected based on current condition and parameters. In this application, the bulk of the time is spent in evaluating the equation G. Therefore, reducing the number of G evaluation is a sufficient indication of speed independent of experimental environments.

As shown in Figure 11.6, the rule-based execution decreases the number of invocation to equation G, and the percentage decrease is annotated for each temperature value. It results in an average 11.33% computational saving. As the problem becomes more complex (the computational costs of G increase), the computational saving will be more significant.

11.3.3 Accord-WS: A Programming System for Autonomic Service-Based Applications

Accord-WS is the service-based prototype of Accord. It builds on the WS-Resource specifications [10] and the Web service specifications [32]. Key components of the prototype include: (1) the formulation of autonomic services that extend WS-Resources with specifications and mechanisms for self-management and (2) a distributed runtime infrastructure to enable decentralized and dynamic compositions of these services.

11.3.3.1 Defining Autonomic Services

An autonomic service (shown in Figure 11.7(a)) consists of (1) a WS-Resource [10] providing functionalities and stateful information, (2) a coordination agent sending and receiving interaction messages for the associated WS-Resource, and (3) a service manager that manages the runtime behaviors of the WS-Resource and its interactions with other autonomic services. Applications can be developed as compositions (possibly dynamic and

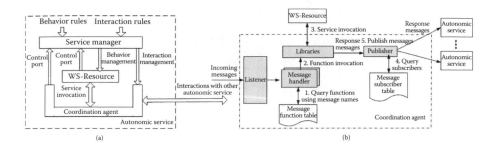

FIGURE 11.7
An autonomic service in Accord-WS.

opportunistic) of these autonomic services. Each managed WS-Resource is extended with a control port specified as a document in Web Services Descriptive Language (WSDL) consisting of sensors and actuators for external monitoring and control of its internal state. The control port can be exposed as part of the service port or as a separate document to the service manager.

The coordination agent acts as a programmable notification broker [32] for the associated WS-Resource. As shown in Figure 11.7(b), a coordination agent consists of four modules that work in parallel: (1) a **listener** module that listens to the incoming messages from other autonomic services, (2) **message handlers** that process the messages using functions defined in the **message function table**, (3) **libraries** that provide functions for processing messages (e.g., translating message formats and combining messages) and invoking the associated WS-Resource and getting response messages, and (4) a **publisher** that sends the response messages to the subscribers. The coordination agent exposes sensors and actuators to the service manager that allows the manager to query and modify its **message function table** and **message subscriber table**. The service manager can dynamically reconfigure the coordination agent by changing the **message function table** to select functions to process messages, and by changing the **message subscriber table** to add and delete subscribers.

The service manager performs (1) functional management using sensors and actuators exposed by the associated WS-Resource based on behavior rules defined by users or derived from application requirements and objectives, and (2) interaction management using sensors and actuators exposed by the coordination agent based on interaction rules derived from application workflows.

11.3.3.2   The Accord-WS Runtime Infrastructure

The runtime infrastructure consists of the Accord portal/composition manager, peer service managers, and other supporting services as shown in Figure 11.8. This infrastructure can facilitate workflow execution and dynamic composition.

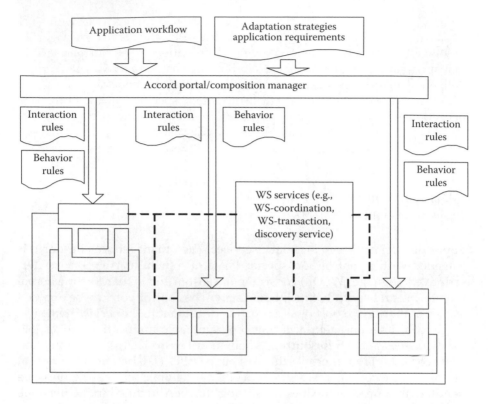

FIGURE 11.8
The runtime framework, with the dashed lines representing the interactions among managers and the solid lines representing the interactions among WS-Resources.

Workflow Execution: To execute an application workflow that may be defined by a user or generated by an automated workflow generation engine, for example [11], the composition manager first discovers and locates the relevant WS-Resources, instantiates a coordination agent for each of the WS-Resources, and further instantiates a service manager for each WS-Resource and coordination agent pair to enable service behavior and interaction adaptations. Coordination agents interact with their associated WS-Resources using messages in Simple Object Access Protocol (SOAP). Service managers are located within the same memory space with their associated coordination agents, and they interact with each other through pointers. The communications among service managers are based on sockets.

The composition manager decomposes the application workflow into interaction rules and injects them into corresponding service managers, which then configure associated coordination agents to dynamically establish publication/subscription relationships and manipulate interaction messages. Specifically, a service manager configures the **message function table** by associating the messages that this autonomic service subscribes to with functions for processing them. Similarly, it also configures the **message subscriber table** by associating the messages that this autonomic service produces with a list of subscribers. These operations are performed by the service manager by invoking the actuators provided by the coordination agent. Further, service managers configure the associated WS-Resource based on the behavior rules defined by users or generated from application requirements.

Compared with the centralized composition specified using BPEL4WS, decentralized composition enables direct interactions among involved services, and therefore avoids unnecessary messages and relieves the bottleneck caused by the centralized unit. Further, the decentralized composition can fully exploit parallelism, since autonomic services without data and control dependencies can automatically proceed in parallel.

Dynamic Composition: Application workflows need to be changed accordingly when business logic or user requirements change. In most cases, these changes affect only a part of the workflow. Workflow decomposition discussed above can benefit the dynamic composition of autonomic services by constraining the modification to the associated part of the workflow without affecting the rest of the application. In Accord, dynamic composition is enabled by adding, deleting, or modifying interaction rules in service managers, which automatically reconfigures the associated coordination agents accordingly.

11.3.3.3 An Illustrative Application: The Adaptive Data-Streaming Application

The application consists of the gyrokinetic toroidal code (GTC) fusion simulation that runs for days on a parallel supercomputer at the National Energy Research Scientific Computing Center (Berkeley, California) and generates multiterabytes of data. These data are analyzed and visualized live, while the simulation is running, at the Princeton (New Jersey) Plasma Physics Laboratory (PPPL). The goal is to stream data from the live simulation to support remote runtime analysis and visualization at PPPL while minimizing overheads on the simulation, adapting to network conditions, and eliminating loss of data. The application consists of the Simulation Service (SS) that executes in parallel on 6K processors of Seaborg an IBM SP machine at NERSC and generates data at regular intervals, the Data Analysis Service (DAS) that runs on a 32-node cluster located at PPPL to analyze and visualize data, the Data Storage Service (DSS) that archives the streamed data using the Logistical Networking backbone and builds a Data Grid of storage services located at the Oak Ridge (Tennessee) National Laboratory and PPPL, and the Autonomic Data Streaming Service (ADSS) that manages the streaming of data from SS (at NERSC) to DAS (at PPPL) and DSS (at PPPL/ORNL).

Scenario 1: Self-optimizing behavior of ADSS: ADSS selects the appropriate blocking technique, orders blocks in the buffer, and optimizes the size of the buffer(s) used to ensure low-latency, high-performance streaming and minimize the impact on the execution of the simulation. The adaptations are based on the current state of the simulation, more specifically the data generation rate, the network connectivity and the network transfer rate, and the nature

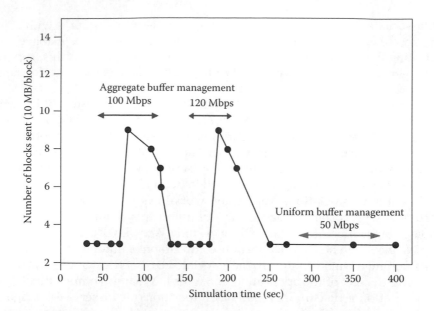

FIGURE 11.9
Self-optimization behaviors of ADSS.

of data being generated in the simulation. ADSS provides (1) Uniform Buffer Management suited for the simulations generating data at a small or medium rate (50Mbps), (2) Aggregate Buffer Management suited for high data generation rates, i.e., between 60 and 400 Mbps, and (3) Priority Buffer Management that orders data blocks in the buffer based on the nature of the data.

To enable adaptations, ADSS exports two sensors, "DataGenerationRate" and "DataType", and one actuator, "BlockingAlgorithm", in XML as part of its control port. As shown in Figure 11.9, ADSS switches to aggregate buffer management during simulation time intervals of 75 sec to 150 sec and 175 sec to 250 sec, as the simulation data generation rate peaks to 100Mbps and 120 Mbps during these intervals. The aggregation is an average of seven blocks. Once the data generation rate falls to 50Mbps, ADSS switches back to the uniform buffer management scheme, and constantly sends three blocks of data on the network.

Scenario 2: Self-configuring/self-optimizing behavior of ADSS: The effectiveness of the data transfer between the simulation service at NERSC and the analysis/visualization service at PPPL depends on the network transfer rate, which depends on data generation rates and/or network conditions. Falling network transfer rates can lead to buffer overflows and require the simulation to be throttled to avoid data loss. One option to maintain data throughput is to use multiple data streams. Of course, this option requires multiple buffers and hence uses more of the available memory. Implementing this option requires the creation of multiple instances of ADSS. In this scenario, ADSS monitors the effective network transfer rate, and when this rate dips

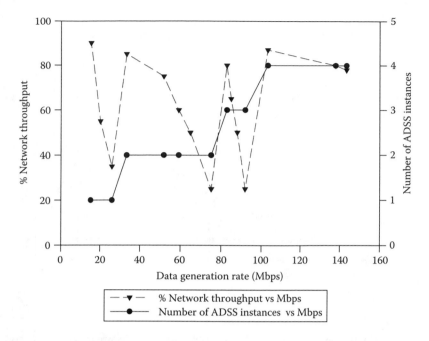

FIGURE 11.10
Effect of creating new instances of the ADSS service when the % network throughput dips to below the user-defined 50% threshold.

below a threshold, the service causes another instance of the ADSS to be created and incorporated into the workflow. Note that the maximum number of ADSS instances possible is predefined. Similarly, if the effective data transfer rate is above a threshold, the number of ADSS instances is decreased to reduce memory overheads.

Figure 11.10 shows that the number of ADSS instances first increases as the network throughput dips below the 50% threshold (corresponding to data generation rates of around 25 Mbps in the plot). This causes the network throughput to increase to above 80%. Even more instances of ADSS services are created at data generation rates of around 40 Mbps, and the network throughput once again jumps to around 80 Mbps. The ADSS instances increase until the limit of 4 is reached.

Scenario 3: Self-healing behavior of ADSS: This scenario addresses data loss in the cases of extreme network congestion or network failures. These cases cannot be addressed using simple buffer management or replication. One option in these cases to avoid loss of data is to write data locally at NERSC rather than streaming. However, these data will not be available for analysis and visualization until the simulation is complete, which could take days. Writing data to the disk also causes significant overheads to the simulation. Therefore, the data need to be streamed to the storage at ORNL, and later transmitted from ORNL to PPPL. As a result, ADSS needs to be dynamically composed with DAS at PPPL or at ORNL.

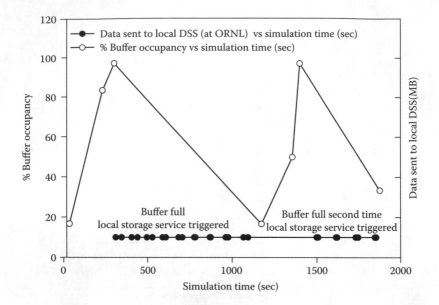

FIGURE 11.11
Effect of switching from the DSS at PPPL to the DSS ORNL in response to network congestion and/or failure.

The buffer size is exposed as a sensor to indicate that when a predefined capacity is reached, data streaming needs to be switched to ORNL. Accordingly, the data streaming destination variable needs to be exposed as an actuator, which could be modified at runtime when rules are fired. The resulting dynamic composition behavior is plotted in Figure 11.11. The figure shows that as the buffer(s) get saturated, the data streaming switches to the storage at ORNL, and when the buffer occupancy falls below 20% it switches back to PPPL. Note that while the data blocks are written to ORNL, data blocks already queued for transmission to PPPL continue to be streamed. The figure also shows that at simulation time 1500 (x-axis), the PPPL buffers once again get saturated and the streaming switches to ORNL. If this persists, the steaming would be permanently switched to ORNL.

11.4   Conclusion

The scale, heterogeneity, dynamism, and uncertainty of emerging wide-area Grid computing environments require that Grid applications be capable of detecting and dynamically responding during execution to changes in both the state of execution environment and the state and requirements of the applications. This places unique and challenging equipments on Grid programming systems. However, existing programming systems either make very strong assumptions about the behavior of the entities, their interactions, and the underlying system, or have limited support for runtime adaptation, which prevents them from meeting the requirements.

This chapter described the Accord programming system that extends existing programming systems to enable the development of self-managing Grid applications wherein the behaviors, organizations, and interactions of elements and applications are sensitive to the dynamic state of the system and the changing requirements of the applications, and adapt to these changes at runtime. The Accord programming system (1) enables the definition of autonomic elements that encapsulate functional and nonfunctional specifications as well as policies (in the form of high-level rules) and mechanisms for self-management, (2) enables the formulation of self-managing applications as dynamic compositions of autonomic elements, and (3) provides a runtime infrastructure for consistently and efficiently enforcing policies to enable autonomic self-managing functional, interaction, and composition behaviors based on current requirements and state and execution context.

The design and evaluations of prototype implementations of Accord based on component-and service-based programming models were presented. The component-based prototype of Accord extends the Common Component Architecture to enable self-management of component-based scientific applications. This prototype supports both function and performance oriented adaptation, enables dynamic composition by replacing components at runtime, and provides consistent and efficient rule execution for intra- and intercomponent adaptation behaviors. The service-based prototype of Accord extends the Axis framework to support self-managing service-based applications and enables runtime adaptation of service, service interactions, and dynamic service composition. Self-managing application scenarios enabled by each of the prototype implementations of Accord were presented.

Acknowledgments

The research presented in this paper is supported in part by the National Science Foundation via grants numbers ACI 9984357, EIA 0103674, EIA 0120934, ANI 0335244, CNS 0305495, CNS 0426354, and IIS 0430826, and by the DOE SciDAC CPES FSP.

References

1. A. Abrahams, D. Eyers, and J. Bacon. An asynchronous rule-based approach for business process automation using obligations. In *3rd ACM SIGPLAN Workshop on Rule-Based Programming (RULE02)*, 323–345, Pittsburgh, PA, 2002.

2. M. Agarwal and M. Parashar. Enabling autonomic compositions in grid environments. In *4th International Workshop on Grid Computing (Grid 2003)*, 34–41, Phoenix, AZ, 2003. IEEE Computer Society Press.

3. B. A. Allan, R. C. Armstrong, A. P. Wolfe, J. Ray, D. E. Bernholdt, and J. A. Kohl. The CCA core specification in a distributed memory SPMD framework. *Concurrency Computation*, 14(5):323–345, 2002.

4. D. Beazley and P. Lomdahl. Controlling the data glut in Large-Scale Molecular-Dynamics Simulations. *Computers in Physics*, 11(3), 1997.

5. F. Berman, R. Wolski, H. Casanova, W. Cirne, H. Dail, M. Faerman, S. Figueira, J. Hayes, G. Obertelli, J. Schopf, G. Shao, S. Smallen, N. Spring, A. Su, and D. Zagorodnov. Adaptive computing on the grid using AppLeS. *IEEE Transactions on Parallel and Distributed Systems*, 14(4):369–382, 2003.

6. P. Boinot, R. Marlet, J. Noy, G. Muller, and C. Consel. Declarative approach for designing and developing adaptive components. In *15th IEEE International Conference on Automated Software Engineering*, 111–119, 2000.

7. J. Bosch. Superimposition: A component adaptation technique. *Information and Software Technology*, 41(5):257–273, 1999.

8. J. Chen, D. Silver, and M. Parashar. Real time feature extraction and tracking in a computational steering environment. In *Proceedings of the 11th High Performance Computing Symposium (HPC 2003)*, Orlando, FL, 2003.

9. *Common Object Broker Resource Architecture (CORBA)*. http://www.corba.org.

10. I. Foster, J. Frey, S. Graham, S. Tuecke, K. Czajkowski, D. Ferguson, F. Leymann, M. Nally, I. Sedukhin, D. Snelling, T. Storey, W. Vambenepe, and S. Weerawarana. *Modeling Stateful Resources with Web Services*. http://www-128.ibm.com/developerworks/library/ws-resource/ws-modelingresources.pdf, 2004.

11. J. Woo Kim and R. Jain. Web services composition with traceability centered on dependency. In *38th Hawaii International Conference on System Sciences*, 89, Hawaii, 2005.

12. B. Kohn, E. Kraemer, D. Hart, and D. Miller. An agent-based approach to dynamic monitoring and steering of distributed computations. In *International Association of Science and Technology for Development (IASTED)*, Las Vegas, Nevada, 2000.

13. H. Liu. *Accord: A Programming System for Autonomic Self-Managing Applications*. PhD thesis, ECE Dept., Rutgers University, 2005.

14. H. Liu, V. Bhat, M. Parashar, and S. Klasky. An autonomic service architecture for self-managing grid applications. In *Proceedings of the 6th IEEE/ACM International Workshop on Grid Computing (Grid 2005)*, 132–139, Seattle, WA, 2005.

15. H. Liu, L. Jiang, M. Parashar, and D. Silver. Rule-based visualization in the Discover Computational Steering Collaboratory. *Journal of Future Generation Computer System, Special Issue on Engineering Autonomic Systems*, 21(1):53–59, 2005.

16. H. Liu and M. Parashar. Enabling self-management of component-based high-performance scientific applications. In *14th IEEE International Symposium on High Performance Distributed Computing (HPDC-14)*, 59–68, Research Triangle Park, NC, 2005.

17. H. Liu and M. Parashar. Rule-based monitoring and steering of distributed scientific applications. *International Journal of High Performance Computing and Networking (IJHPCN)*, 3(4):272–282, 2005.

18. H. Liu and M. Parashar. Accord: A programming framework for autonomic applications. *IEEE Transactions on Systems, Man and Cybernetics, Special Issue on Engineering Autonomic Systems*, 2006.

19. H. Liu, M. Parashar, and S. Hariri. A component-based programming framework for autonomic applications. In *1st IEEE International Conference on Autonomic Computing (ICAC-04)*, 10–17, New York, 2004.

20. *PAPI: Performance Application Programming Interface.* http://icl.cs.utk.edu/projects/papi.

21. M. Parashar, H. Liu, Z. Li, V. Matossian, C. Schmidt, G. Zhang, and S. Hariri. Automate: Enabling autonomic grid applications. *Cluster Computing: The Journal of Networks, Software Tools, and Applications, Special Issue on Autonomic Computing*, 9(2):161–174, 2006.

22. S. Parker and C. Johnson. An integrated problem solving environment: The SCIRun computational steering environment. In Proceedings of 31st Hawaii International Conference on System Sciences, 147–156, 1998.

23. *PCL: The Performance Counter Library.* http://www.fz-juelich.de/zam/PCL.

24. S. R. Ponnekanti and A. Fox. SWORD: A developer toolkit for Web service composition. In *International WWW Conference*, 83–101, Honolulu, Hawaii, 2002.

25. J. Ray, N. Trebon, R. C. Armstrong, S. Shende, and A. Malony. Performance measurement and modeling of component applications in a high performance computing environment: A case study. In *18th International Parallel and Distributed Processing Symposium (IPDPS04)*, 95–104, Santa Fe, NM, 2004.

26. S. M. Sadjadi and P. K. McKinley. Transparent self-optimization in existing CORBA applications. In *1st International Conference on Autonomic Computing*, 88–95, New York, 2004.

27. C. Szyperski. *Component Software Beyond Object-Oriented Programming*. Component Software Series. Addison-Wesley, Boston, MA, 2nd ed. 2002.

28. *TAU: Tuning and Analysis Utilities.* http://www.cs.uoregon.edu/research/paracomp/tau/tautools

29. N. Trebon, J. Ray, S. Shende, R. C. Armstrong, and A. Malony. *An Approximate Method for Optimizing HPC Component Applications in the Presence of Multiple Component Implementations*. Suffix SAND2003-8760C, Sandia National Laboratories, 2003.

30. E. Truyen, W. Joosen, P. Verbaeten, and B. N. Jorgensen. On interaction refinement in middleware. In *5th International Workshop on Component-Oriented Programming*, 56–62, 2000.

31. C. Ururahy, N. Rodriguez, and R. Ierusalimschy. Alua: Flexibility for parallel programming. *Computer Languages*, 28(2):155–180, 2002.

32. WS-BrokeredNotification 1.0 specification. ftp://www6.software.ibm.com/software/developer/library/ws-notification/WS-BrokeredN.pdf.

12

A Self-Configuring Service Composition Engine

Thomas Heinis, Cesare Pautasso, and Gustavo Alonso

CONTENTS

In this chapter we present an architecture for process execution which features an autonomic controller. The controller provides self-tuning, self-configuration and self-healing capabilities targeted at automatically configuring a distributed service composition engine. The system has been designed so that its components can be dynamically replicated on several nodes of a cluster, thereby achieving basic scalability. Thanks to the autonomic controller, the engine reacts to variations in the workload by altering its configuration in order to achieve better performance. The autonomic controller is also able to heal the system in case of failures. In order to illustrate the benefits of the approach, we present a set of experiments which evaluate the ability of the engine to configure itself as the load changes.

12.1   Introduction

Web service integration has been significantly simplified by using process management languages and associated middleware tools [11, 22, 9]. These languages and tools provide useful abstractions to express the business protocols and conversation rules involved [18, 2, 20]. Additionally, and in this context more importantly, they also provide powerful execution environments which can be used to efficiently develop integrated systems out of the composition of distributed services [12, 1, 17].

The most crucial requirement when using process support systems to solve large-scale integration problems is scalability. Popular composite services published on the Web [8] have the potential of being invoked in parallel by large numbers of clients, leading to surges of requests arriving at unpredictable times. Assuming that the service providers participating in the business process do not pose a significant bottleneck, such events affect the underlying process execution infrastructure. Every time a client requests a service, a process instance has to be created and, for all subsequent interactions with the service, its state needs to be kept up to date.

Although using distributed process execution environments can provide the necessary scalability to handle large workloads, it is difficult to determine the configuration of such systems *a priori*, especially when facing unpredictable workloads. In order to deal with such highly variable volumes of processes, dynamic reconfigurability is a very important requirement [21]. With it, the size of a running system can be adapted to keep the balance between servicing its workload with optimal performance and ensuring efficient resource allocation.

In what follows, we present how we applied these principles to JOpera [13], a distributed execution engine for Grid and Web service composition featuring autonomic capabilities. Most existing execution engines fail to provide dynamic mechanisms to tackle scalability issues by only providing means to statically and manually distribute the workload across different sites (e.g., [3]). Instead, our approach dynamically determines the configuration based on the actual workload and autonomically reconfigures the engine [7, 16].

In addition to the basic ability to run service compositions in a distributed environment, thereby achieving scalability in the face of potentially high workload surges, the engine also features an autonomic controller. This component (1) monitors the current workload and state of the system, (2) heals the system in case of failures, and (3) reconfigures the system according to the workload. To do so, the controller acts upon different policies which can be set to work toward different goals (e.g., limit the amount of resources allocated to the system or optimize the response time and throughput of the system).

The remainder of the chapter is organized as follows: Section 12.2 first gives an overview of how service compositions are executed in JOpera and then describes how the functionality of the system can be replicated and distributed across a cluster. In Section 12.3 we present the autonomic controller which configures the system according to the current workload. This section explains the different components of the autonomic controller and introduces the policies upon which they act. In Section 12.4 we provide a performance evaluation of the autonomic capabilities of the engine. We briefly present related work in Section 12.5 before drawing conclusions in Section 12.6.

12.2   Background

This section discusses the architecture of JOpera by first showing how service compositions are executed and then describing how components of the engine can be replicated and distributed across a cluster.

12.2.1   Running a Service Composition with JOpera

In the context of JOpera, Web services are composed as processes. Processes comprise a number of tasks linked by a control flow and a data flow graph. A task in general represents the invocation of a remote service or a local application, such as a Web service, a Grid service, a UNIX shell script, a Java application, or several other kinds of services [14]. Common to all tasks is that they can take input parameters and may return results as output parameters. The data flow defines what data are copied between the tasks, i.e., which output parameters of a certain task are copied into the input parameter of a following task. The control flow, on the other hand, defines the partial order of the execution of tasks.

Figure 12.1 illustrates how JOpera executes processes. In order to execute a process, a client or another process sends a request to the engine application programming interface (API). The engine API puts these requests in the **process queue**. The navigator takes a request from this queue and will, upon the start of the process execution, create a new process instance and begin with the execution of the process. To execute the process, the navigator takes into account the current state of the process, which comprises the results of previous task executions as well as its control flow in order to determine the tasks which need to be executed next. The tasks which are ready to be executed are put into the **task queue**.

Execution of tasks is carried out by the dispatcher component. This component dispatches, hence the name, messages to remote service providers outside the execution engine by using the corresponding adapters. JOpera features adapters to invoke different kinds of services [14]. Upon finishing a service execution, the dispatcher encapsulates the results in a task completion event and puts it in the **event queue**. By doing so, the dispatcher informs the

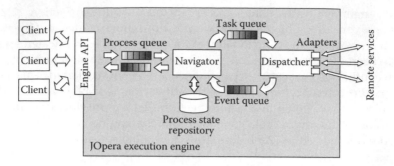

FIGURE 12.1
The JOpera process execution engine executes processes by looping between the navigator and dispatcher components.

navigator about the task completion. The navigator consumes this event and updates the state of the process, which is stored in the process state repository. This closes the process execution loop, as the navigator will then once again identify the tasks which need to be executed next.

The concern of executing tasks and navigating over a process has been separated because the former may take significantly longer than the latter. Executing tasks will therefore not block further navigation. This is essential in case of parallel tasks, where the engine would be blocked while executing a task before being able to schedule the next parallel task. The dispatcher furthermore executes tasks in parallel running threads so that slow tasks do not delay the execution of faster ones.

12.2.2   Distributed Engine Architecture

Decoupling the navigator from the dispatcher by means of asynchronous communication allows the replication of process navigation and task execution functionality. Replicating the dispatcher enables the management of large numbers of parallel task executions. Likewise, replicating the navigator leads to a larger process navigation capacity.

In order to achieve scalability, the navigator and dispatcher components are distributed on a cluster and use two globally accessible queues, the task queue and the event queue, to communicate. The components use these two queues to exchange messages. The navigators thus enqueue task execution request messages, which are subsequently dequeued by dispatchers, in the task queue. Dispatcher components execute these tasks and then enqueue task completion event messages which include the execution results in the event queue. These messages will be consumed by navigators, who will then again navigate over the according processes. The state of process instances is also kept in a global process state repository accessible by all navigators.

The asynchronous interaction between the navigator and dispatcher components allows dynamic reconfiguration at runtime because the number of either component can be increased or decreased without having to stop the whole engine. For this purpose the engine features a management API through which commands that start or stop components can be issued in order to configure each node.

In order to avoid problems related to potential bottlenecks in the access to the centralized queues, the distributed implementation of JOpera uses several levels of caching between the global queues and the components producing and consuming messages. For this purpose, in addition to the two global queues, each of the components also has a local queue in which messages are cached. Messages are sent as close to the consumer as possible. Thus, if a navigator sends a message to itself (e.g., when a process calls another process), the according event message is written to its local memory cache. In case a dispatcher wants to send a task completion event message to the navigator working on the process the task belongs to, it will write the message in the queue nearest to the navigator, i.e., its local queue. If this is not possible, the message will be delivered into the global event queue.

Messages to be dequeued by a component are pre-fetched into its memory cache from either the global or the local queue in order to increase the throughput. This way, the messages are immediately available when a component is able to process them.

To avoid a similar bottleneck regarding the access to the global process state repository, navigators only work on a locally cached copy of the process instance state. This, however, requires that event messages belonging to a certain process instance are delivered to the navigator which locally caches this instance's state. For this purpose, a global routing table associating each process instance with the navigator working on it is maintained so that dispatchers can find the responsible navigator. More specifically, the routing table contains a mapping between the ID of a process instance and the address of the navigator in charge of working on this process instance. This mapping is created as soon as the navigator starts working on the process instance and is removed when the execution of the process instance is finished. With the added complexity of maintaining a routing table, these optimizations help to reduce the load on the queues as well as on the process state repository [15].

12.3 Autonomic Controller

In this section we give an overview of the autonomic controller, which consists of the self-healing, self-tuning, and self-configuring components. The autonomic controller as well as the interactions between its components, the distributed engine, and the configuration information registry are depicted in Figure 12.2.

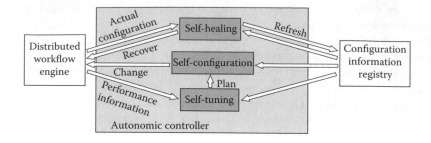

FIGURE 12.2
The autonomic controller architecture, including the interactions between its components.

In brief, the self-tuning component reacts to changes in the distributed workflow engine and defines a new configuration plan, which is passed to the self-configuration component, which in turn will implement the suggested plan. The self-healing component acts autonomously by recovering failures in the distributed workflow engine imposed by failed dispatchers or navigators.

12.3.1   Self-Healing Component

The self-healing component is concerned with ensuring that the information stored in the configuration registry reflects the actual configuration of the system. The configuration information registry stores information about what component is running on which node. Components register themselves in the registry when starting and unregister themselves when stopping. However, the configuration stored in the registry can differ from the actual configuration because of external events such as failures or partitions.

In order to keep the registry synchronized, the self-healing component periodically monitors all nodes in the configuration, queries these to see what component they are running, and compares this information with the stored configuration.

A difference between the actual configuration and the stored configuration detected by the self-healing component is considered to be a failure. In case of a failure the self-healing component will recover affected tasks as well as processes and make sure that the information about the actual configuration is synchronized again. Depending on the component that has failed, the failure is treated differently.

Dispatcher Failure: The tasks which have been running on the failed dispatcher need to be reexecuted because they are lost. To do so, the self-healing component queries the state of the executing process, thereby analyzing what tasks have been executing on the failed component. Task execution requests for these tasks are then put into the task queue for later reexecution.

Navigator Failure: In case a navigator fails, the state of the execution of the processes that were assigned to it is still available in the global process state repository. This is because navigators only work on a locally cached copy of the state.

The self-healing component therefore only needs to remove the entries for these processes in the routing table which point to the failed navigator. By doing so, all pending event messages can be routed through the global queue until a different navigator becomes available to process them. In order to reproduce events lost at the failed navigator, the new navigator will query all dispatchers that are executing tasks according to the state of the recovered process. If any of these tasks is no longer running, the event indicating the finished task is assumed to be lost and the task needs to be reexecuted.

12.3.2 Self-Tuning Component

If either the configuration stored in the configuration information registry changes, due to an external failure or actions taken by the self-healing component, or the performance of the engine changes significantly, the self-tuning component develops a new configuration plan for the engine. The **information policy** defines what performance information indicators the self-tuning component considers in order to determine that the engine is running with a suboptimal configuration. The **optimization policy** defines how to arrive at an optimal configuration of the cluster based on the current performance.

If the performance of the engine is considered to be suboptimal, the self-tuning component defines a new configuration plan on how to partition the cluster in dispatchers and navigators and passes this plan on to the self-configuration component. The self-configuration component will implement the new plan asynchronously. The planning and implementation steps are separated because the time required to develop a new plan is generally significantly shorter than implementing it. This way, it is not necessary for the self-tuning component to wait for the implementation of the plan to complete before evaluating the current configuration and developing a new plan. If a new plan becomes available and the previous one is not yet fully implemented, implementation is aborted and the new plan will be put into effect.

12.3.3 Self-Configuration Component

This component is concerned with the implementation of the configuration plan developed by the self-tuning component. To do so, the self-configuration component gets the current configuration of the engine from the configuration information registry, compares it with the configuration plan, and chooses the nodes to be reconfigured according to the **selection policy**. To do so, this component stops and starts navigators and dispatchers on remote nodes.

12.3.3.1 Starting Components

Starting a component on a remote node is done by sending a command to its JOpera management API. A component can be started only as long as no other component is running.

12.3.3.2   Stopping Components

Navigator Components: When a navigator is stopped, the state of the processes the navigator is working on needs to be migrated, and event messages associated with these processes need to be redirected. The state of a process is migrated by flushing the local cache into the global process state repository. The pending events which have been delivered to the navigator but have not yet been processed are transferred to the global event queue. Events triggered by dispatchers working on tasks belonging to the processes which are to be migrated will also be redirected to the global event queue. This allows a different navigator to pick up the process state and to continue working on it.

Dispatcher Components: Stopping dispatchers is more difficult than stopping navigators, because they are executing tasks. These tasks may involve the execution of local applications or the invocation of remote service providers. It is therefore not possible to interrupt the invocation of all kinds of tasks without losing intermediate results. This is the reason why we employ the *stop* method, which will prevent the dispatcher from taking new tasks from the task queue but will not interrupt the execution of running tasks. Once all tasks have executed successfully, the dispatcher will be stopped. While this method avoids discarding intermediate results of task executions, it has the disadvantage that as long as tasks are running, the node will not be available for starting a different component.

12.4   Measurements

The goal of the measurements is to evaluate how the autonomic process execution platform automatically changes the configuration of the system in response to different workload conditions. In the experiments, the configuration (number of dispatchers and navigators) and the performance indicators used by the controller have been sampled at regular intervals of 1 second, which is reasonable considering the dynamics of the system.

12.4.1   Benchmark and Testbed Description

Given the lack of benchmarks for evaluating the performance of autonomic workflow engines, we have used workloads consisting of processes defined to benchmark existing workflow systems [5] for the experiments described in this paper. These processes are based on the TPC-C order-entry benchmark [4] and are meant to test the workflow execution engines' capability to execute different workflow patterns (i.e., parallel splits and synchronization and multiple instances [19]). The control flow of the process is depicted in Figure 12.3. The process takes as input a number of items to be ordered from different stores, as well as the payment method. In more detail, given a number of items, this process first checks the credit card (if the customer chooses this payment method) and then in two parallel subprocesses notifies

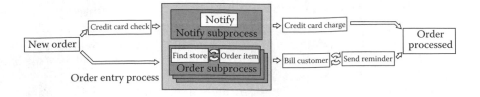

FIGURE 12.3
The order process used to evaluate the system orders a number of items from different stores for the customer and subsequently charges him.

the customer and orders the items from other stores. In the subprocess which orders the items, the first task finds a store which can deliver the item and the second task orders it from there. This subprocess is executed in parallel for each item ordered. Following the parallel execution, either the credit card is charged or the customer is billed (and reminded if he pays late). The process will subsequently be finished. Depending on the number of items ordered, the process consists of more tasks which need to be executed in parallel. Each of these tasks lasts 10 seconds on average (as defined in [5]).

For the experiments, JOpera has been deployed on a cluster of 15 nodes. Each node is a 1.0GHz dual P-III, with 1 GB of RAM, running Linux (Kernel version 2.4.22) and Sun's Java Development Kit version 1.4.2.

12.4.2 Static Configuration

The JOpera execution engine can be configured statically, meaning that the number of dispatchers and navigators is fixed. As we are going to show, using a static configuration may not always lead to optimal performance when the workload changes.

To illustrate the sensitivity of the optimal configuration with respect to the workload and to motivate the need for dynamic reconfiguration, we have carried out a series of experiments where the engine was configured with all possible configurations in a 15-node cluster and have run the two different workloads described in the previous section with each of the configurations. Both workloads simulate 1000 orders paid for by credit card. In the case of the first workload, the customers order 1 item (W_1) and in the case of the second, 20 items (W_{20}).

The results of these experiments are shown in Figure 12.4. The results clearly indicate that the optimal configurations for the two workloads are not the same. Workload W_1 is executed fastest with a configuration between 5 and 7 dispatchers and between 7 and 5 navigators, whereas workload W_{20} is executed fastest using 9 or 10 dispatchers and 6 or 5 navigators, as can be seen in the speedup comparison in Figure 12.4.

A static configuration of the engine, therefore, has two problems. First, a static configuration does not suit all kinds of workloads. If the engine is

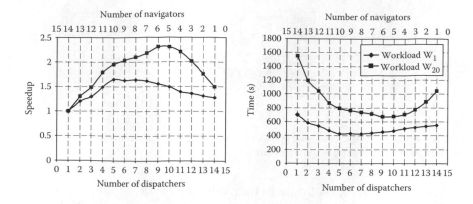

FIGURE 12.4

Speedup relative to the slowest configuration and total time achieved using all possible configurations in a 15-node cluster running workloads W_1 and W_{20}.

configured to service one particular workload, this configuration may not be suitable for other workloads. Hence reconfiguration at runtime may improve the handling of different kinds of workloads. Second, static configurations can potentially lead to inefficient resource allocation, as resources that are no longer required will not be released and can therefore not be used for other purposes.

12.4.3 Self-Configuration

In this experiment we have combined the two workloads from the previous experiment, W_1 and W_{20}, in order to demonstrate the reconfiguration capabilities of the system when the workload characteristics change. We have defined the **information policy** to take into consideration the task queue size and the process queue size. The **optimization policy** has been set to use all available resources so that the workload is executed fastest. Once all resources are in use, this policy will let the controller trade dispatchers for navigators and vice versa depending on the current workload. Excessive growth in the task queue will lead to a configuration with an increased number of dispatchers. In case the process queue grows, it will lead to a configuration with an increased number of navigators. Growth in both queues, however, will not lead to a configuration change. In case both queues shrink, neither dispatchers nor navigators will be stopped, as this would slow down the system. The **selection policy** has been defined to select nodes such that a minimum reconfiguration overhead is induced.

Figure 12.5a shows the size of the process queue over time. This gives a good overview of the rate at which processes are queued to be started (the curve grows) and instantiated and executed by navigators (the curve drops). It also directly reflects the workload which is applied to the system, which—in this experiment—consists of three peaks with varying characteristics: a

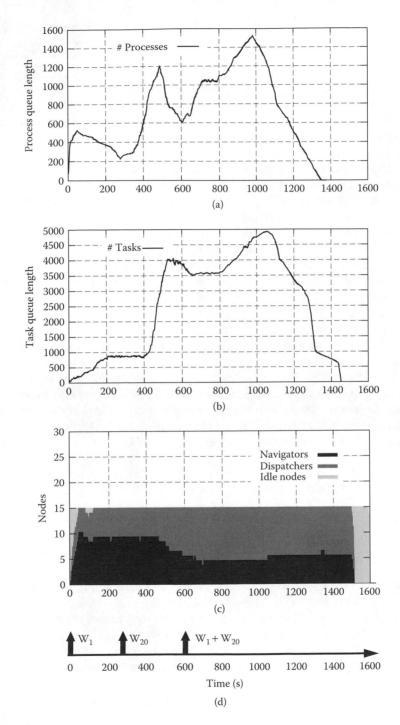

FIGURE 12.5
Autonomic controller reconfiguring the system as workloads with different characteristics arrive.

first peak with the characteristics of W_1, a second peak representing W_{20}, and a last peak consisting of a mixture between W_1, and W_{20}, as depicted in Figure 12.5d. The first peak occurring at $t = 0$ consists of 1000 processes representing the 1000 customers ordering 1 item at the same time. The number of navigators starts to increase rapidly as the number of processes waiting in the process queue increases. The navigators subsequently start to identify the tasks which need to be executed. Because of the control flow of the process used—two consecutive tasks at the beginning—the task queue will grow only mildly, as can be seen in Figure 12.5b. Figure 12.5c shows how the controller configures the system accordingly by allocating only up to 5 dispatchers because of the modest growth of the task queue, while using the rest of the nodes to run navigators. The navigators continue taking processes out of the process queue which will shrink continuously. The configuration will change as the second peak hits the system at $t = 276s$. The second peak again consists of 1000 processes. Customers, however, order 20 items this time. Figure 12.5a shows that in response to this peak, the number of processes waiting in the process queue grows as the new processes are fed into the system.

While these processes begin their execution, the size of the task queue starts to increase slowly. About 100 seconds later, at $t = 395$, the task queue starts to grow at a higher rate. This can again be explained by considering the different characteristics of the processes of the second peak, which execute 20 tasks in parallel. Thus, more dispatchers are required, as there are now more tasks waiting to be executed. Between $t = 463$ and $t = 622$, the controller reconfigures the system by decreasing the number of navigators and by increasing the number of dispatchers because of the surge in the task queue size. Reconfiguring the system to use more nodes as dispatchers will lead to a shrinking task queue.

At $t = 605s$, the third peak hits the system. This peak consists of 500 processes of 1 item ordered and 500 processes of 20 items ordered. Upon this peak being fed into the system, the process queue will again start to grow, and soon after, the task queue will be growing too. The task queue will be growing longer, leading the controller to increase the number of dispatchers. The system will settle at an intermediate configuration which is between the two optimal configurations for W_1 and W_{20}.

After $t = 995s$ the process queue will start to shrink steadily and so will the task queue after $t = 1053s$. The rate of shrinkage in the task queue, however, will decrease at $t = 1318s$. This is because of the structure of the process, which executes the two tasks at the end sequentially. This implies that as soon as the first of these two tasks of a process instance is executed, the second will be put into the queue, leading to this brief stall in the shrinkage of the task queue.

All processes have been started after $t = 1372s$ and thereby all the contained tasks have been enqueued into the task queue. The number of tasks in the queue steadily decreases thereafter. At $t = 1459s$, all tasks have been executed and the controller stops the dispatchers as they become idle. Shortly afterward, the number of navigators reaches zero because the self-configuration component also stops these components as they become idle.

12.4.4 Discussion

In domains which are characterized by unpredictable workloads, it is very difficult to assess the workload characteristics in advance and to tune the system configuration accordingly. In our first experiments (Figure 12.4) we have been able to illustrate that different workload characteristics require different configurations in order to be executed with optimal turnaround time. Finding such a configuration for a given workload is difficult, and misconfiguring the system will lead to degraded performance.

This motivates the addition of the autonomic controller, which adapts the configuration to changing workload characteristics. In our self-configuration experiment we demonstrated that the autonomic controller is able to reconfigure the workflow engine with respect to the changing workload characteristics. When combining the two workloads used for the experiment depicted in Figure 12.4 to emulate a system exposed to changing workload characteristics, the autonomic controller shifted the configuration between the optimal static ones.

In the experiment shown, the benefit of reconfiguration outweighs the cost of reconfiguration (essentially the cost of stopping dispatchers): using reconfiguration, the combined workload is executed in 1492s, whereas it takes 1783s to execute the same combined workload with one of the optimal configurations for W_{20} (9 dispatchers and 6 navigators) and 1991s with one of the optimal configurations for W_1 (5 dispatchers and 10 navigators).

12.5 Related Work

To the best of our knowledge, very little research contributions have been published toward applying autonomic computing principles in the context of distributed workflow engines.

Decentralizing workflow execution is typically done to enable workflows to be executed across several companies without having to use centralized infrastructures [3]. Decentralizing workflow execution in this manner potentially enables improved scalability but also has the downside that the global view over the process is lost. Also, the scalability and reliability problems are only implicitly solved by shifting the problem to nodes executing parts of the process.

Nevertheless, various tools (e.g., GOLIAT [6]) have been proposed to deal with the problem of configuring a distributed workflow engine. Such tools consider the expected characteristics of workloads and make predictions about the performance of the engine when using a given configuration. They thereby interactively give the system administrator advice on how to statically configure the system in order to achieve the desired level of performance. Autonomic computing [10] techniques can instead be used so that the process of optimizing the system configuration does not rely on manual (and static) configuration steps executed by a system administrator.

12.6 Conclusion

In the scope of the work presented here, we have designed an autonomic workflow engine featuring self-tuning, self-configuring, and self-healing capabilities and have evaluated its performance. Using autonomic computing principles, the engine running on a cluster is able to automatically reconfigure itself, taking into account the current workload. As workflow management systems are increasingly used in application domains where unpredictable changes in the workload are common, e.g., process-based Web service orchestration, the ability of a system to configure and heal itself is an important contribution. Scalability and reliability for workflow engines has already been tackled in the past by the distribution of the engine's functionality. However, only little attention has been paid to the need of properly configuring such systems. Doing so is a time-consuming and error-prone task which becomes very difficult in case of unpredictable workloads, as we have been able to illustrate. With our work we have shown that by applying autonomic computing principles, configuration and maintenance of such a distributed system can be greatly simplified.

The experiments presented in this chapter indicate that the autonomic controller of the JOpera engine can reconfigure the system to service workloads of unpredictable and changing characteristics. The controller is also able to heal failures occurring in the distributed engine and to adapt the configuration accordingly [7].

References

1. Boualem Benatallah, Marlon Dumas, Quan Z. Sheng, and Anne H. H. Ngu. Declarative composition and peer-to-peer provisioning of dynamic Web services. In *Proceedings of the 18th International Conference on Data Engineering (ICDE 2002)*, 297–308, San Jose, CA, 2002.
2. Fabio Casati and Ming-Chien Shan. Dynamic and adaptive composition of e-services. *Information Systems*, 26:143–163, 2001.
3. Girish Chafle, Sunil Chandra, Vijay Mann, and Mangala Gowri Nanda. Decentralized orchestration of composite Web services. In *Proceedings of the 13th World Wide Web Conference*, pages 134–143, New York, 2004.
4. The Transaction Processing Performance Council. *TPC-C, an online transaction processing benchmark*. http://www.tpc.org/tpcc

5. Michael Gillmann, Ralf Mindermann, and Gerhard Weikum. Benchmarking and configuration of workflow management systems. In *CoopIS*, 186–197, 2000.
6. Michael Gillmann, Wolfgang Wonner, and Gerhard Weikum. Workflow management with service quality guarantees. In *Proceedings of the ACM SIGMOD Conference*, 228–239, Madison, Wisconsin, 2002.
7. Thomas Heinis, Cesare Pautasso, and Gustavo Alonso. Design and Evaluation of an autonomic workflow engine. In *Proceedings of the 2nd International Conference on Autonomic Computing*, Seattle, WA, June 2005.
8. Thomas Heinis, Cesare Pautasso, Oliver Deak, and Gustavo Alonso. Publishing persistent Grid computations as WS resources. In *Proceedings of the 1st IEEE International Conference on e-Science and Grid Computing*, Melbourne, Australia, 2005.
9. Jens Hündling and Mathias Weske. Web services: Foundation and composition. *Electronic Markets*, 13(2), 2003.
10. IBM. Autonomic computing: Special issue. *IBM Systems Journal*, 42(1), 2003.
11. Frank Leymann. Web services: Distributed applications without limits. In *Proceedings of the International Conference on Business Process Management (BPM 2003)*, 123–145, Eindhoven, Netherlands, 2003.
12. Frank Leymann, Dieter Roller, and Marc-Thomas Schmidt. Web services and business process management. *IBM Systems Journal*, 41(2):198–211, 2002.
13. Cesare Pautasso. JOpera: *Process Support for More Than Web Services*. http:// www.jopera.org
14. Cesare Pautasso and Gustavo Alonso. From Web service composition to megaprogramming. In *Proceedings of the 5th VLDB Workshop on Technologies for E-Services (TES-04)*, 39–53, Toronto, Canada, August 2004.
15. Cesare Pautasso and Gustavo Alonso. JOpera: a toolkit for efficient visual composition of Web services. *International Journal of Electronic Commerce (IJEC)*, 9(2):104–141, Winter 2004/2005.
16. Cesare Pautasso, Thomas Heinis, and Gustavo Alonso. Autonomic execution of service compositions. In *Proceedings of the 3rd IEEE International Conference on Web Services*, Orlando, FL, July 2005.
17. German Shegalov, Michael Gillmann, and Gerhard Weikum. XML-enabled workflow management for e-services across heterogeneous platforms. *VLDB Journal*, 10(1):91–103, 2001.
18. Wil M. P. van der Aalst. Process-oriented architectures for electronic commerce and interorganizational workflow. *Information Systems*, 24(8):639–671, December 1999.
19. W. M. P. van der Aalst, Arthur H. M. ter Hofstede, B. Kiepuszewski, and A. P. Barros. Workflow patterns. *Distributed and Parallel Databases*, 14(3):5–51, July 2003.
20. K. Vidyasankar and Gottfried Vossen. A multi-level model for Web service composition. In *Proceedings of the Second International Conference on Web Services (ICWS2004)*, 462–469, July 2004.
21. K. Whisnant, Z. T. Kalbarczyk, and R. K. Iyer. A system model for dynamically reconfigurable software. *IBM Systems Journal*, 42(1):45–59, 2003.
22. Liang-Jie Zhang and Mario Jeckle. The next big thing: Web services collaboration. In *Proceedings of the International Conference on Web services (ICWS-Europe 2003)*, 1–10, Erfurt, Germany, 2003.

13

Dynamic Collaboration in Autonomic Computing

David M. Chess, James E. Hanson, Jeffrey O. Kephart, Ian Whalley, and
Steve R. White

CONTENTS

13.1 Introduction

The ultimate goal of autonomic computing is to build computing systems that manage themselves in accordance with high-level objectives specified by humans [15]. The many daunting challenges that must be addressed before this vision is realized fully can be placed into three broad categories [14] pertaining to:

- **Self-managing resources**, the fundamental building block of autonomic systems, which are responsible for managing their own behavior according to their individual policies;

253

- **Autonomic computing systems**, which are composed of interacting self-managing resources; and

- **Interactions between humans and autonomic computing systems**, through which administrators can express their objectives, monitor the behavior of the system and its elements, and communicate with the system about potential actions that might be taken to improve the system's behavior.

This chapter concerns an important aspect of the second of these broad challenge areas: how dynamic collaboration among self-managing resources can give rise to system-level self-management. Research on individual self-managing databases, servers, and storage devices, which significantly predates the coinage of the term *autonomic computing*, has resulted in a steady flow of innovation into products from IBM and other vendors over the years. One can be confident that useful innovations in self-managing resources will continue indefinitely. However, isolated, siloed work on individually self-managing resources, while important, will not by itself lead to self-managing computing *systems*. A system's ability to manage itself relies on more than the sum of the individual self-management capabilities of the elements that compose it. We believe that dynamic collaboration among self-managing resources is a crucial extra ingredient required to support system-level self-management.

Why is dynamic collaboration among self-managing resources essential? The primary driver is the ever-growing demand for truly dynamic e-business. For example, IBM's On Demand initiative is a call for more nimble, flexible businesses that can "respond with speed to any customer demand, market opportunity, or external threat" [19]. Customers want to be able to change their business processes quickly. They want to be able to add or remove server resources without interrupting the operation of the system. They want to handle workload fluctuations efficiently, shifting resource to and from other parts of the system as needs vary. They want to be able to upgrade middleware or applications seamlessly, without stopping and restarting the system. In short, dynamic, responsive, flexible e-business requires for its support a dynamic, responsive, flexible information technology (IT) infrastructure that can be easily reconfigured and adapted as requirements and environments change.

Unfortunately, today's IT infrastructures are far from this ideal. Present-day IT systems are so brittle that administrators do not dare to add or remove resources, change system configurations, or tune operating parameters unless it is absolutely necessary. When systems are deployed, relatively static connections among resources are established by configuring each of the resources in turn. For example, once it is decided to associate a specific database with a particular storage system, that connection may exist for the lifetime of the application. When changes such as software patches or upgrades are made, it may take days or weeks to plan and rehearse those changes, thinking through all of the possible ramifications to ensure that nothing breaks, and practicing them on a test system devoted to the purpose. The natural result of such justifiable conservatism is a ponderous IT infrastructure, not a nimble one.

How do we make IT infrastructures more nimble? We believe that, beyond making self-managing resources more individually adaptable, we must endow self-managing resources with the ability to dynamically form and dissolve service relationships with one another. The basic idea is to establish a set of behaviors, interfaces, and patterns of interaction that will support such dynamic collaborations, which are an essential ingredient for system-level self-management.

In this chapter, we propose a set of requirements, identify a set of standards designed to help support those requirements, and show that systems that partially respect these requirements can indeed exhibit several important aspects of self-management. Section 13.2 sets forth our proposed requirements, which focus on supporting dynamic collaboration among autonomic elements. Next, in Section 13.3, we discuss standards and the extent to which they do and do not support those requirements. The next four sections discuss the present status and possible future of two systems that exhibit dynamic collaboration to achieve system-level self-management. The first system, described in Section 13.4, is a datacenter prototype called Unity that heals, configures, and optimizes itself in the face of failures, disruptions, and highly varying workload. Unity achieves these self-managing capabilities through a combination of algorithms that support the behavioral requirements, design patterns that introduce new resources such as registries and sentinels that assist other resources, and well-orchestrated interactions. We discuss the benefits that Unity derives from its use of a subset of the requirements and standards. Then, in Section 13.5, we speculate on the extra degree of self-management that could be attained were Unity to more fully implement the proposed requirements and standards. The second system, described in Section 13.6, focuses more specifically on a particular interaction between two commercially available system components: a workload manager that allocates resources on a fine-grained scale (e.g., central processing unit (CPU) and memory share) and a resource arbiter that allocates more coarse-grained resources such as entire physical servers. We discuss how the requirements and standards support dynamic collaboration between the workload manager and the resource arbiter, and in Section 13.7 we speculate about how their fuller implementation could yield further benefits. We close in Section 13.8 with a summary of our recommendations on requirements and standards, and speculations about the future.

13.2 Requirements

The dynamic nature of autonomic systems has profound implications. We do not believe that such dynamism can be achieved with traditional systems management approaches, in which access to resources tends to be idiosyncratic and management tends to be centralized. On the contrary, our approach is to use a **service-oriented architecture** [13], so that there is a uniform means of representing and accessing services, and to make the resources themselves more **self-managing** [24]. Rather than having a centralized database as the primary source of all information about the resources—what kinds of resources they are, what version they are, etc.—we make the resources themselves **self-describing**. Every resource can be asked what kind of resource it is, what version it is, and so on, and it will respond in a standard way that can be understood by other resources. Similarly, resources respond to queries about their capabilities by describing what kinds of services they are capable of offering. We believe that the advantages that come with making resources self-describing will be at least as great as those that have come from self-description mechanisms such as reflection in programming languages [17].

There is, of course, a role for centralized information. Having a database containing descriptions of all available resources is useful, but having it be the primary source of this information is brittle. As soon as a change is made to a resource, the database is no longer correct. We avoid this problem by **making resources responsible for reporting information** about themselves in these centralized databases. When a new resource comes on line, or as it changes in relevant ways, it informs the centralized database of the change. This keeps the database as up-to-date as possible, without requiring manual updates or periodic polling of all of the resources.

As an essential part of self-management, **resources must govern their actions according to policies**. These policies will typically specify what the resource should do rather than how it should do it. They will specify a performance goal, for instance, rather than a detailed set of configuration parameters that permit the resource to achieve that goal. It is then left to the self-managing resource to determine how best to attain that goal.

Resources will typically use the services of other resources to do their jobs and, as the system changes over time, which resources are being used will change as well. A database may need a new storage service to handle a new table. A new router may need to be found to take over from a router that failed. **Resources must be capable of finding and using other resources** that can provide the services that they need. **They must be capable of forming persistent usage agreements** with the resources that provide them service. Existing work on service composition, such as [20], will be directly relevant to the challenges here.

There will also be cases where **some resource must determine the needs of other resources**, rather than their capabilities. This is clearly necessary at design time, where both manual design tools and automatic deployment planners must be able to ensure, with a high probability of success, that once the system is in operation, all of the necessary resources will be able to obtain the services that they need to function. Similarly, at runtime, a resource's ability to report its needs can provide essential help for system configuration and problem determination.

Owing to the fact that the nature of the services, and the quality of service required, may not be obvious at design time, **resources must be capable of dynamically entering into relationships in which the quality of service is specified** subject to terms that detail constraints on usage, penalties for non-compliance, and so forth. That is, we extend the service-oriented architectural concept of dynamic binding to resources with service-level agreements between the resources themselves. The resource providing the service is required to do so consistent with the terms of the service-level agreement that it has accepted. The consumer of the service may be similarly required to limit the size, frequency, or other characteristics of requests so that the provider can offer the required service level.

In effect, agreements between resources become the structural glue that holds the system together. They replace the static configuration information of today's systems with a mechanism to create similar structures dynamically. Setting up such a structure between two resources involves two steps. In the first step, the resource that wishes to consume a service must find and contact a potential provider. The consumer asks the provider for the details of the agreements that it offers. It then selects one that fits its needs and asks for that specific agreement. If the provider agrees, the agreement is put into place. The second step is for the consumer to use the service of the provider as usual.

It will sometimes be useful for a provider to offer a **default agreement** to any consumer that is allowed to access it. This would be typical, for instance, in a "Yellow Pages" service, which provides a "best effort" response to any queries that it gets. If there is a default agreement, it is not necessary to create an agreement as a separate step. Rather, consumers and providers behave as if they had already put an agreement into place. Default agreements do not give the provider the opportunity to say "no," so it is possible for the provider to get more requests than it can satisfy. Explicit, nondefault agreements permit the provider to determine which agreements it can satisfy, and put limits on its clients accordingly.

While some agreements will be stand-alone bilateral arrangements between two resources, others will involve larger collections of resources. In this latter case, when multiple resources need to be involved in an agreement, there is a need for **resource reservations**. These allow the entity setting up the larger whole to reserve all of the necessary resources before actually forming the agreement.

Just as resources can know their own identity and properties, they can know into what agreements they have entered. Just as resources can report their existence and life-cycle state to a centralized resource database, they can report their agreements to a centralized relationship database and keep this database up-to-date as new agreements are made and old ones ended.

The dynamic restructuring of the system made possible by agreements gives rise to several interesting problems. As in traditional systems, **security** is important—autonomic systems should limit access to those entities that are authorized to access them. The automated nature of autonomic systems makes this even more important. Consider a life-cycle manager that is responsible for moving a small set of resources through their life-cycle states in preparation for maintenance but, because of a bug, sends a "shut down" command to every resource in the system. Authorization control is an important means of limiting such problems. In addition, because of the dynamic nature of autonomic systems, these authorizations need to be dynamic as well, so that the set of authorized resources can change at runtime. To some extent this dynamism of authorization is already present in today's systems, where authorization is typically given to named groups at deployment time, and group membership may change dynamically. However, we expect that the requirements for dynamic authorization in fully autonomic systems will go beyond what can be provided by group membership alone.

In an autonomic system, system management resources can be asked to manage different resources over time. There are likely to be a variety of discipline-specific managers: performance managers, availability managers, security managers, and so on. These disciplines are not orthogonal; increasing availability by mirroring requires extra resources that could have been used to increase performance. In traditional systems, conflicts between the objectives of the various managers are resolved at design time. A single performance manager is assigned to a group of resources. A decision is made ahead of time to trade off performance for availability. In an autonomic system, however, **conflict avoidance, detection, and resolution** must be done at runtime, hopefully without requring the intervention of human administrators.

As the system plans for changes, it must be possible to project the impact of those changes. One approach would be to have a centralized model of the entire system and use it to anticipate the results of change. Such a model would have to possess a rich understanding of the resources themselves and their current operating state. Resource models used in this system model would have to be kept synchronized as the resources themselves are updated, or change their life-cycle state or operating characteristics. The approach that we take instead is to require the resources themselves to **respond to queries about the impact of hypothetical changes** that we might make to them. This eliminates the need for a separate, external model and allows resource designers to handle the queries, and the associated internal processing, in whatever way they want.

13.3   Standards

It is easy to both overestimate and underestimate the importance of standards for computing systems. Having a standard for interchanging a particular class of information is important for interoperability, but it provides little or no help with the problem of actually producing or exploiting that information. On the other hand, given any pairwise interaction between computing resources, it is generally easy to design a method for transferring the information required by the interaction between those resources; no standards are required. But *ad hoc* pairwise interaction design does not scale well, and if autonomic computing is to have a real impact, the heterogeneous resources that form autonomic systems must conform to open, public standards. In this section we review the standards most relevant to the requirements discussed in Section 13.2, both to evaluate their suitability and to find what is missing.[1]

13.3.1 Emerging Web Service Standards

In keeping with our overall service-oriented architecture of autonomic computing systems, all the standards we consider are applicable to systems constructed according to the Web services architecture [23]. The fundamental specification for the description of Web services is the Web Service Description Language (WSDL) [11]. WSDL enables a Web service to characterize the set of messages it is capable of processing—i.e., to describe its own interfaces. WSDL is a highly flexible XML dialect supporting an extremely wide range of interaction styles, from generic messaging to a strongly typed RPC(remote procedure call)-like invocation. It also has the advantage of being independent of the particular transport protocol used, thereby permitting designers to select the protocol most appropriate for their needs.

By itself, WSDL does not make a clear distinction between Web service types and instances. Thus, for example, one common implementation pattern is for a given URL to be associated with a particular service interface as defined by a given WSDL file, but for each message sent to that URL to be handled by a newly instantiated runtime instance of the service. This is insufficient for the needs of long-lived stateful resources. Web Services Addressing (WS-Addressing) [6] fixes this problem by defining a standard form for messaging addresses of *stateful* resources—i.e., the logical "endpoint" address to which messages intended for a given resource instance should be sent.

The long-lived, stateful, active nature of autonomic resources also implies a number of other needs, such as publishing and accessing a resource's state information, publishing and subscribing to events, etc. The Web Services Resource Framework (WSRF) [34, 36, 35, 37, 33] is a constellation of specifications designed specifically to meet these needs in a standardized way. Standardized access to a stateful resource's published state information is given by Web Services Resource Properties (WS-ResourceProperties), which supports both get and set operations for a resource's properties as well as complex XPath-based queries. Web Services Resource Lifetime (WS-ResourceLifetime) provides support for long-lived, but not immortal, resources. The need for registries of resources is met by the Web Services Service Group specification (WS-ServiceGroup).

---

[1] Many of the specifications discussed in this section are still under development, and many are faced with competing specifications that provide more or less similar features. While it is impossible to predict the ultimate form the set of standards will take, it is reasonable to assume that it will not fundamentally differ from what is described here.

Sometimes resources need to expose additional information about their interfaces that may not be expressed using the specifications mentioned so far. For example, it may be the case that a certain property of a resource never changes once set, or may be changed by the resource but not by outside entities, or may be settable by others, and so on. Web Services Resource Metadata [25] provides a standard means by which such additional information may be attached to interface operations and properties.

The essential infrastructure for enabling resources to publish and subscribe to asynchronous events is provided by the Web Services Notification (WS-Notification) specificiations. [30, 31, 32] This is particularly important for autonomic computing, in which resources must be able to react quickly to externally applied changes in their environment, and to report changes in their own state information.

Web Services Security (WS-Security) [26] refers to a collection of related security protocols. For the needs of autonomic systems, we recognize a particular subset of these as being especially useful: resources need to be able to identify the sender of a message and to authenticate that identity; they need to determine the authorization level of the sender, i.e., whether they should pay attention; and they need to be able to communicate without the risk that messages will be intercepted by unauthorized parties. The basic WS-Security standards, defining how data are signed and encrypted and how particular security algorithms are named, are now well established. The higher-level standards, which are required for truly interoperable dynamic security configuration, are still in process.

Built on WSRF, Web Services Distributed Management (WSDM) [28, 29, 27] is another constellation of specifications that meets certain essential needs of autonomic systems. WSDM at present consists of two major parts: MUWS (Management Using Web Services) and MOWS (Management of Web Services). At the core of the WSDM specifications is the notion of a *capability*, which is a URI (Uniform Resource Identifier [5]) that identifies a particular, explicitly documented set of operations, resource properties, and functional behavior. Much of WSDM's attention is given to the manageability capabilities of resources—i.e., the ways in which resources may be managed by traditional systems management architectures. These capabilities are clearly useful to managers in an autonomic system. More importantly, the general notion of capabilities, and the WSDM representation of them, is clearly important as a means of identifying what resources are capable of doing.

Arguably the simplest WSDM capability is *identity*, which is something both essential in autonomic systems and conspicuously absent from the base Web Services and WSRF standards. The notion of resource identity turns out to have subtleties we do not have space to discuss here, but at minimum it should be a unique identifier for a given stateful resource that persists across possible changes in service interface and resource address. Also, since some resources may in fact possess multiple resource addresses, a resource's identity may be used to determine whether or not two different endpoint addresses point to the same resource.

The Web Services Policy Framework (WS-Policy) [3, 7, 4] is a constellation of specifications that partially support autonomic resources' needs for communicating policies. It is primarily an envelope specification, leaving the detailed expression of a policy's content for definition elsewhere.

Web Services Agreement (WS-Agreement) [2] provides the necessary generic support for agreements between autonomic resources, including expression of terms regarding levels of service to be supplied and the conditions under which the agreement applies. It focuses on the outer or envelope format of an agreement document and on the simplest possible mechanisms for agreement creation and management. It leaves open the detailed content of the agreement's terms, recognizing that it will frequently be resource specific.

WS-Agreement specifies a format for agreement *templates*, which allow entities to announce a particular set or class of agreements which they are willing to consider entering into. Resources may use this feature to express their default agreements and, more generally, to describe their own collaborative capabilities. It also supports the expression of the ways in which an agreement may be monitored and the actions that are to be taken when the terms of the agreement are violated. These, taken together with the basic notion of dynamically establishing service commitments at the heart of WS-Agreement, supply important interfaces to help resources avoid, detect, and manage conflicts.

13.3.2 Opportunities

There remain several basic requirements of collaborating autonomic resources that are not obviously met by the Web service standards reviewed above: describing their specific capabilities, expressing policy content, describing the services they need, supporting resource reservations, and responding to queries about the impact of hypothetical changes. These represent opportunities for further development of standardized interfaces.

The notion of capabilities is both extremely important and highly underdeveloped. The basic approach taken by WSDM, in which a capability is a URI representing a human-readable description of behavior, seems correct as far as it goes. But clearly, if resources are to be able to describe their own capabilities in ways that permit automated matching of service providers with service consumers, something richer than opaque identifiers would be of great value. There are a number of ongoing efforts to design interoperable ways of expressing the semantics of Web services, including ways of describing hierarchical ontologies of service capabilities [1].

As noted above, WS-Policy leaves the detailed expression of a policy's content largely unrestricted. Autonomic computing requires a semantically rich policy language that will at least support general "if-then" rules and expression of goals and, equally importantly, detailed ontologies for naming the entities to which the policies apply and describing their salient properties.

In order to describe the services that a resource requires (rather than what it provides), a fairly simple specification should be sufficient. In essence, such a description consists of a list of the interfaces and capabilities that a particular resource expects to find in its environment. The former would be references to WSDL documents, and the latter would be capability URIs. It is important to note that this sort of metadata is required both at design time and at run time; if a system is designed without accurate knowledge of what its basic runtime requirements will be, it is small comfort to know that the problem will eventually be detected during operation.

For reservations, WS-Agreement *could* be used, since a reservation is in fact an agreement to provide something at a future time. But WS-Agreement by itself does not descend to the necessary level of detail; additional, reservation-specific content is required and must therefore also be standardized.

The final item in our list of unmet requirements, estimating the impact of hypothetical changes, is also the least well developed of them. The usage scenarios specifically require that the resources involved *do not* treat a query as an actual request—i.e., that the recipient does not actually carry out the request but only describes what would happen *if* it were to carry it out—which precludes the use of WS-Agreement. The description of the hypothetical changes may refer to the interface elements that would be used in making the change—e.g., the query can take the form "What would happen if I were to invoke operation X on you?", where the operation in question is defined in the resource's WSDL—but no such standard exists at present. A standardized representation for the consequences of the change, such as the impact estimates used in the example systems we present in this chapter, will also ultimately be required.

13.4 The Unity System

In the next four sections, we will consider two systems that have important autonomic features and at least partially meet the requirements of dynamic collaboration; for each system we will consider both its current state and ways in which it could be enhanced to meet more of the requirements.

The system described in [9, 21] is a prototype autonomic system called Unity that illustrates some of the features of dynamic collaboration. (In this chapter we concentrate on the aspects of the system that best illustrate dynamic collaboration; a more complete description is in [9], and [21] includes qualitative data on the system's performance.)

The resources in the Unity system dynamically describe, discover, and configure themselves, form agreements, and exchange impact estimates in order to allocate resources in ways likely to enable the system to meet its goals. Conflicts are avoided either by a centralized utility calculation or (in a variant not described in the original paper but covered briefly below) by simple cooperative utility estimates.

The IT scenario that the Unity system addresses involves resource allocation between multiple applications. A finite pool of resources must be allocated between two or more applications, where each application provides some service for which there is a time-varying level of external demand. The performance of each application depends on the demand being placed on it and the amount of resource allocated to it.

Each application is governed by agreements, along the lines described in [16], which specify the rewards or penalties associated with various possible behaviors of the system. In the Unity system, agreements are expressed in terms ranging from application behavior (for instance, transactions per unit time) to real numbers representing the utility (the "goodness") of a particular level of behavior. The overall success of the system depends on the performance of each application relative to the governing service-level agreement.

The management resources of the system must cooperate in order to optimize the overall system performance relative to the set of service-level agreements in effect. They do this by discovering resources and forming and maintaining relationships using defined interfaces. To acheive these goals, they make use of other system resources that, while not directly involved in management themselves, provide services and capabilities that the management resources require.

Here we will describe the main components of the system as illustrated in Figure 13.1, and briefly describe how the figure illustrates the principles of dynamic collaboration.

The heart of the Unity system's runtime discovery is the *service registry*. Based in this case on the Virtual Organization Registry defined in [18], its function is analogous to Yellow Pages registries in multi-agent systems (see for instance [12]) and very similar to registries based on WS-ServiceGroup as described above (that specification was not available at the time the system was originally planned).

When each resource in the system comes up, it is provided with little initial configuration information beyond the specific task it is to perform and the address of the service registry. Each resource both registers its own capabilities with the registry and queries the registry to find the addresses of the other resources from which it will fetch further information (such as policies) and with which it will form agreements to accomplish its goals.

The syntax and semantics of registry information and queries are very simple: Each resource creates an entry in the registry containing its address (in the form of a Web services endpoint reference) and one or more capability names. Capability names in this system are simple atomic strings, similar to the capability URIs used in the MUWS specifications.

This scheme differs from that used in Universal Description Discovery and Integration technical models (UDDI tModels) [22], in that no hierarchy is represented directly. If one capability is logically a specialization of another, any resource offering the more specialized capability must explicitly register as providing both the more specialized capability and the more general one. This is also the approach taken in [28, 29].

To query the registry, a resource supplies a capability name, and the registry returns a list of addresses of all the resources that have registered as providing

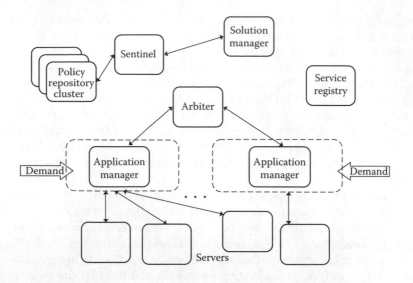

FIGURE 13.1

An abstract schematic of the Unity system. The arbiter component resolves conflicting resource requirements among the application managers, as described in the text. The solution manager uses the sentinel to monitor and heal failures in the cluster of policy repositories. All components access the service registry and policy repository (interaction arrows not shown to avoid clutter), as described in the text.

that capability. There is an implicit correspondance between capability names and service interfaces: Any resource registering itself as having a particular capability is responsible for implementing the corresponding interface. The registry also provides a notification mechanism, by which resources can sign up to be informed when new resources register themselves as providing particular capabilities.

Each application in Unity is represented by an *application manager* resource, which is responsible for the management of the application, for obtaining the resources that the application needs to meet its goals, and for communicating with other resources on matters relevant to the management of the application. In order to obtain the services that it needs, each application manager consults the registry to find resources that have registered as being able to supply those services.

In the system as implemented, the services in question are provided by individual server resources. Each server registers itself as a server with the registry. The management resource that allocates servers to applications is the *arbiter*. It registers itself as able to supply servers, so the application managers can find it. It then consults the registry to determine which servers are available to be allocated. The underlying design scales easily to multiple heterogeneous server pools managed by multiple arbiters and distinguished by the capability names that each type of server exposes and each arbiter queries the registry for.

Another important component of the system is the *policy repository*, which holds the service-level agreements governing the desired behavior of the system, and other system configuration information. All resources in the system, including the management resources, obtain the address of the policy repository from the registry and contact it to obtain the policies that apply to them and to subscribe to changes to those policies.

Policies are scoped according to a flat scheme similar to that used to name capabilities in the registry: Each resource belongs to a scope that corresponds to the task it is performing, and each policy in the repository belongs to one or more scopes. The policy repository is the primary channel through which human administrators control the system.

The ubiquitous use of the registry and the policy repository enables the system to self-assemble at runtime, without requiring manual configuration but still according to human constraints as expressed in the system policies.

In operation, the arbiter is responsible for managing the resource pool, by controlling which resources are assigned to which application. It does this by obtaining from each application manager an estimate, in terms of utilities as determined by the service-level agreements, of the impact of various possible allocations and calculating an optimum (or expected optimum) allocation of the available resources. It is a key responsibility of each application manager to be able to predict how an increase or decrease in the resources allocated to the application would impact its ability to meet its goals.

The arbiter is not concerned with how the individual application managers make their predictions. It simply uses the results of the predictions to allocate resources. Conversely, the individual application managers do not need to know anything about the activities of the other applications. They need only use their local knowledge of conditions within the application, and their local models of probable future behavior, to make the most accurate predictions they can.

As well as allowing for dynamic reaction to changes at operation time, the comparatively loose and dynamic collaborative management enabled by the service registry, the policy repository, and the direct expression of utilities also has advantages in terms of adding new features and functions. These advantages are illustrated by two additions that were made to the system. In the first (described in more detail in [9]), a *sentinel* resource was added to the system. The sentinel can be asked to monitor other resources for liveness and report when a monitored resource stops responding. A *solution manager* resource was also added, responsible for increasing the reliable operation of the infrastructure as a whole (rather than the utility of the applications); it locates a sentinel via the registry and enlists it to monitor for failures in any member of a cluster of policy repositories. When notified of a failure by the sentinel, the solution manager starts a new replacement instance of the policy repository to replace the failed one. Adding these features to the system required no changes to the existing resources, which continued to bind and operate as they had before.

The second modification involved giving the application managers the ability to communicate with each other directly about the hypothetical utilities of various server allocations and to allocate resources properly in the absence of a functioning arbiter. Again the organization of the system as dynamically bound services allowed us to make this change without fundamentally rearchitecting the existing system.

13.5   Improving the Unity System

The Unity system was designed to explore many of the requirements of dynamic collaboration for autonomic computing, but as implemented it does not include all of the requirements listed in Section 13.2. It does not implement significant intercomponent security, for instance, or default agreements, both of which would have obvious advantages for the system.

Perhaps the most significant feature that it does not explore is rich service-level agreements between resources in the system. The most expressive service-level agreements in Unity are logically outward facing, concerned with the rewards and penalties that accrue to the system as a whole with the behavior of the applications. The agreements between the Unity resources are represented as simple, named atomic relationships, without the internal structure of a service-level agreement. Increasing the richness of the interresource agreements by adding structured and parameterized agreement terms and utilities (rewards and penalties), based on explicit quality of service measures, would allow the resources of the system to make better-informed decisions about both their internal operations and their dynamically changing relationships.

Similarly, the policies and agreements governing the behavior of the solution manager are extremely simple; the human administrator must specify relatively low-level details such as the number of replicated copies of the policy repository that should be present in the system. As described in [10], we have designed modifications to the system that would allow the administrator to specify policies in higher-level terms (such as effective availability) and automatically derive the more detailed policies from those. The general problem of deriving detailed IT policies from higher-level business policies is one of the major challenges of autonomic computing.

13.6   Node Group Manager and Provisioning Manager

The second example we will consider, illustrated in Figure 13.2, is described more fully in [8]. (Again we concentrate on the dynamic collaboration features of the system; more details and qualitative performance data can be found in [8].)

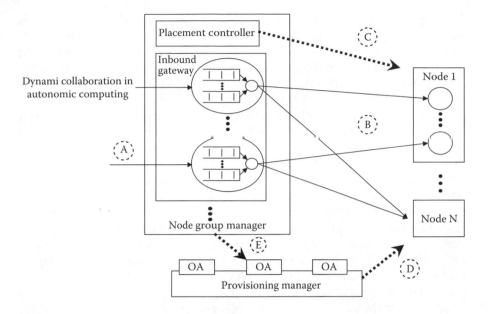

FIGURE 13.2
Overall system structure (described in detail in the text). Incoming requests (A) are queued by the gateway and dispatched (B) to application servers when resources are available to service them. On a longer time scale, the Placement Controller (C) determines which application servers should run in which nodes. In the longest time scale, the provisioning manager determines (D) which nodes to assign to which node groups by consulting its objective analyzers, which receive estimates (E) from the node group manager (for simplicity, only one node group is shown here). Adapted from O. Chess, G. Pacifici, M. Spreitzer, M. Steinder, A. Tantawi, and I. Nhalley. Experience with collaborating managers: Node group manager and provisioning manager. In Proceedings of the Second International Conference on Autonomic Computing, 2005, pp 39–50, (IEEE).

This system consists mainly of two management resources, each with some autonomic characteristics, which collaborate to achieve a level of overall system self-management. While this system as implemented contains fewer dynamic collaboration features than the Unity system, it has the advantage of being based on commercially available management software used in real datacenters rather than on research prototypes. After describing the system as it exists, we will consider how the emerging manageability standards could be used to increase the level of dynamism it exhibits.

The first manager in this system, a node group manager implemented in a middleware application server, uses modeling and optimization algorithms to allocate server processes and individual requests among a set of server machines ("nodes") grouped into node groups. It also estimates its ability to fulfill its service-level objectives as a function of the number of nodes potentially available to each node group.*

---

* The node group manager in question is a version of WebSphere® Extended Deployment, and the provisioning manager is a version of IBM Tivoli® Intelligent Orchestrator.

All of the nodes in a given node group share a set of properties, such as installed operating system, network connectivity, support libraries, and so on. The node group manager is responsible for directing and balancing the traffic within each node group and for allocating server processes to the nodes within the group, but it is not able to carry out the provisioning actions necessary to move nodes from one node group to another. Additionally, the node group manager has only a local view of the set of node groups for which it is responsible. It cannot make higher-level decisions about the allocation of nodes between the demands of its own node groups and those of other processes in the larger datacenter.

The second manager in the system, a provisioning manager, implements another layer of management above the node group manager. It has the knowledge necessary to move nodes from one node group to another through potentially time-consuming provisioning actions, and it can balance the competing demands of multiple managers operating at the level of the node group manager. These may be instances of the node group manager, or of other managers that can provide the required performance estimates. On the other hand, the provisioning manager does not have the node group manager's real-time knowledge of the traffic within each node group.

The two management layers are thus complementary. The two managers perform a very basic form of "collaboration," in that the estimates produced by the node group manager allow the provisioning manager to more effectively allocate the resources that it provisions, and the actions of the provisioning manager give the node group manager more servers to work with.

The provisioning manager utilizes plugins called "objective analyzers" (OAs) to determine the utilization of the various systems it is managing. In operation, the provisioning manager periodically queries the node group managers and each of the other managers at the same level, using the objective analyzers to convert the information from those managers into a form that the provisioning manager understands and can compare across the managers. Specifically, from each of those managers, the provisioning manager requests data estimating, for each potential level of resources that might be allocated to the manager's application, the probability that the application's service-level agreement will be breached if the application is given that level of resources.

This is analogous to the information provided by the application managers to the arbiter in the Unity system described above, except that it is expressed in terms of probabilities of service-level agreement breach rather than in terms of resulting utilities. This is a small but interesting difference. In Unity, the application managers, rather than the arbiter, are aware of the actual agreements in effect. Since each application manager is aware in more detail of the behavior of its application, service-level agreements evaluated by application managers can potentially reflect knowledge of detailed application behavior that is not normally accessible to a central arbiter. On the other hand, if the central arbiter is aware of individual service-level agreements, it can potentially make richer tradeoffs among them. We are currently experimenting with variants of this system to explore the practical effects of where this information is maintained.

The system described in this section displays dynamic self-management at operation time. In the next section, we will outline how other aspects of the system, such as discovery and binding, could be made similarly dynamic.

13.7 Improving the Node Group and Provisioning Managers

In light of the discussion in Section 13.3, it is instructive to consider an idealized version of the node group and provisioning manager discussed in Section 13.6. How might a future version of that system make fuller use of the architecture of dynamic collaboration and the emerging Web services standards, and what advantages would that future version offer over today's?

13.7.1 Discovery

One of the most obvious advantages of widespread adoption of the WS-Addressing and MUWS standards will be the enablement of dynamic binding. In the current system, the provisioning manager must be manually configured with information about the node group manager. It must be told where the node group manager is, what the nodes in the system are, and what the initial node allocation is. Dynamic binding and discovery would enable the provisioning manager to simply detect these properties of the system and configure itself appropriately.

Therefore, the provisioning manager would be notified of the arrival of the node group manager by a service registry. The registry would be based in WS-ServiceGroup and would use WS-Notification to alert interested parties about changes to the registry membership. This would enable the system to configure itself automatically in terms of the binding between the managers, eliminating an unnecessary manual step. It would also allow other system resources, not designed to work with these specific products, to take part in the system, either as a replacement for one of the managers or as third parties taking advantage of the information in other ways. One simple example of this is a system visualization tool that would present the content of the registry to a human administrator.

13.7.2 Negotiation and Binding

When the provisioning manager receives notification from the registry that a new node group manager has entered the system, the provisioning manager would then contact the node group manager to determine whether or not to enter into a management agreement with it. The node group manager would expose agreement templates conforming to the WS-Agreement specification that describe the kinds of agreements into which it is willing and able to enter.

WS-Agreement also provides a simple way for the provisioning manager to propose an agreement, and for the node group manager to respond. How the two managers determine internally whether or not to collaborate is of course outside the scope of WS-Agreement. The content of the agreement indicates that the node group manager will provide the provisioning manager with breach probability estimates, and the provisioning manager will in turn control the set of servers with which the node group manager works. WS-Agreement does not specify a content language in which to express this. Further standardization work is required to establish the detailed discipline-specific conventions required here.

There are several reasons why the provisioning manager and the node group manager might decide not to enter into a management agreement. The node group manager might already have entered into such an agreement with an alternate provisioning manager. Or, the provisioning manager might not be able to manage the type of nodes that the node group manager needs. For the purposes of our scenario, however, we assume that they do enter into a management agreement.

13.7.3   Data Gathering

At this point, the provisioning manager and the new node group manager are bound together. They have a management agreement in place that permits the provisioning manager to make allocation changes to the nodes used by the node group manager. Now the two managers must proceed to actually manage the overall system.

In the current implementation, the provisioning manager polls the node group manager periodically, requesting information concerning the current performance of the node group manager's overall system. In the future system, the node group manager could make these performance statistics available via WS-ResourceProperties, and the provisioning manager would subscribe via WS-Notification to receive updates when these statistics change.

When the provisioning manager decides to change the allocation of the nodes under the control of the node group manager, it would then tell the node group manager to start or stop using a set of nodes. During the formation of the initial agreement, the provisioning manager would have examined the node group manager's manageability characteristics, as specified by the MUWS standard, to ensure that it has the interfaces corresponding to those operations. Here again, the MUWS standard tells us how to expose and access information about manageability characteristics, but it does not give us a specific URI corresponding to the interface that we need. Futher standardization work is required at the discipline level to establish a convention here.

The case where the provisioning manager decides to remove nodes is illuminating. In this case, it would be helpful for the provisioning manager and the node group manager to collaborate on the decision as to which nodes to remove. In the current implementation, for instance, it is often the case that certain nodes under the control of the node group manager require less effort to remove than other nodes. A more detailed exchange of information about the expected impacts of various possible changes would allow the system as a whole to optimize itself even more effectively.

Again, any of the interactions described here could be implemented via *ad hoc* and nonstandard languages and protocols. But by using open interoperability standards such as those of the Web services, resources from different vendors can be used easily, programs that were not specifically designed to work together can collaborate, and new functions not originally anticipated can be composed from existing building blocks.

13.8 Conclusions

While neither system described in this chapter is fully self-managing, both support the central thesis: Dynamic collaboration is an essential ingredient of system self-management. In the Unity datacenter prototype, system components register themselves (along with a very simple description of their capabilities) to a registry, enabling components to find the services they need. A very simple form of negotiation ensues, resulting in an agreement that forms the basis for a relationship that persists until it is no longer required. Even the relatively rudimentary mechanisms for discovery, resource description, negotiation, and agreement make it possible for Unity to assemble itself and exhibit a type of self-healing. Experience with the second system shows that by adding a thin layer of collaborative capability to two commercially available components that were not originally designed to work together, the resulting system can do an effective job of coordinating optimization of system resources at two levels of granularity.

Thus an encouraging lesson can be drawn from experimental observations of the two systems: A modest degree of system-level self-management can be achieved without fully observing all of the requirements listed in Section 13.2. This suggests that although many challenges remain, good progress toward the ultimate vision of autonomic computing can be made long before all of those challenges are met.

Moreover, this chapter's analysis of the strengths and shortcomings of the two experimental systems and the existing body of standards suggests several useful avenues for further work in standards and technology that would bring about a greater degree of system self-management. Many of these center around improving the richness of the dynamic interactions among self-managing resources. Both systems employed very simplistic methods for describing resource needs and capabilities, consisting of a flat mapping between URIs and human-readable descriptions of capabilities and requirements. This very simple language supported some degree of flexibility, such as the ability to add new types of system components without changing the existing ones. Yet it seems likely that richer semantics, accompanied by correspondingly more sophisticated reasoning algorithms, would enable better, more flexible matching between needs and capabilities, and would also form a basis for much more sophisticated forms of negotiation and agreement among resources. Indeed, the negotiation and agreement employed in both systems was quite rudimentary. As suggested in Section 13.5, increasing the richness of the description of agreements from a simple named atomic relationship to a full-fledged WS-Agreement document would be a tremendous step forward. However, even this would not provide the ultimate solution, as the standards and technologies required to support negotiation of such agreements do not yet exist.

As a final observation, the standards mentioned here, and related standards too numerous to list, are helpful in improving interoperability and in supporting a degree of dynamic collaboration. Yet, quite generally, they fall short of what is required in the long term because many of them are essentially envelope or data-container standards. In other words, given a description of an interface, a capability, a possible agreement, or an event description, they tell one how to communicate that description in a self-defining way. Much more work is needed to pin down exactly how to represent capabilities—for example, the fact that a given node group manager is able to produce a particular sort of breach probability estimate, or that a particular router is able to handle a specific packet throughput—in ways that fit within these envelopes. Unifying existing discipline and resource-specific standards with the relevant Web services standards and devising new ones where they are needed will be among the most important drivers of further progress in autonomic computing.

References

1. R. Akkiraju, J. Farrell, J. Miller, M. Nagarajan, M. Schmidt, A. Sheth, and K. Verma. *Web Service Semantics—WSDL-S.* Technical report, A joint UGA-IBM Technical Note, April 2005.
2. Alain Andrieux, Karl Czajkowski, Asit Dan, Kate Keahey, Heiko Ludwig, Jim Pruyne, John Rofrano, Steve Tuecke, and Ming Xu. *Web Services Agreement Specification (WS-Agreement).* http://www.ggf.org/Meetings/GGF12/Documents/WS-AgreementSpecification.pdf, 2004.
3. Siddharth Bajaj, Don Box, Dave Chappell, Francisco Curbera, Glen Daniels, Phillip Hallam-Baker, Maryann Hondo, Chris Kaler, Dave Langworthy, Ashok Malhotra, Anthony Nadalin, Nataraj Nagaratnam, Mark Nottingham, Hemma Prafullchandra, Claus von Riegen, Jeffrey Schlimmer, Chris Sharp, and John Shewchuk. *Web Services Policy Framework (WS-Policy).* ftp://www6.software.ibm.com/software/developer/library/wspolicy.pdf, 2004.

4. Siddharth Bajaj, Don Box, Dave Chappell, Francisco Curbera, Glen Daniels, Phillip Hallam-Baker, Maryann Hondo, Chris Kaler, Ashok Malhotra, Hiroshi Maruyama, Anthony Nadalin, Mark Nottingham, David Orchard, Hemma Prafullchandra, Claus von Riegen, Jeffrey Schlimmer, Chris Sharp, and John Shewchuk. *Web Services Policy Attachment (WS-PolicyAttachments)*. ftp://www6.software.ibm.com/software/developer/library/ws-polat.pdf, 2004.

5. T. Berners-Lee, R. Fielding, and L. Masinter. *Uniform Resource Identifier (URI): Generic Syntax*. http://www.ietf.org/rfc/rfc3986.txt, 2005.

6. Don Box, Erik Christensen, Francisco Curbera, Donald Ferguson, Jeffrey Frey, Marc Hadley, Chris Kaler, David Langworthy, Frank Leymann, Brad Lovering, Steve Lucco, Steve Millet, Nirmal Mukhi, Mark Nottingham, David Orchard, John Shewchuk, Eugne Sindambiwe, Tony Storey, Sanjiva Weerawarana, and Steve Winkler. *Web Services Addressing (WS-Addressing)*. http://www.w3.org/Submission/ws-addressing, 2004.

7. Don Box, Maryann Hondo, Chris Kaler, Hiroshi Maruyama, Anthony Nadalin, Nataraj Nagaratnam, Paul Patrick Claus von Riegen, and John Shewchuk. *Web Services Policy Assertions Language (WS-PolicyAssertions)*. ftp://www6.software.ibm.com/software/developer/library/ws-polas.pdf, 2002.

8. D. Chess, G. Pacifici, M. Spreitzer, M. Steinder, A. Tantawi, and I. Whalley. Experience with collaborating managers: Node group manager and provisioning manager. In *Proceedings of the Second International Conference on Autonomic Computing*, 2005.

9. D. Chess, A. Segal, I. Whalley, and S. White. Unity: Experiences with a prototype autonomic computing system. In *Proceedings of the First International Conference on Autonomic Computing*, 2004.

10. David M. Chess, Vibhore Kumar, Alla Segal, and Ian Whalley. Work in progress: Availability-aware self-configuration in autonomic systems. In Akhil Sahai and Felix Wu, editors, *DSOM*, volume 3278 of *Lecture Notes in Computer Science*, 257–258. Springer, 2004.

11. Erik Christensen, Francisco Curbera, Greg Meredith, and Sanjiva Weerawarana. *Web Services Description Language (WSDL) 1.1*. http://www.w3.org/TR/wsdl, 2001.

12. E. H. Durfee, D. L. Kiskis, and W. P. Birmingham. The agent architecture of the University of Michigan digital library. In *IEEE/British Computer Society Proceedings on Software Engineering (Special Issue on Intelligent Agents)*, February 1997.

13. H. He. *What Is Service-Oriented Architecture?* http://webservices.xml.com/pub/a/ws/2003/09/30/soa.html, 2003.

14. Jeffrey O. Kephart. Research challenges of autonomic computing. In *Proceedings of the 27th International Conference on Software Engineering*, 15–22, 2005.

15. Jeffrey O. Kephart and David M. Chess. The vision of autonomic computing. *Computer*, 36(1):41–52, 2003.

16. A. Leff, J. T. Rayfield, and D. Dias. Meeting service level agreements in a commercial grid. *IEEE Internet Computing*, July/August 2003.

17. Pattie Maes. Concepts and experiments in computational reflection. In *OOPSLA '87: Conference Proceedings on Object-Oriented Programming Systems, Languages and Applications*, 147–155, New York, 1987. ACM Press.

18. J. Nick, I. Foster, C. Kesselman, and S. Tuecke. *The Physiology of the Grid: An Open Grid Services Architecture for Distributed Systems Integration*. Technical report, Open Grid Services Infrastructure WG, Global Grid Forum, June 2002.

19. Stephen Shankland. *IBM: On-Demand Computing Has Arrived.* http://news. zdnet.com/2100-3513_22-5106577.html, 2003.

20. B. Srivastava and J. Koehler. Web service composition—current solutions and open problems. In *ICAPS 2003,* 2003.

21. Gerald Tesauro, David M. Chess, William E. Walsh, Rajarshi Das, Alla Segal, Ian Whalley, Jeffrey O. Kephart, and Steve R. White. A multi-agent systems approach to autonomic computing. In *AAMAS,* 464–471. IEEE Computer Society, 2004.

22. *Introduction to UDDI: Important Features and Functional Concepts.* http://uddi. org/pubs/uddi-tech-wp.pdf, 2004.

23. W3C Web Services Architecture Working Group. *Web Services Architecture.* http://www.w3.org/TR/ws-arch/, 2004.

24. Steve R. White, James E. Hanson, Ian Whalley, David M. Chess, and Jeffrey O. Kephart. An architectural approach to autonomic computing. In *First International Conference on Autonomic Computing,* 2004.

25. *Web Services Resource Metadata 1.0 (WS-ResourceMetadataDescriptor).* http://www.oasis-open.org/committees/download.php/9758/wsrf-WS-ResourceMetadataDescriptor-1.0-draft-01.PDF, 2004.

26. *Web Services Security: SOAP Message Security 1.0 (WS-Security 2004).* http:// docs.oasis-open.org/wss/2004/01/oasis-200401-wss-soap-message-security-1.0.pdf, 2004.

27. *Web Services Distributed Management: Management of Web Services (WSDM-MOWS) 1.0.* http://docs.oasis-open.org/wsdm/2004/12/wsdm-mows-1.0. pdf, 2005.

28. *Web Services Distributed Management: Management Using Web Services (MUWS 1.0) Part 1.* http://docs.oasis-open.org/wsdm/2004/12/wsdm-muws-part1-1.0.pdf, 2005.

29. *Web Services Distributed Management: Management Using Web Services (MUWS 1.0) Part 2.* http://docs.oasis-open.org/wsdm/2004/12/wsdm-muws-part2-1.0.pdf, 2005.

30. *Web Services Base Notification 1.3 (WS-BaseNotification).* http://www.oasis-open.org/committees/download.php/13488/wsn-ws-base_notification-1.3-spec-pr-01.pdf, 2005.

31. *Web Services Brokered Notification 1.3 (WS-BrokeredNotification).* http:// www.oasis-open.org/committees/download.php/13485/wsn-ws-brokered_notification-1.3-spec-pr-01.pdf, 2005.

32. *Web Services Topics 1.2 (WS-Topics).* http://docs.oasis-open.org/wsn/2004/06/ wsn-WS-Topics-1.2-draft-01.pdf, 2004.

33. *Web Services Base Faults 1.2 WS-BaseFaults).* http://docs.oasis-open.org/ wsrf/wsrf-ws_base_faults-1.2-spec-pr-02.pdf, 2005.

34. *Web Services Resource 1.2 (WS-Resource).* http://docs.oasis-open.org/wsrf/wsrf-ws_resource-1.2-spec-pr-02.pdf, 2005.

35. *Web Services Resource Lifetime 1.2 (WS-ResourceLifetime).* http://docs.oasis-open.org/wsrf/wsrf-ws_resource_lifetime-1.2-spec-pr-02.pdf, 2005.

36. *Web Services Resource Properties 1.2 (WS-ResourceProperties).* http://docs.oasis-open.org/wsrf/wsrf-ws_resource_properties-1.2-spec-pr-02.pdf, 2005.

37. *Web Services Service Group 1.2 (WS-ServiceGroup).* http://docs.oasis-open.org/wsrf/wsrf-ws_service_group-1.2-spec-pr-02.pdf, 2005.

14

AutoFlow: Autonomic Information Flows for Critical Information Systems

Karsten Schwan, Brian F. Cooper, Greg Eisenhauer, Ada Gavrilovska, Matt Wolf, Hasan Abbasi, Sandip Agarwala, Zhongtang Cai, Vibhore Kumar, Jay Lofstead, Mohamed Mansour, Balasubramanian Seshasayee, and Patrick Widener

CONTENTS

14.1 Introduction

Distributed information-intensive applications range from emerging systems like continual queries [5, 32], to remote collaboration [35] and scientific visualization [52], to the operational information systems used by large corporations [38]. A key attribute of these applications is their use in settings in which their continued delivery of services is critical to the ability of society to function. A case in point is the operational information system running the 24/7 operations of the airline partner with which our research center has been cooperating [38]. Another example is the use of telepresence for remote medicine or diagnosis. Other well-known critical settings include information flows in financial applications: online data collection, processing, and distribution in automotive traffic management: and more generally, the rich set of information capture, integration, and delivery services on which end users are increasingly reliant for making even routine daily decisions.

The objective of the AutoFlow project is to better meet the critical performance requirements of distributed information flow applications. In this context, multiple technology developments provide us with new ways of meeting these requirements. One is the ubiquitous use of middleware to extend application functionality across the distributed, heterogeneous communication and computational infrastructures across which they must run. Second, while it is not easy or desirable to rewrite applications to make better use of system and network infrastructures, middleware provides a basis on which it becomes possible to customize and extend underlying systems and networks to better meet the needs of the many applications written with these widely used software infrastructures. In other words, middleware can exploit the increasingly open nature of underlying systems and networks to migrate selected services "into" underlying infrastructure. Third, the increasing prevalence of virtualization technologies, on computing platforms and in the network, is providing us with the technical means to safely "extend" existing computation elements on the fly. In particular, virtualization allows us to better control or isolate extended from nonextended elements of the vertical software stacks used by applications and middleware, and to create "performance firewalls" between critical vs. noncritical codes. An outcome of these developments is that vertical extension and the associated control across the entire extended software stack are possible without compromising a system's ability to simultaneously deliver services, critical and noncritical to many applications and application components.

The AutoFlow project exploits these facts to create middleware and open-system infrastructures that jointly implement the following functionalities to meet future applications' criticality and high performance needs:

"Vertical" and "Horizontal" Agility — The AutoFlow middleware presented in this chapter uses information flow graphs as a precise description of the overlay networks used by critical applications' distributed information flows.

Based on these descriptions, applications can dynamically create new services, which are then deployed by middleware as native binary codes to the machines where they are needed. The outcome for these high performance codes is "horizontal agility," which is the ability to change at runtime both what data are streamed to which overlay nodes and where operations are applied to such data. A simple example for real-time scientific collaboration is the runtime augmentation of server functionality to better meet current client needs [53]. Additional benefits of using horizontal agility are described in Section 14.4.2. Furthermore, middleware can also migrate certain services "vertically," that is, for suitable services, middleware can use dynamic methods for system and network extension to realize more appropriate service implementations than those available at application level. Experimental results demonstrating the benefits derived from such vertical service migration appear in Section 14.4.4. The importance of both horizontal and vertical agility is demonstrated with critical information flows for high performance and enterprise applications in Sections 14.4.3 and 14.4.4.

Resource-Aware Operation — The dynamic nature of distributed execution platforms requires applications to adjust their runtime behavior to current platform conditions. Toward this end, AutoFlow uses cross-layer "performance attributes" for access to platform monitoring information. Such attributes are used by application- or model-specific methods that dynamically adapt information flows. Sample adaptation methods implemented for real-time exchanges of scientific data with AutoFlow's publish-subscribe communication model include (1) dynamic operator deployment [53] in response to changes in available processing resources and (2) runtime adjustments of parameterized operators to match data volumes to available network bandwidths [22]. For distributed query graphs, techniques for dynamically tuning operator behavior to maximize the utility of information flows with available network resources are described in [32]. Section 14.4.3 presents results documenting the benefits derived from the network-aware operation of AutoFlow applications.

Utility-Driven Autonomic Behavior — End users of distributed information systems desire the timely delivery of quality information content, regardless of the dynamic resource behavior of the networks and computational resources used by information flows. To meet application needs, AutoFlow uses application-specific utility functions to govern both the initial deployment of information flows and their runtime regulation, thereby enabling the creation of application- or domain-specific autonomic functionality. A utility metric used in this chapter uses application-stated importance values for sending different elements of a high-volume, scientific information flow to best use the limited network bandwidth available across international network links.

Scalability through Hierarchical Management — AutoFlow scales to large underlying platforms by using hierarchical techniques for autonomic management, as exemplified by its automatic flow-graph partitioning algorithm presented in [32]. One way in which this algorithm attains scalability is by making locally rather than globally optimal deployment decisions, thereby limiting the amount of nonlocal resource information maintained by each node. Scalability results presented in this chapter justify hierarchical management with microbenchmarks evaluating the deployment time for a publish-subscribe implementation using a hierarchical approach.

An additional benefit derived from the general nature of *Information Flow Graphs* is their role as a uniform basis for creating the diverse communication models sought by applications, including publish-subscribe [49], information flows implementing continuous queries [5], and domain-specific messaging models like those used by our airline partner [12].

The AutoFlow project leverages a multiyear effort in our group to develop middleware for high-end enterprise and scientific applications. As a result, our AutoFlow prototype uses multiple software artifacts developed in prior work. AutoFlow's resource and network awareness are supported by the monitoring methods described in [25].

High performance and the ability to deal with large data volumes are derived from its methods for dynamic binary code generation and the careful integration of multiple software layers explained in [14], as well as by its binary methods for data representation and runtime conversion for heterogeneous systems [6]. Performance and scalability are due to a separation of overlay-level messaging from the application-level models desired by end users and, as stated above, its hierarchical methods for the efficient deployment of large information flow graphs to distributed systems [32, 33].

While leveraging previous work, the AutoFlow project makes several novel research contributions. First, the AutoFlow middleware uses a formalized notion of information flows and, based on this formalization, provides a complete set of abstractions and a small set of primitives for constructing and materializing different application-level autonomic messaging models. Specifically, information flows are described by *Information Flow Graphs*, consisting of descriptions of sources, sinks, flow operators, edges, and utility functions. *Sources*, *sinks*, and *flow operators*, which transform and combine data streams, constitute the vertices of the flow-graph, while the *edges* represent typed information flows between vertices. Once a flow graph has been described, its deployment (i.e., the mapping of operators to physical nodes and edges to network links) creates an overlay across the underlying physical distributed system. Automated methods for deployment and runtime reconfiguration are based on resource awareness functionality and on *utility functions* that act as a vehicle for encoding user and application requirements. The information flow graph serves as the abstraction visible to applications and provides the concrete representation on top of which to construct domain-specific messaging models like publish-subscribe, for example [14]. Second, in contrast to our previous work on pub-sub [13], scalability and high performance for AutoFlow applications are attained by separating resource awareness functionality placed into AutoFlow *underlays* [32] from the information flow abstractions in the *overlay*, and from the *control plane* used to realize different messaging models. Third, by using utility-driven self-regulation,

AutoFlow flows deployed across heterogeneous, dynamic underlying distributed platforms can continuously provide high-performance services to end-user applications.

Experimental results presented in Section 14.4 demonstrate AutoFlow's basic capabilities. A low send/receive overhead, 5.45μ sec and 14.37μ sec for 10KB messages, respectively, makes the case for high-performance applications, closely matching the performance attained by well-known high-performance computing infrastructures like MPICH. Dynamic reconfiguration is enabled by low instantiation/deletion overheads for overlay components, 0.89μ sec and 0.03μ sec for local (i.e., in the same address space) operations. Experimental results in Section 14.4.3 demonstrate the importance of horizontal agility, using information about the structure of elements in data flows for real-time scientific collaboration to filter flows so as to best meet individual client needs. Section 14.4.4 uses enterprise data to illustrate the performance advantages gained from using a specialized "information appliance" (a network processor able to interpret and process enterprise data) to manipulate application data. Finally, hierarchical deployment and reconfiguration are evaluated in Section 14.4.2.

14.2   Target Application Classes

In order to motivate this chapter, we will first describe some sample applications in which autonomic information flows will be used. Figure 14.1 depicts two classes of information flow applications — real-time scientific collaboration and an airline's operational information system.

14.2.1   Collaborative Visualization

Our first example application is the collaborative visualization of a molecular dynamics (MD) simulation. We note that MD simulations are of interest to computational scientists in a wide variety of fields, from pure science (physics and chemistry) to applied engineering (mechanical and aerospace engineering). Our particular visualization application is geared to deliver the *events of interest* (see Figure 14.2) to participating collaborators, formatted to suit the rendering capabilities at their ends. More details about this application are available in [52].

Collaborative real-time visualization uses a many-to-many information flow, with data originating in a parallel MD simulation, passing through operators that transform and annotate the data, and ultimately flowing to a variety of clients. This real-time data visualization requires synchronized and timely delivery of large data volumes to collaborators. For example, it is unacceptable when one client consistently lags behind the others. Similarly, predictable

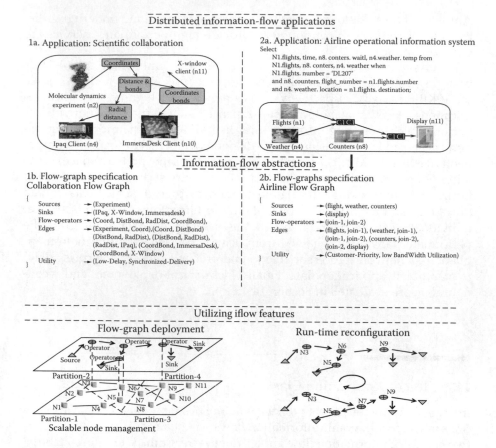

FIGURE 14.1
Implementation of different applications with AutoFlow.

delivery latency is important, since there may be end-to-end control loops, as when one collaborator drives the annotation of data for other collaborators. Finally, autonomic methods can exploit a number of quality/performance tradeoffs, as end users may have priority needs for certain data elements, prefer consistent frame rates to high data resolution (or vice versa), etc. Thus, this online visualization application has rich needs for autonomic behaviors.

14.2.2 Operational Information System

The other information flow in Figure 14.1 represents elements of an *operational information system* (OIS) providing support for the daily operations of a company like Delta Air Lines (see [38] for a description of our earlier work

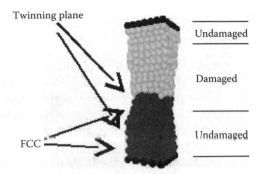

FIGURE 14.2

Events of interest: Sample molecular dynamics data that show different events of interest. For this single simulation of a block of copper being stretched, on the left we see attributes a physicist might want to highlight, while the right side shows the higher-level synthesis a mechanical engineer may want to see.

with this company). An operational information system provides continuous support for an organization's daily operations. We implement an information flow motivated by the requirement to feed overhead displays at airports with up-to-date information. The overhead displays periodically update the weather at the "destination" location and switch over to seating information for the aircraft at the boarding gate. Other information displayed on such monitors includes the names of wait-listed passengers, the current status of flights, etc. We deploy a flow graph with two *operators*, one for selecting the weather information (which originates from the weather station) based on flight information, and the other for combining the appropriate flight data (which originate from a central location like Delta's transaction processing facility) with periodic updates from airline counters that decide the wait-list order, etc. Thus, the three *sources* can be identified as the weather information source, the flight information source, and the passenger information source. They are then combined using the operators to be delivered to the *sink* — the overhead display.

Here, SQL (structured query language) like operators translate into a deployed flow graph with sources, sinks, and operators. Such an OIS imposes the burden of high event rates on underlying resources, which must be efficiently utilized to deliver high utility to the enterprise. Utility-driven autonomic methods for managing event flows may take into account, for example, that a request pertaining to seat assignment for a business-class customer may be given a higher priority because it reflects higher returns for the business. Similarly, other factors like time to departure destination, etc. can drive the prioritized allocation of resources to the deployed information flows. A detailed discussion of the utility-driven deployment of such a flow graph can be found in [33].

14.3   Software Architecture

Insights from our earlier work with the ECho publish-subscribe infrastructure have led us to structure AutoFlow into the three software layers.

First, the *Control Layer* is responsible for accepting information-flow composition requests, establishing the mapping of flow-graph vertices to the physical platform represented by the underlay, and handling reconfigurations. It has been separated from the messaging layer because it must be able to implement different application-specific methods for flow-graph deployment and reconfiguration, each of which may be driven by a different utility function. By providing basic methods for implementing such semantics, rather than integrating these methods into the messaging layer, AutoFlow not only offers improved control performance compared with the earlier integrated ECho pub-sub system developed in our work, but it also gives developers the freedom to implement alternative messaging semantics. In ongoing work, for example, we are creating transactional semantics and reliability guarantees like those used in industrial middleware for operational information systems.

The second layer is the *Messaging Layer*, responsible for both data transport and the application of operators to data. It consists of an efficient messaging and operator module, termed "Stones," and its Web service-enabled distributed extension, termed "SoapStones." A high-performance implementation of information flows can make direct use of Stone functionality, using it to implement data and control plane functions, the latter including deploying new Stones, removing existing ones, or changing Stone behavior. This is how the ECho pub-sub system is currently implemented, essentially using additional message exchanges to create the control infrastructure needed to manage the one-to-one connections it needs for high-volume information exchanges. An alternative implementation of ECho now being realized with AutoFlow provides additional capabilities to our pub-sub infrastructure. The idea is to use AutoFlow's overlays to replace ECho's existing one-to-one connections between providers and subscribers with overlay networks suitably mapped to underlays. Finally, the purpose of SoapStone is to provide ubiquitous access to Stone functionality, leveraging the generality of the Simple Object Access Protocol (SOAP) protocol to make it easy for developers to implement new control protocols and/or realize the application-level messaging protocols they require. Not addressed in this chapter but subject of our future work are the relatively high overheads of SoapStone (due to its use of the SOAP). As a result, it is currently used mainly for initial flow-graph deployment and for similarly low-rate control actions, and higher rate control actions are implemented directly with the Stone infrastructure.

The third layer is the *Underlay Layer*. It organizes the underlying hardware platform into hierarchical partitions that are used by the deployment infrastructure. The layer also implements scalable partition-level resource awareness, with partition coordinators subscribing to resource information from other nodes in the partition and utilizing it to maintain the performance of deployed information flows. Its separation affords us the important ability to add generic methods for dynamic resource discovery, underlay growth and contraction, underlay migration, and the vertical extension of the underlay "into" the underlying communication and computation platforms.

The focus of this chapter, of course, is the autonomic functionality in Auto-Flow. To this end, we next describe the control layer functions used for runtime reconfiguration of AutoFlow applications.

14.3.1   Control Layer — *Composition, Mapping, and Reconfiguration*

The *Control Layer* implements the abstraction of an information flow graph and is responsible for mapping a specified flow-graph onto some known underlay. Deployment is based on the resource information supplied by the underlay layer and a function for evaluating deployment utility. The application can specify a unique utility function local to a flow graph, or a global utility formulation can be inherited from the underlay layer. The control layer also handles reconfigurations to maintain high utility for a deployed information flow.

14.3.1.1   *Information Flow Graph*

The information flow graph is a collection of vertices, edges, and a utility function, where vertices can be sources, sinks, or flow operators:

- A `source` vertex has a static association with a network node and has an associated data-stream rate. A source vertex can be associated with one or more outgoing edges.

- A `sink` vertex also has a static association with a network node. A sink vertex can have at most one incoming edge.

- An `operator` vertex is the most dynamic component in our abstraction because its association to any particular network node can change at runtime as the control layer reconfigures the flow graph's deployment. Each operator is associated with a data resolution factor (which is the ratio of the average stream output rate to the average stream input rate), an average execution-time, and an E-Code [14] snippet containing the actual operator code. An operator vertex can be associated with multiple incoming and outgoing edges.

The `utility` of a flow graph is calculated using the supplied utility function, which essentially contains a model of the system and is based on both application-level (e.g., user priority) and system-level (e.g., delay) attributes. The function can be used to calculate the `net utility` of a flow-graph mapping by subtracting the `cost` it imposes on the infrastructure from the

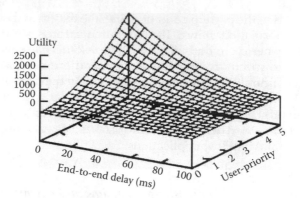

FIGURE 14.3
A sample utility calculation model.

utility value. A sample utility calculation model is shown in Figure 14.3, which depicts a system where end-to-end delay and the user priority determine the utility of the system. The utility model in this scenario can be stated as "High Priority users are more important for the business" and "Less end-to-end delay is better for the business."

The AutoFlow framework also supports a `pin-down` operation, which statically associates an operator with the network node. Pin-down enables execution of critical/nontrivial operators on specialized overlay nodes and/or simply denotes the fact that certain operators cannot be moved (e.g., due to lack of migration support for their complex codes and states). Another interesting feature implemented into the framework is the support for `parameterized "tunable" operators`, which enables remote modification of parameters associated with operator code. For example, a subscription operator might route updates based on a particular predicate, where the control layer supports remotely modifying the predicate at runtime.

14.3.1.2 *Flow-Graph Construction and Mapping*

On close examination of the applications requiring information flow capabilities, we observe that there are two distinct classes of flows.

Basic information flows arise in applications in which only the sources and sinks are known, and the structure of the data flow graph is not specified. For example, in the pub-sub model, there exists no semantic requirement on the flow-graph. AutoFlow can establish whichever edges and merge/split operators may be necessary between publishers and subscribers. This class of "basic" information flows is accommodated by a novel graph construction algorithm, termed `InfoPath`. InfoPath uses the resource information available at the underlay layer to both construct an efficient graph and map the graph to suitable physical network nodes.

Semantic information flows are flows in which the data flow graphs are completely specified, or at least have semantic requirements on the ordering and relationships between operators. For example, SQL-like continual queries over data streams specify a particular set of operations based on the laws of relational algebra. Similarly, in a remote collaboration application, application-specific operators must often be executed in a particular order to preserve the data semantics embedded in scientific work flows. Here, AutoFlow takes the specified flow graph and maps it to physical network nodes using the PathMap mapping algorithm. This algorithm utilizes resource information at the underlay layer to map an existing data flow graph to the network in the most efficient way. The InfoPath and PathMap algorithms are described in detail in [32, 31].

14.3.1.3   Reconfiguration

After the initial efficient deployment has been produced by PathMap or InfoPath, conditions may change, requiring the deployment to be reconfigured. AutoFlow maintains a collection of configuration information, called the IFGRepository, that can be used by the control layer when reconfiguration is necessary. The underlay uses network awareness to cluster physical nodes, and the IFGRepository is actually implemented as a set of repositories, one per underlay cluster. This allows the control layer to perform local reconfigurations using local information whenever possible. Global information is accessed only when absolutely necessary. Thus, reconfiguration is (usually) a low-overhead process.

The AutoFlow framework provides an interface for implementing new reconfiguration policies based on the needs of the application. Our current implementation includes two reconfiguration policies: the Delta Threshold Approach and the Constraint Violation Approach. Both approaches take advantage of the IFGRepository and the resource information provided by the underlay layer to monitor the changes in the utility of a graph deployment. When the change in utility passes a certain threshold (in the Delta approach) or violates application-specific guarantees (in the Constraint approach), the control layer initiates a reconfiguration. The two reconfiguration approaches are described in detail in [31].

A rich set of methods for runtime adaptation controls information flows without reconfiguring their links or nodes. These methods manage the actual data flowing across overlay links and being processed by overlay nodes. Two specific examples are presented and evaluated in Section 14.4 below: (1) the adaptation of the data produced by a scientific data visualization, to meet utility requirements that capture both the quality and the delay of the data received by each visualization client, and (2) improvements in end-to-end throughput for high-rate enterprise data flows by "early" filtering of less important data at an AutoFlow-extended network interface attached to an AutoFlow host.

14.3.2   Messaging Layer — Stones, Queues, and Actions

The messaging layer of AutoFlow is composed of communicating objects, called `stones`, which are linked to create `data paths`. Stones are lightweight entities that roughly correspond to processing points in data flow diagrams. Stones of different types perform data filtering, data transformation, multiplexing and de-multiplexing of data, and transmission of data between processes over network links. Application data enter the system via an explicit submission to a stone but thereafter travel from stone to stone, sometimes crossing network links, until they reach their destination. The actual communication between stones in different processes is handled by the Connection Manager, a transport mechanism for heterogeneous systems, which uses a portable binary data format for communication and supports the dynamic configuration of network transports through the use of attributes. Each stone is also associated with a set of `actions` and `queues`. Actions are application-specified handlers that operate on the messages handled by a stone. Examples of actions include handlers that filter messages depending on their contents and handlers that perform type conversion to facilitate message processing in the stones. Queues associated with stones serve two purposes: synchronizing incoming events to a stone that operates on messages coming from multiple stones and temporarily holding messages when necessary during reconfiguration.

Stones can be created and destroyed at runtime. Stones can then be configured to offer different functionalities by assigning the respective actions. This permits stones to be used as sources/sinks as well as for intermediate processing. Actions assigned to stones can be both typed and untyped. When multiple typed actions are assigned to a single stone, the type of the incoming event determines which action is applied. Some of the actions that can be assigned to stones include:

- An `output action` causes a stone to send messages to a target stone across a network link.

- A `terminal action` specifies an application handler that will consume incoming data messages (as a sink).

- A `filter action` allows application-specified handlers that filter incoming data to determine whether they should be passed to subsequent stones.

- A `split action` allows the incoming messages to be sent to multiple output stones. This is useful when the contents of a single link must be sent along multiple data paths. The target stones of a split action can be dynamically changed by adding/removing stones from the split target list.

- A `transform action` converts data from one data type to another. These actions may be used to perform more complex calculations on a data flow, such as subsampling, averaging, or compression.

Filter and transform actions are particularly noteworthy, as they allow the application to dynamically set handler functions. Handler functions are specified in E-Code, a portable subset of the C language. Dynamic code generation is then used to install and execute the handlers. Not only does this process facilitate dynamic reconfiguration, but it also permits the handlers to be run on heterogeneous platforms. Alternative static implementations of filters and transformers and less portable dynamic methods using Dynamic Link Libraries (DLLs) or similar mechanisms are also available, to accommodate complex stone processing.

The design of stones permits the dynamic assignment of actions and presents a generic framework for messaging. It also allows the dynamic configuration of message handling and transport, and hence offers a suitable base for network overlays.

14.3.2.1 SOAP-Based Overlay Control

The control layer must make calls into the messaging layer to manage stones. However, stones are a general middleware component and may be useful in other infrastructures, an example being the overlay-based implementation of GridFTP described in [8]. For a convenient application programming interface (API) that provides an abstraction of the messaging layer for both AutoFlow and other frameworks, we have developed a Web service front end using SOAP. We call this API SoapStone. The overlay network created with stones can thus be configured and managed through SOAP calls. The SOAP operations for overlay control have been merged with those used for configurations in the control layer, obviating the need for a separate SOAP server for the two layers.

The information flow graph obtained from the control layer and the details of the mapping between the vertices of the flow graph and the corresponding physical network nodes are used to send the appropriate SOAP calls to the corresponding nodes to create and manage stones. Any reconfigurations necessitated by the higher layer during the course of execution can be enacted upon the affected stones through SOAP operations.

14.3.3 Underlay Layer — Network Partitioning and Resource Monitoring

The Underlay Layer maintains a hierarchy of physical nodes in order to cluster nodes that are 'close' in the network sense, based on measures like end-to-end delay, bandwidth, or internode traversal cost (a combination of bandwidth and delay). An example is shown in Figure 14.4. The organization of nodes in a hierarchy simplifies maintenance of the partition structure and provides an abstraction of the underlying system, its administrative domains, and its resource characteristics to the upper layers. For example, when deploying a flow graph, we can subdivide the data flow graph to the individual clusters for further deployment.

We call the clusters *partitions*, although nodes in one partition can still communicate with those in other partitions. Each node in a partition knows about

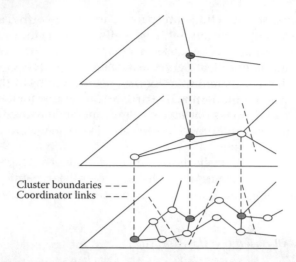

Cluster boundaries – – –
Coordinator links   – – –

FIGURE 14.4
Hierarchical network partitioning.

the costs of paths between every pair of nodes in the partition. A node is chosen from each partition to act as the coordinator for this partition in the next level of the hierarchy. Like the physical nodes in the first level of hierarchy, the coordinator nodes can also be clustered to add another level in the hierarchy. Also, just as in the initial level, all coordinators at a particular level know the average minimum cost path to the other coordinator nodes that fall into the same partition at that level. In order to scalably cluster nodes, we bound the amount of nonlocal information maintained by nodes by limiting the number of nodes that are allowed in each partition.

Horizontal Agility. An important set of underlay responsibilities for autonomic computing is its support for online resource monitoring. To deal with node failures or the addition and deletion of new machines, the underlay layer has functions that handle node Join and Departure requests. The layer also implements a resource-monitoring module, using stone-level data structures that contain per-node network and machine performance data. In particular, each coordinator maintains an IFGRepository of configuration and resource information for its partition. Our current implementation leverages the subscription-based monitoring capabilities from our previous work on the Proactive Directory Service [7], which supports pushing relevant resource events to interested clients. In addition, performance attributes are used to describe information about the underlying platforms captured by instrumented communication protocols, by active network bandwidth measurements, and by the system-level monitoring techniques described in [25]. At the control layer, the coordinator for a particular partition subscribes to resource information from various nodes and intervening links in its partition, aggregates it, and responds to changing resource conditions by dynamically reconfiguring the information-flow deployment.

Vertical Agility. Stones and actions directly support "horizontal" agility. The monitoring support described in the previous paragraph provides system-level resource information to middleware layers. However, none of these abstractions permit the "vertical" agility needed for critical applications. A key attribute of underlays, therefore, is that they can extend "into" the underlying communication and computational platforms. One model for this extension is described in [30], where the C-Core runtime captures the resources available across both general purpose host processors and the specialized communication cores on a host/communication coprocessor pair or more generally, on future heterogeneous multicore platforms. For platforms like these, the underlay extended with C-Core (1) permits the dynamic creation, deployment, and configuration of services onto those cores that are best suited for service execution (e.g., hosts vs. network processors), and (2) monitors and exports the necessary resource utilization and configuration state needed for making appropriate dynamic deployment decisions. In the current design, C-Core provides to the EVPath middleware constructs termed *adaptation triggers*. The application and middleware use these constructs to specify services suitable for mapping to the underlying network. The C-Core infrastructure is responsible for determining, at runtime, the best processing contexts for running such services. By explicitly specifying the application-level services that may be suitable for running "in" the network, the SPLITS (Software architecture for Programmable LIghtweighT Stream handling) compilation support associated with C-Core can statically generate the appropriate representations of such services for network-level execution (i.e., use a different set of code generators) [17].

Specific examples that demonstrate the utility of such "vertical" agility are evaluated in Section 14.4. The first example is drawn from the Delta OIS application. It demonstrates the benefits of deploying filtering handlers "closer" to the network, at the network processor level. The filters extract from the data stream only those Delta events containing information for flights out of the Atlanta airport. Those events are further translated into appropriate formats which can be exchanged with external caterers. Both the filtering handler and the data translation handler can be implemented for network-processor or host-resident execution. Experiments demonstrate that the network near execution of such filtering actions can result in close to 30% performance improvement compared with filter executions on host machines.

The second example further demonstrates the utility of dynamic vertical reconfiguration. Consider the imaging server described in the SmartPointer application (the "Scientific Collaboration" example in Figure 14.1, and also the topic of [52]), which customizes the image data to match the end-users' interests. Our earlier results [17] have demonstrated that networking platforms such as the IXP network processors are capable of performing manipulations of OpenGL image data at gigabit rates. However, depending on current loads on the general purpose computational node (i.e., the host), and the concrete ratio of original image size vs. derived image size, the ability of the host or the communications processor to execute the image manipulation service varies significantly. As a result, it is necessary to monitor platform runtime conditions such as processing loads and application parameters such as image sizes or cropping coordinates. Based on such monitoring, one should then dynamically reconfigure the deployment of the imaging service from a version where all image manipulation is performed at the computational host to another one where the original image is passed to the communications hardware, which then performs all additional image cropping and transmission operations.

14.4 Experimental Evaluation

Experiments are designed to evaluate the autonomic concepts used in the AutoFlow middleware: agility, resource-aware operation, and utility-driven behavior. First, microbenchmarks examine some specific features of the AutoFlow middleware, including the performance of the messaging layer compared with the well-known MPICH (message passing interface) high-performance software and the performance of algorithms implemented with the AutoFlow control layer. We also evaluate the performance of our implementation of the pub-sub communication model. Second, we examine the use of utility functions in evaluating the need to reconfigure an overlay network in the face of changing network resource availability, a form of horizontal agility. Next, we examine several forms of resource-aware operation, including the ability to adjust an application's own bandwidth requirements in response to network resource changes, further adapting application transmission in the context of lower-level network information (such as reported round-trip times [RTTs]). Lastly, the importance of vertical agility is demonstrated with measurements attained on a host/attached IXP platform, emulating future heterogeneous multicore systems.

14.4.1 Messaging Layer: Stone Performance

Microbenchmarks reported in this section are measured using a 2.8 GHz Xeon quad processor with 2MB cache, running Linux 2.4.20 smp as a server. The client machine used is a 2.0 GHz Xeon quad processor, running Linux 2.6.10 smp. Both machines are connected via single-hop 100Mbps ethernet.

Send/Receive Costs: AutoFlow's most significant performance feature is its use of the native data format on the sender side, coupled with dynamically generated unmarshalling code at the receiver to reduce the send/receive cost. "Send side cost" is the time between an application submitting data for transmission until the time at which the infrastructure invokes the underlying network "send()" operation. "Receive side cost" represents the time between the end of the "receive()"operation and the point at which the application starts to process the event. Since these costs are in the range of 0.005ms to 0.017ms, the resulting overheads are quite small compared with the typical

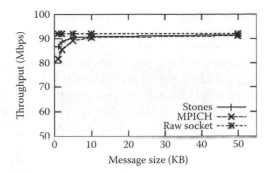

FIGURE 14.5
Throughput using stones.

round-trip delays experienced in local area networks (about 0.1–0.3ms with a Cisco Catalyst 6500 series switch) and negligible for typical wide-area round-trip delays (50–100ms).

`Throughput Comparison against MPICH`: We also compare the throughput achieved for different message sizes using stones with that of raw sockets and MPICH. Figure 14.5 shows that achieved throughput values closely follow the raw socket throughput for packet sizes exceeding 2KB and are almost equal to the values achieved using MPICH. This is very encouraging for AutoFlow applications that target the high-performance domain.

`Stones Instantiation Overheads`: The deployment of a flow graph at the overlay layer consists of creating sources, sink, or filter stones and associating suitable action routines to the stones. Local stone operations require less than a microsecond to complete,[1] while remote SOAPs add just the network overheads associated with the remote operation (typically on the order of a millisecond on local machines).

14.4.2   Utility-Based Reconfiguration

The generator of the Georgia Tech Internetwork Topology Model (GT-ITM) [54] is used to generate a sample Internet topology for evaluating the performance of the control layer, in terms of deployment optimality and reconfiguration benefit. We use the transit-stub topology with 128 nodes for the ns-2 simulation, including 1 transit domain and 4 stub domains. Links inside a stub domain are 100Mbps. Links connecting stub and transit domains, and links inside a transit domain are 622Mbps, resembling OC-12 lines. The traffic is composed of 900 constant bit rate (CBR) connections between sub-domain nodes generated by cmu-scen-gen [42]. The simulation is carried out

---

[1] Some stone operations require portable binary input/output (PBIO) data format registration. Depending upon the options selected for PBIO operation, this may require contact with a network server, increasing time requirements to over a millisecond for the first operation associated with a particular format. Thereafter format information is cached in-process.

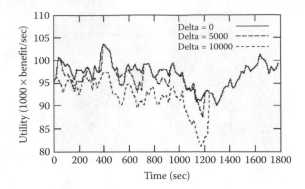

FIGURE 14.6
Utility variation with delta-threshold approach.

for 1800 seconds, and snapshots capturing end-to-end delay between directly connected nodes were taken every 5 seconds. These are then used as inputs to the underlay layer's resource monitoring infrastructure.

Concerning the use of utility for controlling the configuration of distributed overlay networks, it is interesting to compare the performance of two self-optimization approaches: the delta-threshold approach and the constraint-violation approach. The change in utility (where edge utility is determined using the formulation $k \star (c-delay)^2 \star bandwidth_{available} \star bandwidth_{required}$) of a 10-node data flow graph using the delta-threshold approach in the presence of network perturbations is shown in Figure 14.6. The rationale behind the delta-threshold approach is that a reconfiguration is beneficial only when the benefits accrued over time due to reconfiguration surpass the cost of reconfiguration. Hence, pursuing the optimal deployment for smaller gains in utility may not be the best approach. The delta-threshold approach aims to minimize the number of potentially lossy reconfigurations. We note that even for a sufficiently large value of threshold, the achieved utility closely follows the maximum achievable utility, but this is achieved with far fewer reconfigurations (1 with a threshold of 10,000 as compared with 11 with a 0 threshold). Thus, an appropriate threshold value can be used to trade off utility for a lower number of reconfigurations.

Figure 14.7 shows the variation of utility when the constraint-violation approach is used for self-optimization. In this experiment, we place an upper bound on the total end-to-end delay for the deployed data flow graph, and trigger a reconfiguration when this bound is violated. The experiment is driven by real world requirements for delaying reconfiguration until a constraint is violated, because in some scenarios it might be more important to maintain the configuration and satisfy minimal constraints rather than optimize for maximum utility. We note some resemblance in behavior between the delta-threshold approach and the constraint-violation approach. This is because utility is a function of end-to-end delay for the deployed flow graph.

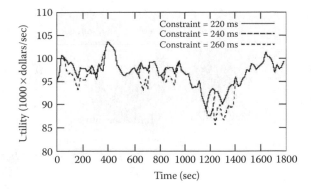

FIGURE 14.7
Utility variation with constraint-violation approach.

However, managing the system by monitoring constraint violations is far easier than optimizing a general utility function. Self-optimization driven by change in utility value is more difficult than one driven by constraint violation, because calculating maximum achievable utility requires knowledge of several system parameters and the deployment ordering amongst various graphs for achieving maximum utility.

14.4.3   Adaptive Downsampling in Congested Networks

In real-time collaboration, one cannot unduly reduce the rates at which data are provided to end users, since that may violate the timeliness guarantee or cause different end users to get "out of sync" with respect to the data they are jointly viewing. A solution is to deploy application-level data filters to down-sample the actual data being sent prior to submitting them to the network transport. These filters can regulate the traffic imposed on the underlying network by "pacing" application-level messages to effectively reduce congestion and maintain better message delivery rates. In contrast to multimedia systems that use statically defined filters specialized for that domain [16], the AutoFlow middleware's API permits clients to dynamically define and then deploy exactly the filters they wish, when and if they need them (also see [14] for more detail). Furthermore, using performance attributes, filters can be controlled to deliver the best-quality data permitted by current network conditions.

The experimental results in Figure 14.8 demonstrate the necessity and effectiveness of middleware-level adaptation through dynamic data downsampling. Here, large cross traffic (250Mbps) is injected as a controlled perturbation into the link from the machine named **isleroyale** at Georgia Tech to the machine named **cruise** at Oak Ridge National Library (ORNL). The network bottleneck is at Georgia Tech's edge router. Permitting the client to

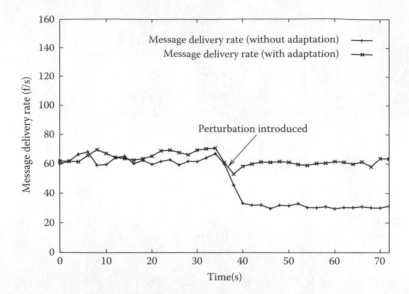

FIGURE 14.8
Adaptive downsampling (ORNL link).

characterize the subset of data most important to it, the client installs a data filter at the server side when congestion occurs, and henceforth receives only "essential" (i.e., as defined by the deployed filter) data at satisfactory speeds. The specific data downsampler used in these experiments removes data relating to visual objects that are not in the user's immediate field of view. That is, the client transfers the current position and viewpoint of the user to the filter (i.e., using attributes), at the server side these values are used to determine what data set the user is currently watching, and that information is then used to transfer appropriately downsampled data to the client. The result is a consequent reduction in the network bandwidth used for data transmission, thereby speeding up data transmission.

This experiment demonstrates the utility of network-initiated data downsampling for maintaining high data rates for limited network bandwidths. By giving end users the ability to define their own filters for implementing data downsampling, similar methods can be applied to other applications, as shown by past work on real-time applications [44, 46], and with services that implement general rather than application-specific compression methods [51].

Other results attained in this research demonstrate some interesting aspects of using dynamic network performance data for controlling middleware-level data movements: (1) If autonomic middleware is given different means of assessing current network behavior, then adaptive methods can use those means to take proactive measures when network capacity changes, one outcome being improved reaction times to network perturbation, and (2) middleware can sometimes cushion applications from undesirable network behavior, such as TCP (transmission control protocol) slow start.

14.4.4   Vertical Agility

This set of experiments illustrates the importance of enabling vertical system agility, that is, the ability of AutoFlow applications to adapt at runtime not only the nodes on which certain actions run, but also "where" on such nodes action execution is carried out. Experiments are conducted using a cluster of eight Dell 530s with dual 1.7GHz Xeon processors running Linux 2.4.18, outfitted with Radisys ENP2611 IXP2400-based programmable communication coprocessors (i.e., network processors, NPs), and interconnected via 1Gbps and 100Mbps links. We have evaluated the viability of executing various application-level services on the IXP NP as compared with host nodes only (i.e., going directly to the host's ethernet link, without an IXP in the path), as well as the performance levels that can be achieved. These evaluations are carried out by implementing different services with handlers that execute jointly on hosts and on their attached IXP2400s.

The results in Figure 14.9 illustrate the importance of offloading certain data translation services from application components executed on standard host onto attached programmable network interconnect cards. The data streams used in this experiment are generated with sequences of demo-replays of representative business data (i.e., data collected for the airline OIS). The

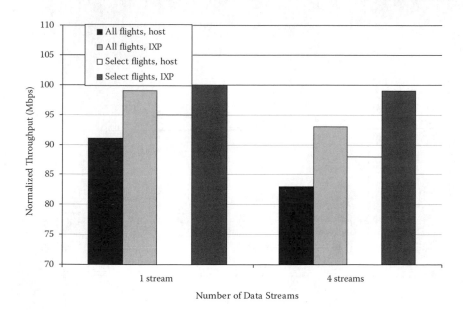

FIGURE 14.9
Data translation/extraction performance on network coprocessor (IXP) or host.

application components executed on the cluster nodes use our version of the Reliable User Datagram Protocol (RUDP) on top of raw sockets [22]. The same protocol is used on the IXP microengines. We use the IXP2400 NP attached to the host via its peripheral component interconnect (PCI) interface to emulate such programmable communication cores. The results represent the performance levels attainable for host- vs. IXP-based execution of rule chains that translate data into the appropriate format (bars marked "all flights"), or translate the data and extract certain information currently needed by the application (bars marked "select flights"). Results demonstrate that the in-network execution of these services results in improved performance, primarily due to CPU offloading and because load is removed from the host's I/O and memory subsystems. In addition, some unnecessary data copying and protocol stack traversals are avoided. Additional results on the utility of execution of application-level services on communications platforms such as the IXP network processor and vertical runtime adaptations appear in [20, 19].

While it may be intuitive that executing filtering functionality "as early as possible," at the network interface, can result in improved service quality, the "vertical" reconfiguration of other services is more sensitive to current workload or platform resources. The results in Figure 14.10 compare the host's vs. the IXP's ability to crop 1MB OpenGL-produced images against application-provided bounding boxes.

The first conclusion from these experiments is that even non-communications-related service components can be considered for vertical deployment "into" the platform, with this example showing performance

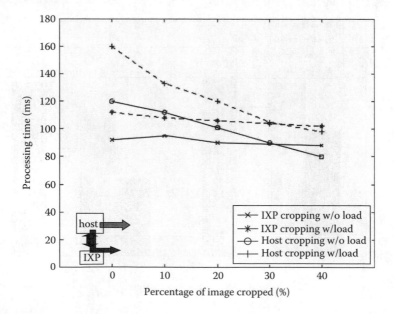

FIGURE 14.10
Host vs. IXP performance in cropping incoming images.

gains reaching up to 40%. This gap increases with the amount of additional load on the host's processing contexts. Hence, performance improvements can be observed if services are redeployed on the platform overlay, particularly since our related work already demonstrates that such reconfigurations have negligible overheads [18].

The second conclusion that we derive is that middleware must consider carefully certain application-specific service parameters when determining the deployment context for such data cropping. Specifically, when cropping is implemented on the host, the CPU crops the image and sends a (possibly much) smaller image to the network device (via the PCI interface), thereby reducing image transfer time. In the IXP implementation, the host transfers the entire image to the NIC, which limits the IXP's cropping performance to essentially the throughput of the PCI interface between host and IXP. Hence, as the cropping window size decreases, the performance of the host implementation starts increasing, whereas the performance of the IXP implementation is dictated by the data rates delivered from the PCI interface and does not change significantly. We note, however, that as with graphics processors associated with host CPUs, future heterogeneous multicore platforms are not likely to experience PCI-based performance limitations.

14.5 Related Work

Publish-Subscribe Infrastructures and Application-Level Multicast. Pub-sub middleware like IBM's Gryphon [49], ECho [14], ARMADA [1], and Hermes [41] are examples of application-level messaging middleware. These systems automatically route information to subscribers via scalable messaging infrastructures, thereby simplifying the development of distributed messaging applications. While many pub-sub middlewares make use of Internet Protocol (IP) multicast [39], AutoFlow (like Gryphon [49]) uses application-level overlays for efficient information dissemination. Other systems that implement application-level message multicast include SCRIBE and SplitStream [10], both focused on peer-to-peer systems. AutoFlow does not currently deal with common peer-to-peer issues, like dynamic peer discovery or frequent peer disconnection, but the basic multicast functionality offered by these systems is easily realized with our abstractions. Other distinguishing features of AutoFlow are its support for dynamic deployment and runtime optimization techniques to adapt information flows to changes in user requirements or resource availabilities.

Distributed Data Stream Processing. Projects like TinyDB [37], STREAM [5], Aurora [9], and Infopipes [27] have been making progress towards formalizing and implementing database-style data stream processing for information flow applications. Some existing work deals with resource-aware distributed deployment and optimization of SQL-like execution trees [2, 47, 32].

AutoFlow goes beyond database-style streams in order to provide a general abstraction for expressing complex information flows, including pub-sub, scientific data flows, and others in addition to SQL-like queries. In addition, AutoFlow facilitates the use of application-specific operators that implement desired runtime quality/performance tradeoffs. AutoFlow also provides self-regulating information flow overlays to deal with runtime resource variations, a capability not present in many existing systems.

Scientific Collaboration. High-speed networks and grid software have created new opportunities for scientific collaboration, as evidenced by past work on client-initiated service specialization [52], remote visualization [36], and the use of immersive systems across the network [15] and by programs like the Terascale Supernova Initiative. In all such applications, scientists and engineers working in geographically different locations collaborate, sometimes in realtime, by sharing the results of their large-scale simulations, jointly inspecting the data being generated and visualized, running additional analyses, and sometimes even directly running simulations through computational steering [24] or by control of remote instruments [40]. Such large-scale collaborations require infrastructures that can support the efficient "in-flight" capture, aggregation, and filtering of high-volume data streams. Resource awareness is required to enable suitable runtime quality/performance tradeoffs. AutoFlow addresses these needs with built-in support for the resource-aware deployment of customized information flow graphs and by supporting dynamic reconfiguration policies that maintain high performance levels for deployed flow graphs.

Self-Configuring Services, Architectures, and Infrastructures. Researchers in the pervasive computing domain believe that with the computing power available everywhere, mobile and stationary devices will dynamically connect and coordinate to seamlessly help people in accomplishing their tasks [21]. Tools like *one.world* provide an architecture for simplifying application development in such environments. While AutoFlow's information flow abstraction is sufficiently rich to deploy flows that accomplish user tasks in mobile environments, the focus of its implementation on high-end systems makes it complementary to much of the work being done in the pervasive computing domain. In contrast, it is straightforward for AutoFlow to manage evolving data sources, as done in systems like Astrolabe, which has the capability to self-configure, monitor, and adapt a distributed hierarchy to manage evolving data sources. An interesting generalization of AutoFlow would be to introduce more complex concepts for automatic service synthesis or composition, an example of the latter being the "service recipes" in projects like Darwin [23]. Finally, infrastructures like AutoFlow will strongly benefit from efforts like the XenoServer project [28], which proposes to embed servers in large-scale networks that will assist in deployment of global-scale services at a nominal cost. Accord [34] is a framework for autonomic applications that focuses on object-based *autonomic elements* that are managed via sets of interaction rules. A composition manager assists in automatically composing sets of objects, which then manage themselves via rule interactions. Accord's more declarative approach to composition fills the same role as the InfoPath's hierarchical con-figuration/reconfiguration algorithms (Section 14.4.2). Its rule-based behaviors are a more formal representation of the filter-based AutoFlow behavior. Autonomia is a Java-based effort to create a general framework consisting of self-managed objects, including XML-based control and management policy specifications, a knowledge repository, and an Autonomic Middleware Service that handles the autonomic runtime behavior.

Utility-Driven Self-Regulation. Adaptation in response to change in environment or requirements has been a well-studied topic. The challenge of building distributed adaptive services with service-specific knowledge and composition functionalities is dealt with in [23]. Self-adaptation in grid applications using the software-architectural model of the system is discussed in [11]. A radically different approach, similar to AutoFlow, for self-adaptive network services is taken by [26], where the researchers propose a bottom-up approach, by embedding an adaptable architecture at the core of each network node. In contrast, AutoFlow's self-regulation is based on resource information and user preferences, the latter expressed with flow-specific utility functions.This utility-driven self-management is inspired by earlier work in the real-time and multimedia domains [29], and the specific notions of utility used in this chapter mirror the work presented in [50], which uses utility functions for autonomic datacenters. Autonomic self-optimization according to business objectives is also studied in [3], but we differ in that we focus on the distributed and heterogeneous nature of distributed system resources.

Vertical Agility in Exploiting Network Coprocessors. The utility of executing compositions of various protocol- vs. application-level actions in different processing contexts is already widely acknowledged. Examples include splitting the TCP/IP protocol stack across general purpose processors and dedicated network devices, such as network processors, FPGA (Field Programmable Gate Array)-based line cards, or dedicated processors in SMP (Symmetric Multiprocessing) systems [43], or splitting the application stack, as with content-based load balancing for an http server [4] or for efficient implementation of media services [45]. Similarly, in modern interconnection technologies, network interfaces represent separate processing contexts with capabilities for protocol offload, direct data placement, and OS (operating system)-bypass [55, 48]. In addition to focusing on multicore platforms, our work differs from these efforts by enabling and evaluating the joint execution of networking and application-level operations on communications hardware, thereby delivering additional benefits to distributed applications.

14.6   Conclusions and Future Work

The AutoFlow project leverages a multiyear effort in our group to develop middleware for high-end enterprise and scientific applications and multiple software artifacts developed in prior work, including Dproc [25], PBIO [14],

ECho [13], IQ-RUDP [22], SplitStream [17], IFlow [32], and SmartPointer [52]. This chapter has examined the autonomic abilities and application benefits resulting from the combined techniques represented by the AutoFlow middleware, demonstrating that it has sufficient base efficiency to be used in high-performance environments and that applications can use it to demonstrate both horizontal and vertical agility to improve or maintain performance in the context of changing resource availability. We demonstrated the benefits of resource-aware adaptation of application behavior, including the ability to adjust an application's own bandwidth requirements in response to network resource changes, further adapting application transmission in the context of lower-level network information.

In ongoing work, we are further enriching the AutoFlow middleware to include support for lossless reconfiguration, fault tolerance at the underlay layer, improvement of the performance of SoapStones, development of extended heuristics for mapping and remapping of overlay networks to available computation and communication resources, management of the evolution of distributed systems over time, and further exploitation of both network coprocessors and potentially in-network processing resources.

References

1. T. Abdelzaher, M. Bjorklund, S. Dawson, W.-C. Feng, F. Jahanian, S. Johnson, P. Marron, A. Mehra, and T. Mitton et al. ARMADA middleware and communication services. *Real-Time Systems Journal*, 16:127–153, 1999.
2. Y. Ahmad and U. Cetintemel. Network-aware query processing for distributed stream-based applications. In *Proceedings of Very Large Databases Conference*, 2004.
3. S. Aiber, D. Gilat, A. Landau, N. Razinkov, A. Sela, and S. Wasserkrug. Autonomic self-optimization according to business objectives. In *Proceedings of the International Conference on Autonomic Computing, ICAC-2004*, 2004.
4. George Apostolopoulos, David Aubespin, Vinod Peris, Prashant Pradhan, and Debanjan Saha. Design, implementation and performance of a content-based switch. In *Proceedings of INFOCOM 2000*, 2005.
5. S. Babu and J. Widom. Continuous queries over data streams. *SIGMOD Record*, 30(3):109–120, 2001.
6. Fabian Bustamante, Greg Eisenhauer, Karsten Schwan, and Patrick Widener. Efficient wire formats for high performance computing. In *Proceedings of Supercomputing 2000*, Dallas, TX, November 2000.
7. Fabian E. Bustamante, Patrick Widener, and Karsten Schwan. Scalable directory services using proactivity. In *Proceedings of Supercomputing 2002*, Baltimore, Maryland, 2002.
8. Zhongtang Cai, Greg Eisenhauer, Qi He, Vibhore Kumar, Karsten Schwan, and Matthew Wolf. Iq-services: Network-aware middleware for interactive large-data applications. *Concurrency and Computation. Practice and Exprience Journal*, 2005.
9. Don Carney, Ugur Cetintemel, Mitch Cherniack, Christian Convey, Sangdon Lee, Greg Seidman, Michael Stonebraker, Nesime Tatbul, and Stan Zdonik.

Monitoring streams — a new class of data management applications. In *Proceedings of the Conference on Very Large Databases*, 2002.

10. Miguel Castro, Peter Druschel, Ann-Marie Kermarrec, Animesh Nandi, Antony Rowstron, and Atul Singh. SplitStream: High-bandwidth multicast in cooperative environments. In *Proceedings of 18th Symposium of Operating Systems Principles (SOSP-18)*, Bolton Landing, NY, 2003.

11. Shang-Wen Cheng, David Garlan, Bradley Schmerl, Joao Sousa, Bridget Spitznagel, and Peter Steenkiste. Software architecture-based adaptation for grid computing. In *Proceedings of High Performance Distributed Computing (HPDC-11)*, July 2002.

12. Delta Technologies, Delta Air Lines. Delta Technologies Messaging Interface (DTMI). Private communication.

13. Greg Eisenhauer, Fabian E. Bustamante, and Karsten Schwan. A middleware toolkit for client-initiated service specialization. *ACM SIGOPS*, 35(2):7–20, April 2001.

14. Greg Eisenhauer, Fabian Bustamente, and Karsten Schwan. Event services for high performance computing. In *Proceedings of High Performance Distributed Computing (HPDC-2000)*, 2000.

15. I. Foster, J. Geisler, W. Nickless, W. Smith, and S. Tuecke. Software infrastructure for the i-way high performance distributed computing experiment. In *Proceedings of the 5th IEEE Symposium on High Performance Distributed Computing*, 562–571, 1997.

16. Armando Fox, Steven D. Gribble, Yatin Chawathe, Eric A. Brewer, and Paul Gauthier. Cluster-based scalable network services. In *Proceedings of SOSP*, 1997.

17. Ada Gavrilovska. *SPLITS Stream Handlers: Deploying Application-Level Services to Attached Network Processors*. PhD thesis, Georgia Institute of Technology, 2004.

18. Ada Gavrilovska, Sanjay Kumar, Srikanth Sundaragopalan, and Karsten Schwan. Platform Overlays: Enabling in-network stream processing in large-scale distributed applications. In *15th Int'l Workshop on Network and Operating Systems Support for Digital Audio and Video (NOSSDAV'05)*, Skamania, WA, 2005.

19. Ada Gavrilovska and Karsten Schwan. Addressing data compatibility on programmable networking platforms. In *Proceedings of Symposium on Architectures for Networking and Communications Systems (ANCS'05)*, Princeton, NJ, 2005.

20. Ada Gavrilovska, Karsten Schwan, and Sanjay Kumar. The execution of event-action rules on programmable network processors. In *Proceedings of Workshop on Operating Systems and Architectural Support for the On-Demand IT Infrastructure (OASIS'04), in conjunction with ASPLOS-XI*, Boston, MA, 2004.

21. Robert Grimm, Tom Anderson, Brian Bershad, and David Wetherall. A system architecture for pervasive computing. In *Proceedings of the 9th Workshop on ACM SIGOPS European Workshop*, 2000.

22. Qi He and Karsten Schwan. IQ-RUDP: Coordinating application adaptation with network transport. In *Proceedings of High Performance Distributed Computing*, July 2002.

23. An-Cheng Huang and Peter Steenkiste. Building self-configuring services using service-specific knowledge. In *Proceedings of the 13th IEEE Symposium on High-Performance Distributed Computing (HPDC'04)*, July 2004.

24. J. E. Swan II, M. Lanzagorta, D. Maxwell, E. Kuo, J. Uhlmann, W. Anderson, H. Shyu, and W. Smith. A computational steering system for studying microwave interactions with space-borne bodies. In *Proceedings of IEEE Visualization 2000*, 2000.

25. J. Jancic, C. Poellabauer, K. Schwan, M. Wolf, and N. Bright. dproc — Extensible run-time resource monitoring for cluster applications. In *Proceedings of International Conference on Computational Science*, 2002.

26. Nico Janssens, Wouter Joosen, Pierre Verbaeten, and K. U. Leuven. Decentralized cooperative management: A bottom-up approach. In *Proceedings of the IADIS International Conference on Applied Computing*, 2005.

27. Ranier Koster, Andrew Black, Jie Huang, Jonathon Walpole, and Calton Pu. Infopipes for composing distributed information flows. In *Proceedings of the ACM Multimedia Workshop on Multimedia Middleware*, October 2001. http://www.cc.gatech.edu/projects/infosphere/papers/acmmm_koster.pdf

28. Evangelos Kotsovinos and David Spence. The xenoserver open platform: Deploying global-scale services for fun and profit. Poster, ACM SIGCOMM '03, August 2003.

29. R. Kravets, K. Calvert, and K. Schwan. Payoff adaptation of communication for distributed interactive applications. *Journal of High Speed Networks*, July 1998.

30. Sanjay Kumar, Ada Gavrilovska, Karsten Schwan, and Srikanth Sundaragopalan. C-Core: Using communication cores for high performance network services. In *Proceedings of 4th Int'l Conf. on Network Computing and Applications (IEEE NCA05)*, Cambridge, MA, 2005.

31. Vibhore Kumar, Zhongtang Cai, Brian F. Cooper, Greg Eisenhauer, Karsten Schwan, Mohamed Mansour, Balasubramanian Seshasayee, and Patrick Widener. Iflow: *Resource-aware overlays for composing and managing distributed information flows*. Submitted to Eurosys-2006, Leuven, Belgium, 2005.

32. Vibhore Kumar, Brian F. Cooper, Zhongtang Cai, Greg Eisenhauer, and Karsten Schwan. Resource-aware distributed stream management using dynamic overlays. In *Proceedings of the 25th IEEE International Conference on Distributed Computing Systems (ICDCS)*, 2005.

33. Vibhore Kumar, Brian F. Cooper, and Karsten Schwan. Distributed stream management using utility-driven self-adaptive middleware. In *Proceedings of the International Conference on Autonomic Computing*, 2005.

34. Hua Liu and Manish Parashar. Accord: A programming framework for autonomic applications. *IEEE Transactions on Systems, Man and Cybernetics*, 2005. Special Issue on Engineering Autonomic Systems, Editors: R. Sterritt and T. Bapty, IEEE Press.

35. LSC. http://www.ligo.org/. LIGO Scientific Collaboration, 2003.

36. K. Ma and D. M. Camp. High performance visualization of time-varying volume data over a wide-area network status. In *Proceedings of Supercomputing 2000*, 2000.

37. S. Madden, M. Franklin, J. Hellerstein, and W Hong. Tinydb: An acqusitional query processing system for sensor networks. *ACM Transactions on Database Systems*, March 2005.

38. Van Oleson, Greg Eisenhauer, Calton Pu, Karsten Schwan, Beth Plale, and Dick Amin. Operational information systems — an example from the airline industry. In *First Workshop on Industrial Experiences with System Software*, 1–10, San Diego, CA, October 2000.

39. Lukasz Opyrchal, Mark Astley, Joshua S. Auerbach, Guruduth Banavar, Robert E. Strom, and Daniel C. Sturman. Exploiting IP multicast in content-based publish-subscribe systems. In *Proceedings of the Middleware Conference*, 185–207, 2000.

40. B. Parvin, J. Taylor, and G. Cong. Deepview: A collaborative framework for distributed microscopy. In *Proceedings of the IEEE Conference on High Performance Computing and Networking*, 1998.

41. Peter Pietzuch and Jean Bacon. Hermes: A distributed event-based middleware architecture. In *Proceedings of the 1st International Workshop on Distributed Event-Based Systems (DEBS'02)*, Vienna, Austria, July 2002.

42. VINT Project. The network simulator — ns-2. http://www.isi.edu/nsnam/ns

43. Greg Regnier, Dave Minturn, Gary McAlpine, Vikram Saletore, and Annie Foong. ETA: Experience with an Intel Xeon Processor as a packet processing engine. In *Proceedings of Hotl 11, 2003*.

44. D.I. Rosu and K. Schwan. FARACost: An adaptation cost model aware of pending constraints. In *Proceedings of IEEE RTSS*, Dec. 1999.

45. S. Roy, J. Ankcorn, and Susie Wee. An architecture for componentized, network-based media services. In *Proceedings of IEEE International Conference on Multimedia and Expo*, July 2003.

46. Lui Sha, Xue Liu, and Tarek Abdelzaher. Queuing model based network server performance control. In *Proceedings of Real-Time Systems Symposium*, Dec. 2002.

47. M. A. Shah, J. M. Hellerstein, S. Chandrasekaran, and M. J. Franklin. Flux: An adaptive partitioning operator for continuous query systems. In *Proceedings of ICDE'03*, 2003.

48. P. Shivam, P. Wyckoff, and D.K. Panda. Can user level protocols take advantage of multi-CPU NICs? In *Int'l Parallel and Distributed Processing Symposium*, 2002.

49. Robert Strom, Guruduth Banavar, Tushar Chandra, Marc Kaplan, Kevan Miller, Bodhi Mukherjee, Daniel Sturman, and Michael Ward. Gryphon: An information flow based approach to message brokering. In *International Symposium on Software Reliability Engineering '98 Fast Abstract*, 1998.

50. W. E. Walsh, G. Tesauro, J. O. Kephart, and R. Das. Utility functions in autonomic systems. In *Proceedings of the International Conference on Autonomic Computing, ICAC-2004*, 2004.

51. Yair Wiseman and Karsten Schwan. Efficient end to end data exchange using configurable compression. In *Proceedings of International Conference on Distributed Computer Systems*, March 2004.

52. Matt Wolf, Zhongtang Cai, Weiyun Huang, and Karsten Schwan. Smart pointers: Personalized scientific data portals in your hand. In *Proceedings of Supercomputing 2002*, November 2002.

53. Matthew Wolf, Hasan Abbasi, Ben Collins, David Spain, and Karsten Schwan. Service augmentation for high end interactive data services. In *IEEE International Conference on Cluster Computing (Cluster 2005)*, September 2005.

54. Ellen W. Zegura, Ken Calvert, and S. Bhattacharjee. How to model an internetwork. In *Proceedings of IEEE INFOCOM*, March 1996.

55. Xiao Zhang, Laxmi N. Bhuyan, and Wu-Chun Feng. Anatomy of UDP and M-VIA for cluster communications. *Journal on Parallel and Distributed Computing*, 2005.

15

Scalable Management — Technologies for Management of Large-Scale, Distributed Systems

Robert Adams, Paul Brett, Subu Iyer, Dejan Milojicic, Sandro Rafaeli, and Vanish Talwar

CONTENTS

Future enterprises are anticipated to comprise globally distributed datacenters hosting and providing on-demand access to shared information technology (IT) infrastructure (compute and storage resources) as a utility. These systems are characterized by complexity and scale of infrastructure and applications; heterogeneity of resources and workloads; and dynamism of underlying computing environments. Management technologies for such environments need to support complete service life-cycle functionalities — allocate, deploy, monitor, adapt. The scale, complexity, and business demands require that the management systems be implemented in an automated manner with minimal or close to no human intervention. This requires the development and use of autonomic capabilities. To be successfully deployed in future large-scale and globally distributed IT systems, such autonomic management systems need to be loose-coupled and decentralized, and have to deal with incomplete knowledge. In this chapter, we propose a new approach to management of large-scale distributed services, based on three artifacts: a scalable publish-subscribe event system, scalable Web service (WS)–based deployment, and model-based management. We demonstrate that these techniques improve the manageability of services. In this way we enable service developers to focus on the development of service functionality rather than on management features.

15.1 Introduction

15.1.1 Scalability Issue

Contemporary computing systems are increasing in scale and broad deployment across the globe. This is true for enterprise and scientific systems, as well as consumer space. Traditional centralized enterprise datacenters are expanding into dozens of geographically dispersed datacenters. Remote operations are contending with even more management complexity while also dealing with the emergence of hundreds of so-called closet computers in small branch offices and home offices. Leveraging computation or data assets in Grid [1] or PlanetLab [2] environments pose similar requirements. As applications and services move out of the datacenter and into distributed installations, a new class of applications and services are coming about which are large-scale, geographically distributed, shared, and heterogeneous.

This has dramatically changed the design assumptions for such systems and applications. Scalability is not limited any more by physical or administrative boundaries — systems span the globe and cross organizations. Availability is not driven only by private networks and corporate policies — many systems are connected over wide-area networks and outside of a given administrative domain. This results in a significant dynamism in terms of unexpected loads, rebooting, and upgrading machines and services.

We claim that as systems continue to grow in size and wide-area deployment, traditional management approaches, such as those currently used by OpenView, Tivoli® [4], and Unicenter® [5], will become less effective. The management systems are moving toward service-oriented architectures [16], as demonstrated by the recent standards, such as Web Services Distributed Management (WSDM) [17] and WS-Management [18]. But, scalability, availability, and dynamism create additional requirements.

15.1.2   Scalable Management Features

The features we consider essential to the new global scalable enterprise are *loose coupling* of the management stack (communication, deployment services, and model-based automation), *decentralization* (distribution, no central point of management), and as a result of the previous two, *dealing with incomplete knowledge*.

To demonstrate the utility of these features to scaled management, we created a scalable, decentralized, distributed service provisioning and management system which includes three significant artifacts: Planetary Scale Event Propagation and Router (PsEPR, pronounced "pepper") [6], an infrastructure for a scalable, publish-subscribe event system which scales significantly better than point-to-point or hierarchical topologies; a WS-based service deployment tool [7] which decouples deployment specification from the dependencies and component models; and finally model-based automation which enables changes to the design of the system at runtime, enabling a higher degree of automation.

These three artifacts enable future application developers to more easily design, develop, and manage distributed applications that have no deployment or management center (decentralized), that are geographically disperse, and that adapt to changing resource availability and workload. We have built these three artifacts on the PlanetLab test bed [2], which has been used for the last several years for deployment and testing of this class of applications. Some key learning from running very large scale applications and services on PlanetLab aligns very well with our goals of decoupling, decentralization, and dynamism.

15.1.3   Autonomic Computing and Scalable Management

Scalable management is closely related to the work in autonomic computing. The two areas share a number of required and recommended behaviors for autonomic computing. For example, in order to accommodate scale, management must be fully automated, i.e., self-managed; it must handle problems locally whenever possible (i.e., the impact of a change in an area should not impact other services at the global scale); and scalable services' behaviors and relationships must be managed so that they meet service-level agreements. Furthermore, in order to accomplish scalable management, underlying systems must implement design patterns such as self-configuration, self-healing, and self-optimization. While in this chapter we do not explicitly address these patterns, the topics of manageability automation, adaptation, performance, and dependencies are critical for large-scale autonomic systems. In addition, scalable management requirements such as decoupling, decentralization, and dealing with incomplete knowledge are also features of autonomic systems.

15.1.4   Motivating Scenarios

To illuminate the required features of scalable management, we present three motivating scenarios: global service health, inventory, and plug-in.

Global Service Health. Consider a service that provides some functionality to people or computers all over the world, runs 24/7, is hosted at hundreds of locations that are geographically separated, and is made up of many interacting components. Somehow, the service must decide which hosts to run on, allocate the resources for those hosts, and then install and configure itself on those hosts. The set of hosts will be constantly changing because of hardware failures, network failures, purposeful reconfiguration of the hardware or network, and malicious activity. Additionally, the number of hosts required by the service can change because of workload or new business requirements. Also, the number of separate components of the application can be constantly changing, and thus the installation and reconfiguration process is continuous.

Running on multiple, geographically disperse locations has the advantage that the service has increased immunity to failure and attack. But, from a management point of view, it is hard to know if the service is running correctly. This service demonstrates the extremes of decentralization and decoupling. So, besides the problems of deploying and configuring a decentralized application, there are problems of management and control.

Global Service Inventory. Consider an installation of computers that spans the globe. This could be all the desktop computers in a multinational corporation or all of the blades in a collection of data centers that have been geographically located around the world. Monitoring and controlling all of these computers becomes difficult at some scale. Manual and semi-automatic management of the systems will seek solutions like running similar applications on all of the computers and limiting the variations in hardware configurations. However, datacenters will only grow and the number of client computers will only increase. This growth will require automated management and control. Because of unreliable monitoring systems and the network, the management and control feature will need to run in multiple locations.

Usual solutions are to centralize management and build hierarchies of managers — clients are managed by low-level management systems, and these low-level managers are managed by other managers, and these managers are controlled by a central manager. It is easy to see that these layers create more complexity and more things to manage. Additionally, managing the managers has the same problems as managing the low-level computers. In a summary, this scenario requires scalable communication that connects managers and other scalable components and automated management/control interface.

Global Service Plug-in. Consider a service that uses several services to perform its function. If there is a need to install this service in a new environment, a number of services that this service depends on may already be running, but some may not. Of the already running services, some of them may be the right version, but the others may be obsolete, and a new version needs to be installed.

Furthermore, the running services, with the right version, need to be verified for correctness of operation prior to installing a new service. Correctness also includes the service-level agreements that need to be guaranteed for the composite service. Once everything is verified and all dependencies have been resolved, the new service needs to be "plugged- in" vis-à-vis existing services, by dynamically connecting new services with existing services. In a summary, this scenario requires service discovery or the updated model of the system, service health monitoring, and loose and recoverable connection between services.

15.1.5 Chapter Outline

The remainder of the chapter is organized in the following manner: Section 15.2 motivates the paper with an analysis of scalability and complexity. In Section 15.3 we present related work in the area. Section 15.4 describes architecture, design, and implementation of the three artifacts. In Section 15.5 we evaluate performance of our solutions, followed by lessons learned in Section 15.6. Summary and future work are presented in Section 15.7.

15.2 Dealing with Scale and Complexity

To further motivate the need to deal with scalability and complexity, we have performed two experiments.

Discrete event simulations of centralized, hierarchical, and decentralized control structures were constructed in order to predict the behavior at large scale of these control structures. These simulations, based on published measurements of the global Internet for response times [9] and packet loss [10], simulate the effects of delays in transmission control protocol (TCP) and losses on the planetary scale command and control structure implemented on top of TCP. Figure 15.1 shows the results obtained for a resource constraint of 100 simultaneous connections at each node, with a 3.5% probability of packet loss and with the TCP recovery strategy [11]. Additionally, these simulations included introducing delays due to resource constraints at each node and connection failures.

Centralized management and control of applications were seen to suffer from significant performance degradation at scale, due to resource constraints and error rates on globally distributed networks.

Hierarchical topologies improve the control and management scalability significantly, but reach limits on very large networks due to the cumulative effect of network failure as the tree depth increases.

Shared routing overlay networks such as PsEPR provide better performance on globally distributed networks by amortizing the overhead of maintaining optimum routing architectures over many applications. Additionally,

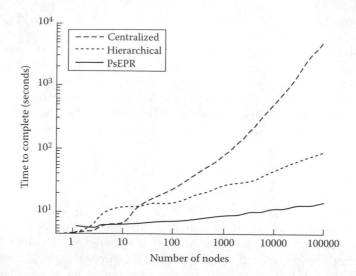

FIGURE 15.1

Comparison of managing approaches: centralized, hierarchical (spanning tree), and publish-subscribe (simulated).

the use of a common command and control structure provides improved resiliency to real-world latencies and error rates.

Relatively speaking, improvements in latency significantly lag improvements in bandwidth [12]. The impact of this is the decreasing efficiency of static, hierarchical communication structures and the increasing performance of decentralized, dynamic structures.

In the second experiment, we have compared the number of required changes as a result of a reconfiguration or failure. We have evaluated the number of changes as a function of service complexity and scale. We looked at system changes in response to dynamic events for a simple application, a JPetStore medium complex application, a local content provider, and a complex application (an airline reservation system running in multiple countries in different languages). Our analysis of system changes in response to dynamic events exhibits the challenges faced in designing automated management services with an ever-increasing complexity of systems.

The dynamic events introduced are those of *application server failures* and *addition of new application servers*. The higher-level dynamic events result in several subsequent changes within the system. This is attributed to the complex interdependencies that exist among the system components. As can be seen from the graph presented in Figure 15.2, the number of needed system changes grow exponentially with an increase in complexity. The problem becomes even more challenging in very large scale systems wherein management services are decentralized and have to make decisions based on incomplete knowledge. The graphs illustrate a very critical problem, that

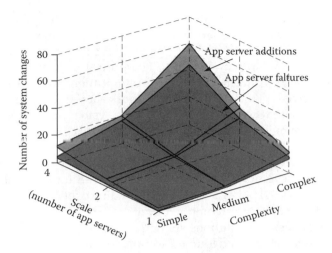

FIGURE 15.2

A number of complex service changes as a function of scale and complexity. The scale is shown with respect to number of application servers which are affected by the dynamic event.

of designing automated management services that can deal with an ever-increasing complexity of the system.

Our approach to handling this complexity is to model applications and services and multiple loosely connected components. The information model infrastructure captures the current state of the system. Whenever dynamic events occur, they are propagated to the information models. The information models are then reasoned upon by adaptation management services to determine the low-level system changes that need to be implemented in the system. These changes are then executed within the system. Other management services that rely on system knowledge — for example, monitoring services — need only to refer to the updated information model to obtain updated current system state.

15.3 Related Work

Our work on scalable management draws a lot of similarities with work in many areas. While we leverage the existing experience, we are different in that our primary focus is on the very large scale, global services. In particular, we base our work on service-oriented architectures, but in order to accomplish the scale, we are required to adopt autonomic techniques.

PsEPR is similar in concept to publish-subscribe systems. These range from Java Message Service to TIBCO Corporation's Rendezvous. PsEPR differs from these by dynamically creating communication points so that event senders and receivers have minimal dependencies. PsEPR's overlay routing is also opaque, thus allowing services to adapt to its structure — for instance, moving computation "close" to data sources. This sort of messaging structure is also being explored in Astrolabe [24]. There are also many other examples of publish-subscribe coordination and communication efforts [25,26,27,28].

In terms of related work in the area of application management systems, several deployment tools exist. The Deployme system for package management and deployment supports creation of the packaging, distribution, installation, and deletion of old unused packages from remote hosts. Kramer et al. describe CONIC, a language specifically designed for system description, construction, and evolution [31]. Cfengine provides an autonomous agent and a middle- to high-level policy language for building expert systems which administer and configure large computer systems [32].

Existing management solutions similarly address functionalities in other areas of our interest, e.g., adaptation to failures and to performance violations ([3],[4],[5]). The effectiveness of these traditional solutions in large distributed systems is significantly reduced by a number of properties of these solutions. These are centralized control, tight coupling, nonadaptivity, and semi-automation. Furthermore, these solutions do not adequately address the needs and characteristics of large-scale distributed services. Most of the tools do not by themselves provide the complete life-cycle management capability necessary in large dynamic systems such as PlanetLab.

In contrast, we are designing our management system by leveraging scalable technologies, some of which are mentioned in this section (e.g., publish-subscribe, decentralized agents and control, decentralized decision making), and extending them to the next level of very large scale global services. We provide solutions for deployment, eventing, and adaptation for services life-cycle management. We also propose higher-level abstractions for service and system descriptions through languages and models, which aid in formally capturing the complex needs of emerging services.

15.4 Architecture, Design, and Implementation

The architecture of our system is presented in Figure 15.3. It consists of the PsEPR [6], on top of which three industry standard packages are running: OASIS (WSDM) defines management interfaces and schemas [17], DMTF *Common Information Model* (CIM) describes how information and state are modeled [19], and *Business Process Execution Language* (BPEL) supports the workflow for services [20]. On top of these components, the deployment service is running as an implementation of the Global Grid Forum (GGF) *Configuration Description, Deployment, and Lifecycle Management* (CDDLM) standard [21]. On top of the stack is the automation engine that automates

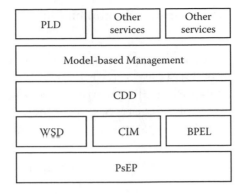

FIGURE 15.3
High-level architecture.

deployment and management of the whole stack. As an example of a managed application, we are using PlanetLab Data Base (PLDB). In the rest of the section, we describe in more detail the PsEPR, deployment, and automation layers.

15.4.1 PsEPR/PLDB

To build loosely coupled, distributed applications, we created the event-based communication system PsEPR. For communication of monitoring and control information, PsEPR creates an overlay network for the distribution of XML messages from a source to one or more receivers (see Figure 15.4).

Our experience with building and managing a large, distributed service [2] leads us to conclude that loose coupling among components (within or between distributed services) is necessary for robust communication — specifically, communication between virtual endpoints, where those endpoints can move (or be transparently redirected) as necessary because

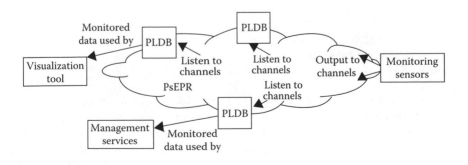

FIGURE 15.4
PsEPR infrastructure and PLDB.

of the ever-changing characteristics of communication bandwidth and availability. Our definition of "loose-coupling" includes:

- location independence of senders and receivers — "location" both in network address space and in physical space;
- service independence — neither the provider nor the user of a service needs to know of the existence of the other;
- state independence — reliability or delivery guarantees are not required;
- connection flexibility — one-to-many and many-to-many communication is easy.

Thus, PsEPR creates an overlay network on the existing Internet that efficiently moves event messages from clients sending the events to clients who have asked to receive the messages.

The PsEPR communication model sends XML-formatted messages over named "channels." Channels are hierarchically named, with channels having sub-channels, etc. A client authenticates itself to PsEPR and sends event messages to any channel. To receive events, a client requests a "lease" on a particular channel — a request to receive messages of a particular type from a channel and its subchannels.

This is similar to a publish-subscribe system where event senders create the messages and receivers "subscribe" to the messages they wish to hear. For instance, a client on a host named "x.example.com" could send heartbeat messages on a channel:

```
con = new PsEPRConnection(credentials);
con.send(heartbeatEvent,
"/example.com/heartbeat/x.example.com/");
```

One or more receivers could be listening to all heartbeat messages for this class of clients:

```
c = new PsEPRConnection(credentials);
les = c.getLease("/example.com/heartbeat/", 120,
    typeHeartbeatEvent);
event = les.receiveEvent();
```

In this simple example, the receiver has asked for all heartbeat events on the "/example.com/heartbeat/" channel and all of its subchannels for the next 120 seconds. Since the sender is sending events addressed to a subchannel of that lease, the receiver will see it along with events sent on channels of other hosts. If the receiver wanted events from only the one host, it could subscribe to that particular subchannel.

Internally, PsEPR is made up of routers which accept events from clients, route the events among the routers, and deliver the events to other clients based on routing tables. The leases from clients are kept in routing tables which filter and direct incoming events to the appropriate connections. Leases have explicit expirations which allow a default routing table cleaning process. Clients that wish to keep receiving events from particular channels must resubscribe before the end of the last lease.

The routers communicate among themselves to pass information on where leases are originating. In this way, the routers implement an ever-changing tree from senders and receivers. The current implementation uses a single-level routing, with connected routers requesting events for channels that their clients have requested through leases. Future work on PsEPR will introduce more complex routing and optimizations to create routing localities.

Measurements of PsEPR show that while PsEPR is less efficient at point-to-point communication and high-volume transfers, its flexibility makes adaptation of changing service configuration simple. Loose-coupling of service components, in the ways that PsEPR makes available, creates more reliable and scalable services and systems.

One service that has been implemented on top of PsEPR is PLDB. Since PsEPR events are transient (they are lost if not received), PLDB is a service that recalls past events that appeared on certain channels.

All PlanetLab nodes have clients which output node state information onto PsEPR channels. Any program wishing to know the current state of a node can listen to that node's channel. But, these clients only put out a tuple of information when the value of that tuple changes. This necessitates some way of finding the last event sent out. Rather than creating a query-like communication to a tuple sender (it's not the sender's problem that the receiver hasn't been listening forever), requests for events from the past are generalized into a service ("PLDB") which listens to channels and remembers the last values for tuples.

PLDB is made up of multiple supervisors who each manage a collection of monitors. Each individual monitor listens to one channel and collects and stores tuples that are seen on that channel. The monitor can also generate tuples based on requests it sees on the channel — some client wishing to see a past tuple value sends a request event and the monitor replays a version of the tuples with time information.

Each supervisor who is managing a group of monitors also listens to the traffic on a set of channels and independently evaluates the number of monitors that are operating on a channel. If there are too few monitors running, it creates a monitor for that channel. If there are too many monitors on a channel, the supervisor can terminate one of its monitors. Heuristics around the number of monitors on the channel, the timing of creation and destruction, and geographical load balancing create an ecosystem of monitors listening to a set of channels.

This creates a complex service which adapts to changing resources and load. Of the hundreds of PlanetLab nodes, some are being rebooted at any time. This means PLDB is continually adding and removing monitors and supervisors to provide the past event query service. This application demonstrated the

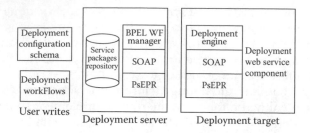

FIGURE 15.5
Deployment components.

advantages of loosely coupled, distributed, event-based services by running without interruption for many months.

15.4.2   Deployment

We built a system for deploying large-scale decentralized services within wide-area infrastructures. The functional requirements for this deployment system are to perform the installation, configuration, activation, and reconfiguration of services. It addresses the challenges of scalable performance, high reliability, and fast recovery time in response to dynamic faults and workload variations. The design of our deployment system builds on the lessons and experiences gained from the SmartFrog Project at Hewlett-Packard [33] and the CDDLM working group at GGF [21].

Figure 15.5 shows the conceptual view of the deployment system components. The key aspects are: *decentralized management components* which describe the deployment actions of a service, a *deployment configuration schema* that describes the configuration information needed during deployment, *deployment workflows* that compose the management components, and a *decoupled communication mechanism* based on Simple Object Access Protocol (SOAP)-PsEPR. The deployment system components are conceptually distributed among a deployment server and deployment target machines.

Application providers or writers typically specify the deployment details of the application service (e.g., steps needed to install the software, the list of dependent packages and services) in README files and manuals. Given the scale and complexity of the systems, there is a need to express the deployment information in a more structured and machine-readable manner, so as to be able to automate the complete deployment process in a repeatable way. Given an application service, an administrator using our proposed approach describes the logic for installation, configuration, and activation of the service as Java methods of a management component. The management components extend well-defined deployment interfaces. The code snippet below shows an example management component. These management components are then distributed to all of the deployment targets.

```
public class GenericRPMInstaller {
  public boolean install(String parameters) {
    ....
// download the packages
    RsyncDownloader downloader = new
    RsyncDownloader(downloadFromDir,downloadToLocation,
      new Integer(downloadBlockSize).intValue());
    downloader.download();

// install the package
  String installCmd = rpmCmd+downloadToLocation+"/"+rpm;
  File file = new File(downloadToLocation);
    .....
p = Runtime.getRuntime ().exec (installCmd,null,file);
    .....
  }
}
```

At the time of deployment, the deployment administrator expresses the configuration information needed during the deployment process in a well-defined deployment configuration schema.

The administrator also describes the various dependencies that the service has with other distributed services and applications as a BPEL workflow. In this workflow, the deployer maps the dependency requirements that the application service provider has specified to the actual instances of the packages and services within the system. For example, an application writer specifies that this application needs an Oracle Data Base (DB). The deployer maps this requirement to an actual Oracle DB available somewhere and specifies that in the BPEL workflow. The BPEL workflow appears as a composition of the management components.

```
<sequence name="main">
  <receive name="receiveInput" partnerLink="client" portType="tns:
  PLDBInstallation-Sequence" operation="process" variable="input"
  createInstance="yes"/>
    .....
  <invoke name="invoke-1" partnerLink="deploymentengine-node-24"
   operation="invokeEngine" portType="nsx24:DeploymentEngine"
   inputVariable="net-xmpp_input"/>
    .....
  <invoke name="invoke-2" partnerLink="deploymentengine-node-15"
   portType="nsx15:DeploymentEngine" operation="invokeEngine"
   inputVariable="net-psepr_input"/>
......
</sequence>
```

The BPEL workflow is provided to a BPEL process manager responsible for orchestrating the deployment actions in accordance with the specified workflow. The BPEL process manager communicates with a *deployment engine* that exists on all of the deployment targets. The deployment engine on a deployment target node is responsible for receiving and processing all of the deployment requests given to that deployment target node. It parses the requests sent through a BPEL engine, locates the appropriate management component responsible for a request, and then invokes the appropriate methods on that component. That method is responsible for executing the deployment actions for the service.

A workflow for a typical complex service would involve multiple management components, some of which are invoked and executed in parallel and others in sequence.

We are implementing our deployment service on the basis of the design presented above. Our initial use case scenario is the deployment of PLDB. The software package for PLDB consists of a tar file for the core software, and a set of dependent libraries that the software needs. The dependencies are expressed as BPEL workflows and supplied to a ActiveBPEL workflow engine. We are creating a library of commonly used deployment components. For example, we have an RPMInstaller component, an RSyncDownloader component, and a Notifier component among others. These components are being written in Java.

These generic components are then reused for the design of the deployment components written for PLDB application. An early version of the deployment engine has been developed in Java and hosted as a Web service within a Tomcat-Axis container on every deployment target (managed client).

A new transport mechanism has been integrated in the Axis stack to enable handling SOAP calls over PsEPR. Extending the Axis stack is just a matter of extending the BasicHandler class. We have called our PsEPR handler PsEPRSender:

```
public void invoke(MessageContext msgContext) throws AxisFault {
    SoapPayload myPA = new SoapPayload(
        msgContext.getRequestMessage().getSOAPPartAsString() );
    PsEPREvent myEV = new PsEPREvent();
    myEV.setPayload(myPA);
    PsEPRConnection  myConn = new PsEPRConnection(credentials);
    myConn.sendEvent(myEV);
}
```

The PsEPR-enabled client is required to set the new transport to the call object. The Axis engine finds the link between PsEPRTransport and PsEPRSender in the client-config.wsdd created from the XML file below:

```
<deployment name="pepr" xmlns="http://xml.apache.org/axis/wsdd/"
        xmlns:java="http://xml.apache.org/axis/wsdd/providers/java">
    <handler name="PsEPRSender"
type="java:soap.pepr.PsEPRSender" />
    <transport name="PsEPRTransport" pivot="PsEPRSender" />
</deployment>
```

The key benefits of our proposed design are (i) a decentralized deployment process through management components and BPEL dependency specification for scalable and reliable deployment, (ii) standards based, high-level interaction (SOAP, WSDL, and BPEL) for increased interoperability and decreased recovery time, (iii) workflow description expressiveness through BPEL language. Overall, we provide automation and lowered management costs through our system.

15.4.3 Model-Based Automation

Large-scale application services and systems provide challenges to the design of automated management systems. For example, a management service responsible for self-adaptation to dynamic changes is required to deal with information and processes that are heterogeneous, of large size, and dynamic. The problem gets compounded further as complexity of the system increases, mandating a knowledge of intricate administrator learnings during key management decisions.

We propose a model-based design of automated management services to deal with the challenges mentioned above. Figure 15.6 shows the conceptual view of the components of our design. In such a design, information models present a structured, formal representation of the information about the IT system and services. The information model provides a set of well-defined modeling classes and schemas to represent information about hardware elements, software services, and their relationships and associated constraints. An example of such schemas and specifications are those

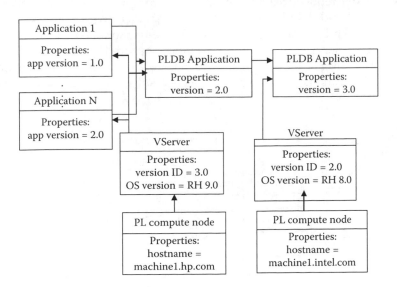

FIGURE 15.6
Conceptual partial view of the PLDB and the underlying infrastructure model.

defined by CIM. We build upon the CIM schemas within our prototype implementation.

The models are stored in model repositories. For scalable management, our design proposes a federation of distributed model repositories, and each individual repository captures the local system information. A well-defined model object manager and interfaces exist to access the information contained in the repositories.

Distributed model repositories present several challenges. First, an appropriate partitioning of the system information is needed which accounts for locality of reference and semantics of the stored information. Second, models need to be kept consistent across the system as a whole and have to deal with partial updates.

Thirdly, the model object managers must support a scalable distributed query mechanism. Further, the model repositories themselves need to be self-adapting to changes, e.g., faults, occurring in the system. As an ongoing effort, we are designing solutions addressing these challenges. We are also extending the model management subsystem to provide support for histories, transactions, and multiple consistency levels.

In a typical usage of models, schemas for the system under consideration are designed. The designed schemas support multiple levels of abstraction of system information. Instances of the designed model are subsequently created and stored in the distributed model repositories. They are initialized with information on current state of the local system for which they are responsible. Thereafter, the instance models are continuously updated to reflect new states of the system.

The model repositories thus capture the complex system information in a structured and distributed manner and together provide a near-real-time view of the entire system (see Figure 15.7). Our model-based decentralized management services rely on the information captured by models during their decision-making processes. At any given time, a particular component of the decentralized management service selects a subset of model repositories to

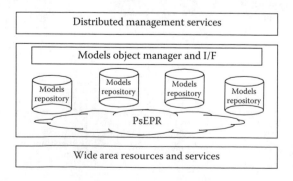

FIGURE 15.7
Model-Based automation component.

obtain current system information. The choice of this subset is statistical and depends on various performance and locality properties.

Once the subset of model repositories is selected, the management service chooses the level of abstraction within the model that is most appropriate to its needs. The decision making is then done using the incomplete system knowledge.

We present two examples of automated management services to illustrate the design. First, consider an adaptation service that determines the set of adaptation actions to be taken in response to dynamic events. Whenever an event occurs in the system, it is propagated to the model repositories, and the model instances are updated to reflect the event. The models at this point have captured the complex current state of the system in a structured, meaningful manner. The adaptation service applies reasoning on this structured information and determines the set of low-level system-wide changes (e.g., the set of redeployment actions) that need to be implemented in the system.

Next, consider a resource allocation management service that processes monitoring data collected in a distributed system. The service needs to know information about the monitoring collectors/reporters, etc. This is a challenge in a complex, dynamic, and heterogeneous system consisting of several hundreds of computing elements, each with its own collecting and reporting infrastructure. With our approach, this complex information is captured in models. The management service is designed to refer to the models only to obtain the information. The model is continuously updated to reflect the new system state even in the presence of dynamic system changes.

We are using OpenPegasus software as the basic infrastructure for storing and retrieving CIM-based models. The implementation of distributed model repositories is currently a work in progress. Our future and ongoing efforts also include prototyping an adaptation service based on our model-based approach within the PlanetLab environment.

15.5   Performance Evaluation

We have performed a few experiments in order to verify the scalability of our management system. We have performed measurements for PsEPR, WS-based deployment, models, and the PLDB.

We were interested in comparing the performance of two types of Web service communication: the synchronous SOAP over HTTP with the asynchronous SOAP over PsEPR. The Web service used in these experiments is a dummy service, which simulates the execution of a deployment operation. We have used Axis as the Web service container for both HTTP and PsEPR scenarios. Axis is easily integrated into Tomcat and also allows the transport layer to be changed. The HTTP Web service was made available through Tomcat, and we have written a PsEPR server to substitute Tomcat for handling PsEPR-SOAP requests.

The PsEPR server gets a lease on a channel and retrieves the SOAP payload received in PsEPR events. The SOAP payload is handed to an Axis engine. The Web service response is returned by the Axis engine to the PsEPR server and then it is sent back to the originator of the call.

We have two clients for originating requests: one executes synchronous communication (HTTP), which means the client sends the request and receives the response in the same thread, and the other executes asynchronous communication (PsEPR), which means the request is sent in a thread and the response is received in an independent thread.

Our network of Web services runs on over 50 nodes of PlanetLab, each one running a few instances of the Tomcat and the PsEPR server, totaling around 65 instances of each server. In each experiment (HTTP and PsEPR), we have measured the time taken to execute the calls to a set of Web services running on a number of nodes. A call consists of sending a request and waiting for a response either synchronously or asynchronously. We run the same experiments increasing the size of the payloads. As described in Section 15.4.2, deployment configuration schemas are distributed to deployment targets. We wanted to verify the impact of using larger SOAP envelopes (simulating more complex schemas) on the performance of Web service calls. Figure 15.8 demonstrates the increase on the time span of a deployment operation when the number of nodes involved in the operation increases and the size of the payload is increased.

We conducted a few experiments to find out the feasibility of WSDM Management Using Web Services (MUWS) for our scalable management solution. We compared Web services using Axis as the service container against MUWS.

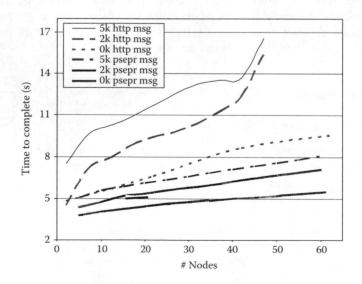

FIGURE 15.8
SOAP over HTTP vs. SOAP over PsEPR.

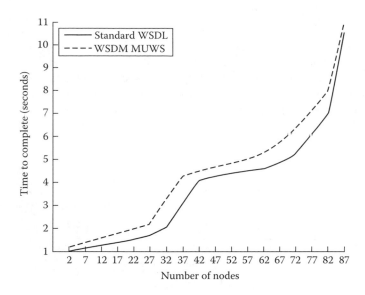

FIGURE 15.9
Scaling management using WSDM.

Our experiment focused mainly on finding out how much overhead both systems incurred when the scale increased. We ran experiments comprising 2 to 87 nodes on PlanetLab. Figure 15.9 shows how the two solutions compared with increasing scale. The experiment consisted of making a synchronous call to a Web service running WSDL and another one running MUWS and waiting for a response. As you can see from the figure, WSDM has more overhead than the WSDL (Web Service Definition Language)-based approach, but the overhead is not significant. So, we believe that the advantage of using WSDM MUWS outweighs the disadvantage of the overhead.

We conducted several experiments that compare SmartFrog-based deployment against our Web services-based deployment solution. The experiments consisted of deploying PLDB in a series of PlanetLab nodes. The nodes were chosen from a geographically dispersed set of locations around the world. We varied the number of deployment nodes from 1 to 30 and found that the Web services-based lightweight solution consistently outperformed the SmartFrog-based deployment solution in terms of deployment time when the scale increased. The results of this test are shown in Figure 15.10. The experiments cemented our belief that although Web service–solution is perceived to perform slower and consume more memory and more of the central processing unit (CPU), the differences are less marked in realistic applications [34]. Moreover, we believe that some of the known advantages of the Web services–based solution such as interoperability and extendability simply outweigh the drawbacks. The purpose of the experiments was not to prove that the Web

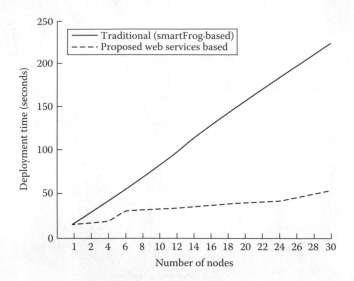

FIGURE 15.10
SmartFrog- vs Web services-based deployment compared on increasing number of targets.

services-based solution is better than the SmartFrog-based solution, but rather how Web services-based deployment is a viable solution for large-scale deployment. It doesn't seem fair to compare SmartFrog, which is feature rich, against our lightweight Web services-based solution. Nonetheless, the results are encouraging and show a pattern that we expect to see in terms of scalability and extensibility. In the future, we plan to improve our solution to have a richer functionality. We also plan to replace the underlying communications stack from HTTP to PsEPR. We believe that using PsEPR for deployment will improve scalability and reliability, while the use of Web services will improve interoperability.

One of the services we have constructed based on our eventing, deployment, and management principles is the PlanetLab database service, a tuple-store service providing management information for PlanetLab. Numerous PLDB monitors running on PlanetLab observe properties like load average, currently installed packages, and kernel checksums which are transmitted via PsEPR to any listening services. PLDB achieves robustness, reliability, and high availability through service replication. A management supervisor monitors health of the tuple-store service and dynamically starts and stops local monitors on a per channel basis in order to maintain robustness, reliability, and availability goals. Figure 15.11 shows a set of PLDB monitors running on a group of channels. When all the monitors on a single node are killed, other node supervisors detect the reduction in redundancy on a per channel basis and automatically create new channel monitors to restore the service to its design parameters.

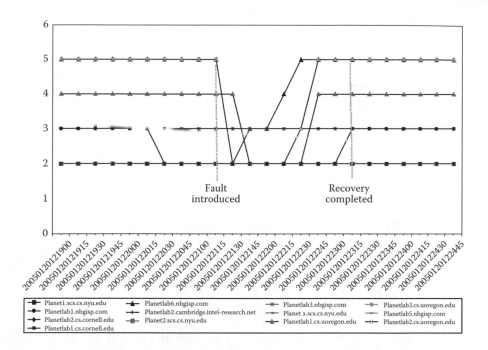

FIGURE 15.11
Flexing PLDBs.

15.6 Lessons Learned

In this section we summarize some lessons learned while exploring scalable management of global services.

- There are trade-offs between performance and reliability for traditional point-to-point communication versus loosely coupled publish-subscribe. Our preliminary simulation as well as real system performance measurement indicated the scalability benefits of the latter. PsEPR enables decoupling at the lowest layer. PLDB's ability to use multiple channel monitors showed that loose coupling (clients finding each other in PsEPR channels) creates easy and transparent reliability from the client perspective.

- Decoupling at the communication layer is not enough. We also need decoupling higher in the stack. This requires an event-driven programming model in the design of management services. While a fully decentralized and decoupled service is ideal to handle scale, it opens a problem in managing it. We thus end up with building a decentralized management solution to manage a decentralized application service, and then we need another decentralized management solution to manage a decentralized management service, and so on. There is still an open question in how to fine-tune the balance between decentralization and ease of management.

- We decoupled the expression of dependencies from the component model, such as in SmartFrog. This enabled us to reason and manage the dependencies through workflows. However, this introduced the need to manage the expressed dependencies (install, update as they change, etc.). There exists a trade-off between improved expressiveness and development time (e.g., of workflows, language). As we develop higher levels of abstractions, such as expressing dependencies for deployment, it enables more degrees of runtime design changes.

- The proposed solution to address the complexity problem is to build solutions that capture complexity in a structured manner, based on models. This way the "effort" needed in dealing with complexity is shifted from "runtime" to "development time." However, there is a trade-off that exists in this shift in terms of maintenance cost, software development effort, and disruption to existing systems design.

- There is a need to architect for global services. Services for reliable global operation are different from applications built for the machine room. Their requirements are different and rely primarily on scalability, complexity, and dealing with incomplete knowledge.

15.7   Summary and Future Work

We have presented a new approach for scalable management, based on decoupling, decentralization, and dealing with incomplete knowledge. We demonstrated design and implementation of three system components that contribute to the architecture of scalable management: a scalable publish-subscribe evening, WS-based deployment, and a model-based management. We have evaluated performance of these components in terms of scalability. All these features are critical for autonomous systems of the future.

In the future, we are going to explore extensions to the WSDM interface for scalable management (multicast management channels) and workflows for managing multiple interdependent components. We are also going to add more features to our management components and make them available as a toolkit to the PlanetLab community. We plan to capture the experience of researchers in the form of best practices for scalable management. One area that we specifically want to focus on is policies and best practices for management of large-scale globally distributed services. Once we have the basic scalable management infrastructure in place and it is used by the PlanetLab users, we shall be able to experiment with different policies and capture and derive the best practices.

Acknowledgment

We are indebted to Martin Arlitt, Greg Astfalk, Sujata Banerjee, Ira Cohen, Puneet Sharma, and William Vambenepe for reviewing the paper. Their comments significantly improved the contents and presentation of the paper. Mic Bowman and Patrick McGeer provided original support and ideas for pursuing this effort. Dongyan Xu shepherded our paper through the review and submission process.

References

1. I. Foster et al. The Physiology of the Grid: An Open Grid Services Architecture for Distributed Systems Integration. Open Grid Service Infrastructure WG, Global Grid Forum, June 22, 2002. http://www.globus.org/research/papers/ogsa.pdf
2. L. Peterson et al. A blueprint for introducing disruptive technology into the internet. In *Proceedings of the First ACM Workshop on Hot Topics in Networking (HotNets)*, October 2002.
3. HP OpenView http://www.managementsoftware.hp.com/.
4. IBM Tivoli, http://www.tivoli.com/.
5. Computer Associates Unicenter, http://www3.ca.com/solutions/solution. asp?id=315.
6. P. Brett et al. A Shared Global Event Propagation System to Enable Next Generation Distributed Services. WORLDS'04: First Workshop on Real, Large Distributed Systems, San Francisco, CA, December 2004.
7. V. Talwar et al., Approaches for service deployment, *IEEE Internet Computing*, vol. 9, no. 2, pp. 70–80, March-April 2005.
8. S. White, et al. An architectural approach to autonomic computing, *Proceedings of the International Conference on Autonomic Computing*, 2–9, May 2004, New York, NY, USA.
9. Jeremy Stribling. *All Pairs Pings for PlanetLab.* http://www.pdos.lcs.mit.edu/~strib/ pl_app
10. *Internet Traffic Report — Global Packet Loss.* http://www.internettrafficreport. com/ 30day.htm.
11. V. Paxson. *RFC-2988: Computing TCP's Retransmission Timer.* http://rfc.net/rfc2988.html, November 2000
12. David A Patterson. Latency lags bandwidth. *Communications of the ACM,* October 2004, 71–75.
13. J. Dunagan et al. Towards a self-managing software patching process using black-box persistent-state manifests. *Proceedings of the International Conference on Autonomic Computing,* 106–113, May 2004, New York.
14. G. Chen and D. Kotz. Dependency management in distributed settings, *Proceedings of the International Conference on Autonomic Computing,* 272–273, May 2004, New York.

15. S. Aiber et al. Autonomic self-optimization according to business objectives, *Proceedings of the International Conference on Autonomic Computing*, 206–213, May 2004, New York.
16. M. N. Huhns and M. P. Singh, Service-oriented computing: Key concepts and principles, *IEEE Internet Computing*, vol. 9, no. 1, 2005, 75–81.
17. OASIS WSDM WG Charter http://www.oasis-open.org/committees/wsdm/charter.php
18. WS-Management. http://msdn.microsoft.com/ws/2005/02/ws-management
19. DMTF CIM, http://www.dmtf.org/standards/cim/
20. OASIS BPEL Working Group Charter: http://www.oasis-open.org/ committees/wsbpel/charter.php
21. CDDLM Charter, https://forge.gridforum.org/projects/cddlm-wg
22. M. Happner et al. *Java Message Service 1.1.* http://java.sun.com/products/jms/docs.html.
23. TIBCO Corp., *TIBCO Rendezvous.* http://www.tibco.com/software/enterprise_backbone/rendezvous.jsp
24. R. van Renesse, K. Birman, and W. Vogels. Astrolabe: A robust and scalable technology for distributed system monitoring, management, and data mining, *ACM Transactions on Computer Systems*, vol. 21, no. 2, 164–206, May 2003.
25. P. R. Pietzuch. *Hermes: A Scalable Event-Based Middleware.* Ph.D. thesis, Computer Laboratory, Queens' College, University of Cambridge, February 2004.
26. P. Wyckoff, S. W. McLaughry, T. J. Lehman, and D. A. Ford. T Spaces, *IBM Systems Journal*, vol. 37, no. 3, 454–474, 1998.
27. M. Narayanan, et. al., *Approaches to Asynchronous Web Services.* http://www-106.ibm.com/developerworks/webservices/library/ws-asoper/
28. IBM Corporation. *MQSeries: An Introduction to Messaging and Queueing.* Technical Report GC33-0805-01, IBM Corporation, June 1995. http://ftp.software.ibm.com/ software/mqseries/pdf/horaa101.pdf.
29. T. De Wolf et al. Towards autonomic computing: Agent-based modeling, dynamical systems analysis, and decentralized control, *Proceedings 1st Int'l Workshop on Autonomic Computing Principles and Architectures*, 2003.
30. K. Oppenheim and P. MCormick. Deployme: Tellme.s package management and deployment system, *Proceedings of the Usenix IVth LISA Conference*, December 2000, New Orleans, 187–196.
31. Jeff Magee, Jeff Kramer, and Morris Sloman. Constructing distributed systems in Conic, *IEEE Transactions on Software Engineering*, 15(6):663–675, June 1989.
32. Mark Burgess. A site configuration engine, *USENIX Computing Systems*, vol. 8, no. 3, 1995, http://www.cfengine.org
33. P. Goldsack et al. *Configuration and Automatic Ignition of Distributed Applications*, 2003 HP Openview University Association conference.
34. N. A. B. Gray, Comparison of Web services, Java-RMI, and CORBA service implementations, *Proceedings of Fifth Australasian Workshop on Software and System Architectures* (ASWEC), April 2004.

16

Platform Support for Autonomic Computing: A Research Vehicle

Lenitra Durham, Milan Milenkovic, Phil Cayton, and Mazin Yousif

CONTENTS

This research is intended to provide autonomic support for platforms. Even though computer systems have increased exponentially in power and sophistication over the decades, much less progress was made in making them easier to manage. An ever-growing number of computer systems being deployed to solve today's business problems, coupled with advanced technologies that result in multiple logical and physical execution structures within each "physical machine" (such as multi-threading, virtualization, and multiple cores) are increasing the cost and complexity of management. Management complexity tends to have an adverse impact on system reliability, efficiency, and adaptability. It also tends to hinder acquisition and deployment of new equipment and services. Autonomic systems [1, 2] represent an approach to managing complexity by making individual system nodes and components self-managing. These systems manage themselves by monitoring and adjusting their operation to varying system conditions and anticipating user needs to allow users to concentrate on accomplishing tasks rather than managing the computer system.

As of this writing, most autonomic systems and prototypes reported in literature seem to be implemented in higher software layers, mostly in user space with perhaps some OS modifications. Our research focus is on platform support for autonomics. Specifically, we are looking at dedicating platform resources and firmware to implement a set of management and autonomic behaviors that are exposed via well-defined interfaces. Our long-term vision is to create platforms with on-board intelligence that makes them discoverable, configurable, self-managing, self-healing, and self-protecting even when the host operating system is not active. We intend such platforms to provide the agile infrastructure to support the evolution of Internet Distributed Computing [3]. We start by creating a separate execution container on the platform in which such autonomic behaviors are implemented and made available for invocation through a Web services based interface, WS-Management in our case [4]. As described in [3], we favor Web services technology because of its machine independence, discovery and self-description capability, and support for run-time binding.

A platform autonomics container may be provided in a variety of ways such as a dedicated micro-controller in the chipset, a plug-in option card, or a dedicated management virtual machine in systems that support virtualization. In any of these cases, the container is expected to have dedicated physical or virtual execution resources, such as processor and memory, supporting a dedicated software execution environment that is isolated and possibly very different from the OS and execution environment of the "host" platform dedicated to execution of user applications. Isolation from the host operating environment and separation of manageability and autonomics functions into a dedicated execution environment provides some fundamental advantages, resulting primarily in increased availability.

Most commercial management solutions operate by placing a software agent in the host execution environment to monitor OS and application health and optionally control its operational state. In autonomic terminology, the management agent implements sensors and effectors for the OS, applications, and — in some instances — the underlying platform hardware. The agent is designed to communicate with a remote console that usually resides at a central management location where it provides information and control interfaces to human operators. In the basic but typical case, local instrumentation data gathered by the agent is sent to the remote console where it is stored in a database, analyzed and, when necessary, used to trigger a control decision that can be completed by the operator or, in some instances, by the management console automation policies. Autonomics may be added to this basic structure in the form of local policy and knowledge engines that reside in the local management agent and close the control loop in response to the changes of the observed local state, thus eliminating the latencies and overhead of a round-trip communication with the remote console and possibly the operator.

A primary weakness of implementing manageability and autonomics in the same execution environment with the applications and OS that they monitor is that malfunctions of the monitored environment may impair the agent's operation and the agent's lifecycle is limited by the environment's lifecycle. Put simply, when the OS crashes it takes the agent down with it. As a result, no management is possible when the OS is not running.

Our platform autonomics approach rectifies this problem by providing a separate execution environment for the autonomic manager that has a different lifecycle and is expected to function in both pre-OS and post-OS states. In the pre-OS state, this is useful for configuration and provisioning actions, and an obvious benefit in the post-OS state is the ability to perform forensic analysis by examining the machine state exactly as it was left by the crash. When coupled with event logging within the manager, this can be a powerful tool in determining the root cause of failures.

With proper design, a platform autonomic manager can be decoupled from the power states of the host processor, so that it is powered even when the host processor is off. This is very useful for performing host power on/off operations, performing hardware setup and configuration, and automating provisioning that facilitates platform self-configuring behavior and attributes. While a separate container provides a number of benefits, it is only as useful as the autonomic functions implemented in it and made available to external entities for use in various phases of host system lifecycle — starting with pre-power, pre-OS states, assisting the OS when it is present, and taking over when it is not.

Here we describe a research project in Intel's Research Lab exploring the nature and value of providing platform support for autonomic computing. Our primary focus is on exploring the execution environment of the autonomic container and in discovering which set of manageability and autonomic behaviors is best suited for platform implementation. We postulate that autonomic systems can provide increased manageability and higher availability by executing in an autonomous execution environment and by having rich sensory information at their disposal. We have built a research platform that combines hardware sensors, such as temperatures at targeted points, processor voltage and frequency, and fan speeds, with soft sensors, post-boot OS level software tools that monitor system and component statistics — such as CPU utilization, memory usage and process queue length — all augmented with ambient (wireless) sensors instrumenting the environment. We provide a common software abstraction and use it to tie sensor information with an authoring tool for autonomic and policy-based management.

16.1   Architecture of the Autonomic Research Platform

The architecture of our autonomic research platform combines hardware sensors with both soft and ambient sensors with the goal of creating a research vehicle for exploring platform support for autonomic computing, as shown in Figure 16.1.

The architecture consists of standard server hardware equipped with on-board platform (wired) sensors. Ambient sensors, wireless in our setup, provide information about the physical environment and communicate with the platform via a physically connected sensor gateway. The runtime execution environment has access to platform sensor information such as processor temperature and fan speed as well as soft sensor data such as memory usage and process queue length. In addition, the platform supports various effectors which may be used to modify the system state and behavior as a result of changes in analyzed sensor data and applicable management policies. The sensors and effectors are consolidated in a software abstraction layer before being exposed to higher level policy authoring tools. The policy operation layer handles the reception of sensor data, evaluation of policies/rules and

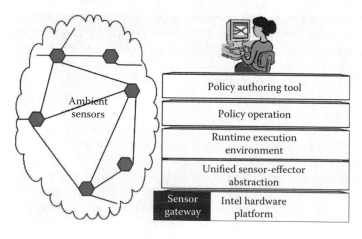

FIGURE 16.1
System architecture.

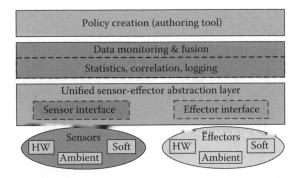

FIGURE 16.2
Architectural components.

actuation of effectors to implement change. The policy authoring tool provides a graphical user interface for policy-based management of the platform. The intent is to use the tool to rapidly implement and test a variety of policies, assess their usefulness, and then concentrate on the most promising ones.

Figure 16.2 shows the major architectural components of the autonomic research platform. The lowest layer contains the sensors and effectors. Our goal with this project was to create and experiment with management policies that used multi-sensor data; thus the architecture supports hardware, soft and ambient sensors and effectors. The next component is the Unified Sensor-Effector Abstraction Layer. This layer abstracts the information presented by the sensors and effectors to present multiple disparate sensor and effectors in a uniform manner to the higher layers, e.g. policy engines and other external tools. By abstracting sensor readings and effector controls into single sensor and effector interfaces, we are able to simplify and speed the development of management tools and enable new sensors and effectors to be incorporated into existing frameworks. The Data Monitoring and Fusion layer preprocesses sensor data and provides functions such as alert notification, correlation of fused data, sensor data formatting, aggregation, filtering, and logging. The final layer is policy creation where the platform autonomic manager evaluates the sensed information and determines the appropriate actions to take. We discuss each of these architectural components in the following subsections.

16.1.1   Sensors

Multi-sensor data can provide a rich base for sophisticated management operations and be utilized to derive context awareness. We want to explore the use of multi-sensor data to aid in autonomic computing systems, specifically platform management. The addition of the ambient sensor data with hardware and soft sensor data can further enhance the management capabilities. For example, combining ambient temperature and airflow sensors with hardware sensors in computing nodes can help localize hotspot origins and obstruction locations.

Due to the differences between sensors types (hardware, software and ambient), the integration of sensors into the autonomic research platform presented various challenges. The ambient sensors integration revealed that the higher levels must treat sensors as asynchronous and no hard guarantees can be met regarding sampling rates as the data rates vary with sensor technology. In the next subsections we discuss how we access the hardware, ambient and soft sensors.

16.1.1.1   *Integrated Hardware Sensors*

The Intelligent Platform Management Interface (IPMI) [5] specification defines platform independent data formats and capabilities. We decided to use IPMI to gain access to the platform hardware sensor information such as voltage, temperature, cooling fan status and speed, chassis intrusion, Error Correction Code (ECC) memory, processor status, and power supply status. We searched for drivers that would provide us access to the IPMI sensors and found an open source software project called OpenIPMI [6]. The OpenIPMI project is an effort to create a full-function IPMI system to allow full access to all IPMI information on a server and to abstract it to a level that will make it easy to use.

OpenIPMI consists of two parts: device driver and user-level library. The device driver goes into the operating system kernel; in our case, Linux kernel, but we found that the OpenIPMI driver included in the Linux version we use was adequate enough (although it was not configured to load by default). The OpenIPMI library allows for creation of a connection to a domain and iteration over the entities in the domain (CPU, memory, fan, etc.) and the sensors belonging to those entities. IPMI is dynamic and asynchronous so everything in OpenIPMI happens through callbacks.

16.1.1.2   *Ambient Mote Sensors*

We chose to use motes [7, 8, 9] as the ambient sensors in our platform since they offer a wide range of sensors such as acceleration, barometric pressure, humidity, light, magnetic field, sound, temperature and vibration. Motes are tiny, low-cost, self-contained computers that use radio links to communicate with each other and are able to self-organize into ad-hoc networks. The sensors provided by the motes offer many possibilities for implementing autonomic policies in a datacenter. A rise in either temperature or humidity can be a sign of a failed cooling system, which can be detrimental to servers. On desktop systems, the first sign of a failing fan or hard drive is often the noise noticed by the system's user. However, in a datacenter with thousands of servers, these warning signs go unnoticed. Administrators rely on monitoring tools and a mote monitoring for increased noise levels could provide a useful backup to these existing monitoring sensors. Constricted airflow, such as a box placed on top of an air vent during a server installation, can lead to thermal problems in the datacenter. Airflow sensors could provide useful information for predicting thermal hotspots in a datacenter.

Power consumption is a crucial component in the design of motes. Battery life of a mote can range from days to years, which is mostly determined by the software it runs. The key to extending the battery life is to keep the CPU asleep and radio turned off as much as possible. To do so, the sampling rate of the motes must be kept low and asynchronous communication, rather than polling, must be used.

16.1.1.3   Soft Sensors

Soft sensors are OS-level post-boot tools that use OS utilities to report system state and status. An example of a potential soft sensor is the /proc file system, and an example of a standard use of the /proc filesystem is the standard UNIX® utility "top." The top utility uses the /proc filesystem to provide source data to provide a rolling display of processor, memory, swap, and process information. We wrote a soft sensor that interfaces with the /proc filesystem to gather data about system state and status. It calculates statistics and provides information on instantaneous load averages, number of active processes, CPU states and percentages (idle, user, kernel), memory states (real, free, swap), and utilizations of individual processes by process ID.

The Process/System soft sensor is initialized by verifying that the /proc filesystem is mounted and reporting the sensors it has available and what granularity it can use to report updated information. The soft sensor runs in the background and continuously polls for sensor information. The Process/System soft sensor updates the sensor data and pushes the data at the prescribed interval. Individual /proc readings are treated as individual sensor inputs to which a policy manager can subscribe. This sensory input can aid load modulators by providing close granularity of the total load and stress being applied to each system; this aids in deciding which systems should be given more work or need to be relieved. It can also aid in swift determination of declining health of individual servers through the monitoring of its memory and CPU.

16.1.2   Effectors

As with sensors, effectors can be "soft" (i.e., controlled by the OS), hardware (e.g., changing fan speeds or processor frequency), or ambient (e.g., increasing air-flow at a certain location). Integrating effectors into the autonomic research platform involved first researching available platform-level effectors and determining which would be appropriate. The investigation began with Dynamic Voltage and Frequency Scaling (DVFS) and hard-drive spin-down, two common power-savings techniques, and later extended to higher-level effectors, such as load redirection and suspending to disk.

Effector testing shows DVFS to be extremely effective for power reduction on server platforms — however, the interaction between frequency, power, and performance is not a simple relationship — it is very workload-dependent.

Suspend to disk is a possible effector better suited for high-level management. Many workloads, such as clusters performing batch jobs have long idle periods. In addition, some workloads have many identical front-end servers which are only needed during peak hours. During idle times, unneeded servers can be suspended for reduced power consumption and better thermals in a datacenter. The next subsections detail the findings for DVFS and suspend to disk effectors.

16.1.2.1　Dynamic Voltage and Frequency Scaling

Intel Enhanced SpeedStep® technology [10] allows a system to dynamically adjust processor voltage and core frequency, which results in decreased power consumption, with the tradeoff of a possible slight performance degradation. Only certain voltage and frequency pairs may be selected and are referred to as P-States (Performance States). The current P-State for a processor is selected via an ACPI call inside the kernel. When selecting a P-State, the kernel chooses only the frequency — the hardware will automatically scale to the appropriate voltage.

Using DVFS required an update of the BIOS to version 1.80, upgrading the Linux kernel to version 2.6.12, and installing the additional kernel modules by statically compiling them into the kernel. We conducted experiments with different workloads that show varying power savings possibilities when DVFS is used. Depending on the workload, we can net an energy savings of between 0% and 17% (see Table 16.1).

DVFS is an effective approach to reducing system power consumption, when a system is not fully utilized. Further analysis is needed to evaluate the relative benefits of DVFS in comparison with Suspend to Disk when it can be used to completely power off a system. If not done properly, combining DVFS and Suspend to Disk can result in a negative energy savings. In order to optimize power using both effectors, a complex model with application-level knowledge is required.

16.1.2.2　Suspend to Disk

Suspend to Disk is very similar to the Windows® operating system hibernate feature. It provides the ability to take a snapshot of the state of a computer's memory and save it to disk, and restore that image later. This allows

TABLE 16.1

DVFS Energy Savings

Test Case	Energy Savings by Using 2.8 GHz Instead of 3.4 GHz
Spinning in Cache	17.3%
Compiling Linux 2.6 Kernel	11.2%
JPEG Compression	10.6%
BZip2 Decompression	10.2%
Reading from Disk	2.3%

you to fully power down a computer at an arbitrary point in time and not need to reload programs and/or reopen documents on power-up. It is much faster than normal shutdown, restart, re-opening programs, re-establishing program state.

Tradeoffs between suspend/resume and shutdown/restart/reinitialize applications include time, energy savings, and potential for system-crash. The time it took our development platform to save system and memory state was 8 seconds. After normal restart, the time to re-open applications and get them to the point before shutdown is highly system variant. The time it takes a system to resume following a Suspend varies depending on combinations of BIOS and OS, but it is generally on the order of 10s of seconds. While suspended, the system is in the same state as if it were powered off completely. On our servers, only the Baseboard Management Controller (BMC) and a few critical components remain running during power-off, and the system power consumption is 10 W to 15 W.

Due to the nature of 3rd party kernel patches, and the potential for unfortunate interaction with other patches/software/configurations on the system, there always remains the possibility that using the suspend/resume feature will result in system crashes or other failed-startup experiences. Suspend to Disk is an effective means of reducing power in a datacenter. However, it relies on a set of machines that are either non-critical, or have redundancy — for instance a cluster of Web servers sitting behind a load-balancer or a cluster performing batch jobs, such as a compile farm.

16.1.3 Unified Sensor-Effector Abstraction Layer

We wanted to provide a common software abstraction for the hardware, software and ambient sensors called the unified sensor-effector abstraction layer (see Figure 16.2). This layer allows for the "sensing" of platform data and "effecting" of desired actions/changes. Hardware sensor information along with information received from the software and ambient sensors is presented to higher layers in a common format. In addition to presenting the sensor data, the unified sensor-effector abstraction layer allows for instrumentation of management functions via common access methods. The layer provides a single point of entry for management of the platform. It provides a means for discovering the available sensors, accessing the sensor data and effector methods and notifying of events. In the next sections we describe the two parts of the unified sensor-effector abstraction layer.

16.1.3.1 *Sensor Abstraction Interface Layer (SAIL)*

SAIL is a system for taking multiple, disparate sensor inputs and presenting the combined readings in a coherent, uniform manner to policy engines and other external tools (e.g. performance monitors). By abstracting sensor readings into a single interface, we are able to simplify and speed the development of management tools and enable new sensors to be incorporated into existing frameworks. The sensor interface specifies the Init, Cleanup, Run,

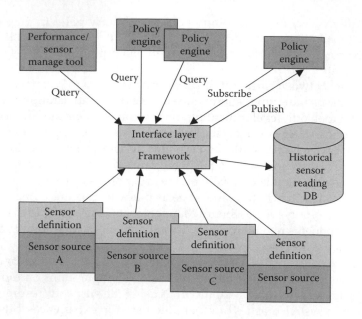

FIGURE 16.3
SAIL components.

Stop, and Version functions. In addition, a RequestSensor function may be implemented that allows higher levels to specify the sensors of interest.

We use the term "sensor source" to describe any hardware device or software package that can provide one or more sensor readings; this can include remote devices attached via a network protocol. SAIL is composed of an Interface Layer, a Sensor Definition Layer, a Historical Sensor Reading Database and a Framework for tying these together (see Figure 16.3). The Interface Layer defines the standard interface provided to the consumer (e.g., policy engines). Through this layer all sensors provide the same capabilities, regardless of data source. The interface allows for management applications to make ad-hoc queries of sensor readings or to subscribe to changes in one or more sensor readings and have the SAIL system publish results when changes occur. The Sensor Definition Layer provides a "plug-in" wrapper definition to tie new sensor sources into the system. For example, if a new hardware sensor with an interface driver is implemented, a thin wrapper matching this definition will allow the new sensors to be added to the framework and accessed through the Interface Layer. The Historical Sensor Reading Database provides the additional functionality of recording past sensor readings for use in trend detection and failure prediction.

The final component is the framework tying all sensors that implement the Sensor Definition Layer together with the Database into the Interface Layer. This is the software layer hidden from the external components that performs the actual sensor reading conversion. Through this package, we are able to both have a sensor source announce the existence and state of new sensors to the system or to have the higher level request specific sensor information.

16.1.3.2 Effector Abstraction Interface Layer

Effectors encapsulate the functionality of sensors in an autonomic research platform. Essentially effectors represent controls in the systems. In order to prevent conflict between different policies trying to control a single effector, they are initialized as. Single Controls where only one policy may have control at any time or Multi-Control where several policies may concurrently have controls at the same time.

The effector interface provides three functions: GetName, RegisterController and ReleaseControl. The GetName function returns the name of the effector which was set during initialization along with the control type. RegisterController and ReleaseControl allow an effector to one or more controllers. For instance, a CPU speed effector should probably have a single controlling policy or the policies may conflict and desire that the CPU speed be set to different values. A meta-policy is placed over a single control interface to handle multiple policies attempting to control an item.

16.1.4 Data Monitoring and Fusion

At this time the project has focused mainly on providing sensor data aggregation in the form of a sensor database. The sensor data is currently stored in a MySQL® database as a convenient expeditious interim implementation. Sensor information can be read from the database in a polling fashion. This can be changed to an event-driven framework with support for stored procedures.

16.1.5 Policy Creation

The research project investigated policies for self-management of a platform taking into account information received from hardware, soft, and ambient sensors. We defined a basic policy class which uses the sensors and effectors and handles the details of updating them. For a policy to use sensors a developer need only define what sensors are used and then request the readings as needed — the updates happen automatically. The Policy class does not directly manage effectors to be consumed by concrete policy implementations; however it does enable a policy to "expose" policies to be consumed elsewhere. In the next subsections we discuss two of the policies we have implemented.

16.1.5.1 Hibernate (Suspend to Disk)

The Hibernate policy will hibernate a server based upon ambient thermal conditions. It uses the Suspend to disk effector found in the effector investigation. The Hibernate package is an example of a single-use effector that is protected by a meta-policy which in turns exposes an effector that multiple other policies can use. The Hibernate policy is an example consumer of the meta-effector. The policy subscribes to ambient mote sensors to determine the ambient temperature for each server. Based on ambient temperature readings, the policy can request (using the effector provided by the meta-policy) four separate hibernate states: Must Hibernate, Hibernate If Possible, Stay Awake If Possible, and Must Stay Awake. The meta-policy receives all of the requests for hibernate states and decides which to implement.

16.1.5.2 *Linux Virtual Server (LVS) Control*

The open source Linux Virtual Server (LVS) project [11] extends the Linux kernel to provide network load balancing. Our system includes one "virtual server," several "real servers," and a feedback and control system (see Figure 16.4). The virtual server distributes, or directs, computing requests from clients to dedicated "real servers." Traffic is routed based on an algorithm which takes into account environmental sensor data and the dynamic load on each real server. The real servers transparently handle the requests and send responses directly back to the requesting client.

The virtual server presents a single point of entry into a computing system that provides a particular service, e.g. Web services, decision support mechanisms, transaction processing, database access, e-mail, etc. Typically, we would expect several thousand clients to access such services simultaneously, and one server would be incapable of handling such a load. Therefore it is necessary to distribute this load to various machines while ensuring that the process is completely transparent to the client. The job of the virtual server is to accept incoming requests and distribute them to the real servers. The algorithm governing which particular physical real server receives a request has significant impact on overall system performance and therefore on power

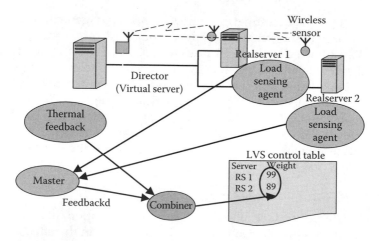

FIGURE 16.4
LVS with thermal feedback.

consumption, heat generation, and ultimately, operating costs and Total Cost of Ownership (TCO).

Each real server is generally a stand-alone server high performance general purpose computing platform. Their purpose is to run applications, receive requests from the virtual server, process such requests, and provide responses directly to the requesting client. The distribution of incoming requests by the virtual server takes into account the number of requests already being serviced at each real server, preset weights that reflect the relative capabilities of each different real server and requirements for persistence in the connections between a client and the real server that services it.

Current feedback mechanisms [12] merely sense the instantaneous load on each real server and modify the distribution algorithm accordingly. The current feedback/control algorithm does not have a mechanism to sense the real time load on each real server and the influence of environmental indicators such as ambient temperature, airflow, and humidity. This limitation results in sub-optimal distribution of load to the real server causing unfavorable thermal distribution, power consumption, and uneven wear and failure rates in the datacenter.

We use sensors to measure the temperature, airflow, and humidity at each real server inlet and exhaust. This is then combined with information about individual system load and is used to adjust the request routing algorithm (see Figure 16.4). The algorithm modifies the weights associated with each real server and thereby changes the real server's probability of receiving the next request. This policy is advantageous in that intelligent distribution of incoming requests and the subsequent load due to handling those requests results in a more uniform operating environment and a more even wear pattern. Greater operating environment uniformity yields lower TCO and cooling costs for the entire datacenter.

16.2   Prototype Environment

This section describes our prototype environment. The autonomic research platform combines hardware, soft, and ambient sensor information with a policy authoring tool to allow for the creation and experimentation with multi-sensor based management policies. Figure 16.5 shows the prototype environment. It consists of a server platform with integrated platform sensors and access soft sensors as well as ambient wireless sensors provided via a wireless sensor gateway physically connected to the server.

The platform provides the container for the autonomic manager that collects and analyzes the platform and environmental sensor data. Based upon knowledge it has gathered and the management policy in place, the platform autonomic manager creates a plan to "effect" the desired management change. The knowledge may be from observations, past experiences, centralized

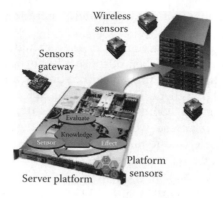

FIGURE 16.5
Autonomic research platform prototype environment.

policies and/or peer updates. In the following subsections, we discuss each of the hardware and software components used in the autonomic research platform.

16.2.1 Software

This section describes the software with which the autonomic research platform interacts. First we discuss the operating system and then we discuss the policy authoring tool used for policy-based management of the platform.

16.2.1.1 Operating System

We have chosen to use Linux for our operating system. We wanted an OS that we could modify to allow us access to the hardware as well as "soft" sensors. When starting this project we knew that IPMI would provide access to the hardware sensors on the server platform but we also needed to consider what IPMI drivers/libraries would be available. Linux and Windows versions of IPMI drivers are available and Linux also has built in IPMI drivers in the kernel that can be enabled. In terms of available effectors, Linux 2.6 and later also provides a cpufreq subsystem that allows clock speed to be explicitly set [10, 13].

16.2.1.2 Policy Authoring Tool

For the policy authoring tool we use the Execution Management System (EMS™) [14] provided by Enigmatec Corporation Limited. EMS is designed with autonomic principles and the reduction of business operational expense in mind. Policies in EMS are operational polices that take the form of "when X occurs do Y." The software describes these policies in a form derived from Event Condition Action (ECA) rules languages [15, 16] called the Event Condition Workflow (EC-W). The procedural part of the ECA rule language is replaced with the invocation of an EMS distributed workflow. The event-condition component is expressed as a set of events that can be sensed while the workflow describes a set of steps that need to occur to reach some defined goal.

Within EMS, the managed element provides the basic logical representation of a management interface. The management interface consists of two parts: sensors that provide information and effectors that are operations to modify the environment. A managed element may not correspond to a single element in the infrastructure but may represent a collection of real-world components that behave as a single item. EMS provides the notion of an autonomic manager that consumes data provided by a managed element and responds appropriately using the effector interfaces provided by the managed element. The autonomic manager can also present its own managed element interface allowing it to be managed by another autonomic manager. Base managed elements represent the integration point between EMS policies and the outside world. EMS services embody the notion of autonomic managers in that an EMS service monitors and controls one or more managed elements including other EMS services.

Figure 16.6 shows an example policy for managing the processor frequency based upon sensor information received about the processor temperature and the CPU utilization. The policy authoring tool provides a graphical representation of the workflow for the sensors and effectors. The tool allows a user to select the sensors and effectors of interest and define the workflow between them using scenarios or use-cases. The policy authoring tool also translates the graphical representation of the management policy into procedural rules for implementation. The policy shown in Figure 16.6 is to receive notification

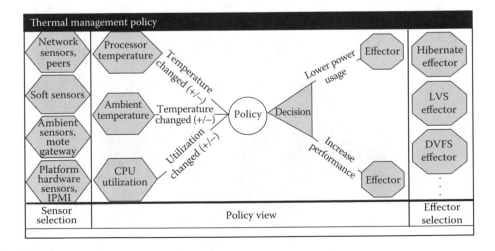

FIGURE 16.6
Policy-based management view.

of either a processor temperature change and/or CPU utilization change from the respective sensors. Thus, the tool must use SAIL to obtain this sensor data. The policy logic decides if power usage should be lowered or if performance needs to be increased, and calls the appropriate effector via the effector abstraction interface layer.

16.2.2 Hardware

This section describes the hardware we used to prototype the autonomic research platform. We begin with a discussion of the ambient sensors and then discuss the server platform we have chosen.

16.2.2.1 Ambient Sensors

We have chosen to use motes for the ambient sensors since they offer a wide range of sensors. Berkeley [7, 8] motes are commercially available via Crossbow® Technology Inc. [17]. The left side of Figure 16.7 shows the Crossbow MPR400 MICA2 series mote we used. This processor board measures $58 \times 32 \times 7$ mm. The Mote Processor Radio (MPR) 400 can operate in two frequency regions: 868-870MHz (up to 4 channels) and 902-928 MHz (up to 54 channels). The motes feature a 4-Mbit serial flash for storing data, measurements, and other user-defined information. A 64-bit serial ID chip is also provided on the MICA2 motes. A 51-pin mote interface connector is used to interface with sensor boards and other interface boards. The motes are designed for battery power and the MICA2 mote form factor is designed to match up with two AA batteries.

Crossbow also provides a variety of sensor boards for use with the Berkeley motes. Their basic board provides light and temperature sensing. It has a precision thermistor and a light sensor/photocell. The next level multi-sensor board adds acoustic and sounder sensing modalities (shown in Figure 16.7). A board with light, temperature, acoustic, sounder, 2-axis accelerometer, and 2-axis magnetometer is also available. The modular and stackable board design of the motes allows for easy switching of sensor boards. For our research, we used the MTS300 sensor board that has a light, temperature, acoustic and sounder sensors. All of the sensors have a power control circuit

FIGURE 16.7
Crossbow mote and Intel® Mote.

with a default setting of off to minimize power consumption of the sensor board. The sensors are turned on by sending the appropriate control signals to the power switches.

The right side of Figure 16.7 shows an Intel® Mote [9]. These motes were built to enhance the original mote technology designed at the University of California at Berkeley [7, 8]. They provide increased CPU performance, more system memory, and a higher bandwidth/reliability radio in a smaller package and are intended for more demanding applications such as vibration and acoustic monitoring, and local data compression and analysis. The Intel Mote hardware is based on a stackable architecture with a backbone interconnect providing power and bidirectional signaling capability. The stackable boards consist of the main board (3 × 3cm in size), power board, and sensor boards. These motes include an Intel Mote specific software layer for Bluetooth support and platform device drivers. The Intel Mote has a 512kB FLASH, 64kB RAM, a Bluetooth 1.1 radio with ~30m range, and an integrated 2.4GHz antenna.

16.2.2.2 Sensor Gateways

As mentioned earlier, the ambient sensors communicate with our autonomic research platform via a sensor gateway. Any Intel Mote connected to the server via USB will act as a gateway or base station to communicate with the other Intel Motes. For the Berkeley motes, we use both the MIB510 [18] or the Stargate [19] sensor gateway; both are provided by Crossbow. The MIB510 (Mote Interface Board) provides an interface for an RS-232 mote serial port and reprogramming port. This serial gateway has an Atmel® ATmega16L processor that can be used to program code to the MICA2 motes. The Stargate uses a 32-bit, 400-MHz Intel XScale® processor (PXA255) [20] and is low-power and small in size. Designed within Intel's Ubiquitous Computing Research project, the Stargate was licensed to Crossbow Technology Inc. for commercial production. The Stargate processor board comes with an embedded Linux distribution and software pre-installed. A daughter card provides interfaces for a RS/232 serial, 10/100 Ethernet, USB Host, and JTAG ports.

16.2.2.3 Server Platform

We selected the SR1400JR2 server platform [21] shown in Figure 16.5 for two reasons. First, onboard platform instrumentation supports IPMI [5] version 1.5 and an Intel® Management Module (IMM) [22] upgrade to the BMC provides support for IPMI version 2.0 with serial features such as console redirection, Serial Over LAN, and terminal-mode command-line interface. The second reason for choosing this platform was its support of effectors such as Intel's demand-based switching (DBS) technology [10]. DBS is an Intel server power management technology that allows for control of processor frequency and voltage throttling. It is based on the Enhanced Intel SpeedStep Technology that allows processors to run at multiple frequency and voltage settings. The Intel® Xeon® processor integrates DBS with Enhanced

Intel SpeedStep technology to dynamically adjust power and lower the processor's power demand. Intel Xeon processors with an 800MHz bus at operating frequencies of 3.40 GHz or higher support DBS with Enhanced Intel SpeedStep Technology.

16.3   An Example Usage Scenario

Current datacenters suffer from a number of inefficiencies including, but not limited to: (i) Provisioning for peak loads, and sometimes for multiple peak loads; (ii) Low average resources utilization, which usually is in the range of less than 25%; (iii) Availability in the range of 99.99% or possibly less, which translates to 53 minutes of downtime. Such downtime costs businesses thousands or millions of dollars; (iv) Lack of autonomic features including inability to self-optimize when runtime workloads vary, inability to self-configure when resources are added/removed and inability to self-heal when failures happen; (v) Datacenter manageability remains ad-hoc with considerable overhead on the datacenter TCO. It is estimated that 70 cents of each dollar is spent on managing the datacenter, with the majority of that going towards operational expenses.

In addition, current datacenter deployments have so far exhibited strong binding between servers, running applications and administrators. The objective of such one-to-one tie is mainly to ease managing servers and applications. It is also due to the lack of technology to deploy otherwise. The vision we have established for future IT datacenters is grid-like, autonomic, and accessed as a utility. Grid-like datacenters allow their resources to be dynamically allocated/de-allocated based on the compute and I/O resources requirements of running applications. Making datacenters autonomic refers to providing management infrastructure to self-optimize, self-configure, self-heal and self-protect datacenters. These self-* features are key to considerably reducing datacenters, TCO through optimizing resources efficiency, increasing datacenter availability and protection, maintaining Quality of Service (QoS) Service Level Agreement (SLA), and enforcing power budgeting schemes. Finally, treating the compute and I/O power of datacenters as a utility allows clients to pay only for resources needed to run their service.

One technique to address most of the datacenter inefficiencies above is through first converting the datacenter servers into a pool of compute and I/O resources, then creating Dynamic Platforms (DP) from the pool. A *DP, created from a collection of compute and I/O resources, includes the ability to allocate, de-allocate, and reallocate resources dynamically and transparently to system software, and as runtime changes (including workload, fault, and service) require.* Achieving this demands considerable manageability infrastructure that dynamically and automatically discovers, configures, and provisions resources, predicts failures and enforces trust. The manageability infrastructure also dynamically reconfigures resources to enforce high-level operational polices. Sensor-rich platforms as described earlier are key ingredients to future data-centers in-general and the manageability infrastructure in-specific and are needed to enforce all self-management aspects of a datacenter.

The management infrastructure is built around the Intel® Active Management Technology framework and contains a number of Capability Modules responsible for various tasks such as resources discovery, static and dynamic provisioning, and power management. Specifically, this management infrastructure: (i) Performs out-of-band (OOB) discovery, inventory and cataloging of resources; (ii) Executes platform-based static provisioning to load various system software images, under various high-level policies; (iii) Manages the enclosure's health through monitoring enclosure- and platform-level events including those related to performance, power and external events; (iv) Collects information about resources in each partition and audits these resources to guarantee the integrity of each platform; (v) Chooses the root of trust that will be used to validate the sets of resources and their accessibility as well as the isolation across platforms; and (vi) Enforces various high-level policies related to power budgeting, Quality of Service (QoS), availability, and serviceability.

16.4   Concluding Remarks

The aspects of self-management (self-configuration, self-optimization, self-healing and self-protection) [2] have been investigated in a variety of autonomic computing research. In [23] the authors identify requirements for self-configuring network services. The Quicksilver project [24] investigates self-organizing and self-repairing peer-to-peer technologies. Self-* Storage [25] research explores the integration of automated management functions into storage architectures. The SLIC (Statistical Learning, Inference and Control) project [26] focuses on automated decision-making, management, and control of complex IT systems. A state-based approach to change and configuration management and support is taken in STRIDER [27]. However, there is virtually no related work on platform support for autonomic computing in industry or academia; the focus has been on the higher software layers, mostly in user space with perhaps some OS modifications.

What is described in this article is a research project in Intel's Research Lab focused on exploring the nature and value of providing platform support for autonomic computing. We provide a separate execution environment for the autonomic manager that has a different lifecycle and is expected to function in both pre-OS and post-OS states. The project extends from the belief that autonomic systems can provide increased manageability and higher availability by executing in an autonomous execution environment and would benefit from having rich sensory information at their disposal. We have described the research platform we built that combines hardware, soft, and ambient sensors. The hardware sensors are used to measure things like the processor temperature and fan speed while the soft sensors provide information such as kernel memory usage and process queue length. This information is combined with data obtained from ambient wireless sensors, Intel and Crossbow motes. We provide a common software abstraction interface to tie the sensor information with an authoring tool for autonomic and policy-based management.

The project is in the initial phase and our future work includes construction of a reference autonomic manager. We intend to use the policy authoring tool, EMS, to rapidly implement a variety of policies and conduct a variety of experiments. We plan to work with academia to use the research platform to explore the platform support needed for their autonomic computing systems and prototypes. With the research platform in place, we will start to answer a number of research questions we have on wireless sensor use, autonomic manager architecture, and policy-based management that may significantly enhance the manageability of future IT infrastructure.

Acknowledgments

We would like to express our thanks to Paul Drews, Rob Erickson, Leo Singleton, and Sandeep Sira for their contributions to the development of the Autonomic Research Platform prototype, to the Intel Research Berkeley Lablet and the Intel Mote sensor team for providing us with motes and technical support, and to Enigmatec for their policy authoring tool, EMS.

References

1. P. Horn, Autonomic Computing: IBM's Perspective on the State of Information Technology, IBM Corp., 2001. www.research.ibm.com/autonomic/manifesto/
2. J. Kephart and D. Chess, The Vision of Autonomic Computing, *IEEE Computer*, 36(1):41–50, 2003. www.research.ibm.com/autonomic/research/papers/
3. M. Milenkovic, et al., Toward Internet Distributed Computing, *IEEE Computer*, 36(5):38–46, 2003. ieeexplore.ieee.org/iel5/2/26966/01198235.pdf
4. R. McCollum (ed.), et al., "Web Services for Management (WS-Management June 2005)," v1, 3rd Ed., 2005. intel.com/technology/manage/downloads/ws_management_june_2005.htm
5. Intel® Intelligent Platform Management Interface website. intel.com/design/servers/ipmi/

6. OpenIPMI website. openipmi.sourceforge.net/
7. S. Hollar, "COTS Dust," Master's Thesis, Univ. of California, 2000. www-bsac.eecs.berkeley.edu/archive/users/hollar-seth/publications/cotsdust.pdf
8. J. Hill, *System Architecture for Wireless Sensor Networks*, PhD Thesis, Univ. of California, 2003. www.jlhlabs.com/jhill_cs/jhill_thesis. pdf
9. Intel® Mote webpage. intel.com/research/exploratory/motes.htm
10. V. Pallipadi, Enhanced Intel SpeedStep® Technology and Demand-Based Switching on Linux*, Intel Corp., 2005. intel.com/cd/ids/developer/asmo-na/eng/195910.htm
11. Linux Virtual Server website. linuxvirtualserver.org/
12. J. Kerr, Using Dynamic Feedback to Optimize Load Balancing Decisions, *Linux.Conf.Au* 2003, Perth, Jan. 2003. redfishsoftware.com.au/projects/feedbackd/lca-paper.pdf
13. Linux kernel CPUfreq subsystem website. kernel.org/pub/linux/utils/kernel/cpufreq/cpufreq.html
14. Execution Management System. Enigmatec Corp., 2004. www.enigmatec.net/download/enigmatec_ems_ white_paper.pdf
15. J. Widom and S. Ceri (eds.), *Active Database Systems: Triggers and Rules for Advanced Database Processing*, Morgan Kaufmann, CA, 1996.
16. G. Papamarkos, A. Poulovassilis, and P. Wood, Event-Condition-Action Rule Languages for the Semantic Web. *First International Workshop on Semantic Web and Databases*, Berlin, Sept. 2003, pp. 309–327. www.cs.uic.edu/ ifc/SWDB/ proceedings.pdf
17. Crossbow Wireless Sensor Network webpage. xbow.com/Products/products details.aspx?sid=3
18. MIB510 Serial Interface Board DataSheet. Crossbow Technology, Inc.xbow .com/Products/Product_pdf_files/Wireless_pdf/MIB510CA_Datasheet.pdf
19. Stargate Datasheet, Crossbow Technology, Inc. xbow.com/Products/Product _pdf_files/Wireless_pdf/6020-0049-02_A _Stargate.pdf
20. Intel® PXA255 Processor website. intel.com/design/pca/prodbref/252780.htm
21. Intel® Server Platforms Quick Reference Sheet, Intel Corp., 2005. intel. com/design/servers/buildingblocks/download/04-760_servplat_quick _reference.pdf
22. Intel® Management Module Installation and User's Guide, Intel Corp., 2004. download.intel.com/support/motherboards/server/sb/immugen.pdf
23. B. Melcher and B. Mitchell, "Towards an Autonomic Framework: Self-Configuring Network Services and Developing Autonomic Applications," *Intel Technology Journal*, 8(4):279–290, 2004. intel.com/technology/itj/2004/volume08issue04/art03_autonomic/vol8_art03.pdf
24. K. Birman, *Bringing Autonomic, Self-Regenerative Technology into Large Data Centers*, Presented at the Workshop on New Directions in Software Technology, St. John, Dec. 2004. www.cs.cornell.edu/projects/quicksilver/public_pdfs/BringingAutonomic.pdf
25. M. Mesnier, E. Thereska, G. Ganger, D. Ellard, and M. Selzer, "File Classification in *Self-* Storage Systems," *First IEEE International Conference on Autonomic Computing*, NY, 2004, pp. 44–51. www.pdl.cmu.edu//PDL-FTP/ABLE/ICAC04.pdf
26. I. Cohen, J. Chase, M. Goldszmidt, T. Kelly, and J. Symons, "Correlating Instrumentation Data to System States: A Building Block for Automated Diagnosis

and Control," *Sixth USENIX Symposium on OSDI*, CA, Dec. 2004, pp. 231–244. usenix.org/events/osdi04/tech/full_papers/cohen/cohen.pdf

27. Y. Wang, C. Verbowski, J. Dunagan, Y. Chen, H. Wang, C. Yuan, and Z. Zhang, "STRIDER: A Black-box, State-based Approach to Change and Configuration Management and Support," *Proceedings of the 17th USENIX Large Installation Systems Administration Conference*, CA, Oct. 2003, pp. 159–172. usenix.org/events/lisa03/tech/wang/wang.pdf

Part IV

Realization of *Self** Properties

17

Dynamic Server Allocation for Autonomic Service Centers in the Presence of Failures

Daniel A. Menascé and Mohamed N. Bennani

CONTENTS

Service centers are used to host several types of application environments and the persistent data associated with them. These centers are typically composed of a large number of servers — including Web, application, and database servers — and host several application environments (AE). Each AE has its own workload with its characteristics defined in terms of transaction types and their workload intensity levels. At any given time, a subset of the servers of the center is allocated to an AE. It is common for Service Level Agreements (SLA) to be established between the owners of the application environment (i.e., the customers of the service center) and the service center management. These SLAs include maximum response time per transaction type, minimum throughput for each type of transaction, and service availability. In some cases, penalties are established for failure to comply with established SLAs.

An adequate number of servers have to be deployed to each application environment to ensure that SLAs are met. A static allocation of servers to application environments is not cost-effective because one would have to provision for the peak load expected by each application environment. A dynamic approach is preferable, since it is not uncommon for the peak in workload intensity of one application environment to occur during a period of low intensity for another application environment. Thus, SLAs can be met more effectively if the service center is able to switch servers among application environments as needed.

This chapter shows how autonomic computing techniques can be used to dynamically allocate servers to application environments in a way that maximizes a global utility function for the center. This utility function is a function of the measured performance metrics (e.g., response time) and their SLAs for each application environment. The technique is shown to successfully reallocate servers when workload changes and/or when servers fail.

17.1   Importance of Autonomic Systems

Autonomic computing, a term introduced by IBM, indicates systems that are self-managing, self-tuning, self-healing, self-protecting, and self-organizing. These systems are also called *self-** systems [1,4,6]. Modern computer systems are quite complex and are composed of many different types of hardware and software components, which are organized in multitiered architectures. The workloads of these systems tend to exhibit large variations and it is very difficult, even impossible, for human beings to constantly tune the large number of system parameters to ensure that a system operates at its best performance level at the current workload intensity level. Sopitkamol and Menascé [12] showed the significant impact of 28 configurable parameters on the performance of an e-commerce site.

To illustrate the importance of configuration parameters on performance, consider the very simple case of a multithreaded server with a fixed number, m, of threads. Arriving requests join a queue for an available thread. Each thread processes one request at a time and, when idle, picks the first waiting request from the common queue. A request being processed by a thread uses the underlying physical resources of the server (e.g., processors and disks). Thus, all running threads compete for physical resources. Consider now the variation of the response time of a request as a function of m depicted in Figure 17.1 for four different arrival rate (L) values: 1 req/sec, 2 req/sec, 3 req/sec, and 3.5 req/sec. The curves of this figure were obtained by solving a two-level analytic performance model. The top-level model is a Markov chain (a birth-death process in particular) whose states k ($k = 0, 1, \cdots$) represent the number of requests in the system (waiting for a thread or using a thread).

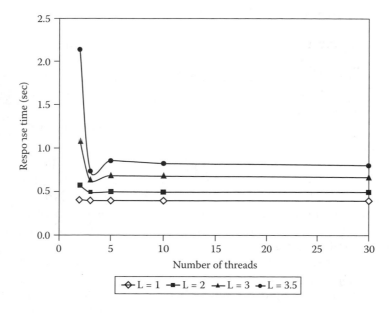

FIGURE 17.1

Response time vs. number of threads for a multithreaded server for various workload intensity levels (in requests/sec).

The birth rate of this Markov chain is the same for all states and is equal to the request arrival rate. The death rate is state dependent. In particular, for all states $k \leq m$, the death rate is equal to the system throughput $X_0(k)$, obtained by solving the lower-level queuing network model, which considers the processors, disks, and their queues. The death rate for states $k > m$ is just $X_0(m)$ because there can be at most m threads in execution. More details about these modeling techniques can be found in [10].

As expected, as the arrival rate increases, so does the response time. The most interesting aspect of the behavior represented by Figure 17.1 is the interplay of two factors: software contention (i.e., waiting for a software thread) and physical contention (i.e., waiting for a processor or disk). As m increases, the waiting time for a software thread tends to decrease. However, at the same time, contention for physical resources increases. Therefore, depending on which factor dominates, the response time decreases or increases as a function of m. Thus, for each workload level, there is an optimal number of threads. An autonomic multithreaded server dynamically adjusts the number of threads as the workload varies to obtain the best possible performance at each load level. An example of an autonomic controller for a multithreaded server is illustrated in [3,7], where additional configurable parameters and metrics other than just response time were considered.

17.2 Autonomic Control Approaches

There are two main approaches to designing autonomic controllers: white box and black box approach. A *black box* approach considers only the inputs and outputs of a system and tries to infer relationships between them without any knowledge of the internal components of a system. For example, consider the multithreaded server of Section 17.1. A black box approach does not take into account that there is a queue for threads nor that threads use processors and disks. The controller observes the output, i.e., the response time, and acts on the controllable parameter, i.e., the number of threads, to reduce the response time. The controller "learns" from observations a relationship between the input and the output for a given value of the parameter. Examples of black box controllers include the machine learning approach used in [13] and the control theoretic approach of [5].

In a *white box* approach, the controller knows the internal details of the system being controlled and can use this knowledge to build a model that relates the output with the input for a given value of the controllable parameter. Our previous work [3,7–9] falls in the category of white box controllers. We developed a technique to design self-organizing and self-tuning computer systems based on the combined use of combinatorial search techniques and analytic queuing network models [10]. In that work we defined the function to be optimized as a weighted average of the deviations of response time, throughput, and probability of rejection metrics relative to their SLAs.

The advantage of black box controllers is that they do not require expertise in model building. On the other hand, they require training, which may be time-consuming and can be effectively applied only if the system operates under the conditions in which it was observed for training purposes.

17.3 Autonomic Service Centers

We consider a service center with M online application environments (AEs) and N servers. Due to failures, a subset of the servers may be operational at any given time. The application environment AE_i has S_i classes of transactions that are characterized by their transaction arrival rate and by their demands on the physical resources. The relevant performance metrics for AE_i are the response times $R_{i,s}$ for each class s of AE_i, which can be obtained by solving an analytic performance model $\mathcal{M}_i$, for AE_i. We represent these response times in the vector $\vec{R}_i = (R_{i,1}, \cdots, R_{i,S_i})$. The value of these metrics is a function (1) of the workload vector $\vec{\lambda}_i = (\lambda_{i,1}, \cdots, \lambda_{i,S_i})$, where $\lambda_{i,s}$ is the arrival rate, in transactions per second (tps), for class s of AE_i, and (2) of the number of servers, n_i, allocated to AE_i. Thus,

$$\vec{R}_i = \mathcal{M}_i(\vec{\lambda}_i, n_i). \tag{17.1}$$

The model $\mathcal{M}_i$ is a multiclass open queueing network [10]. Let $D_{i,s}^{CPU}$ be the service demand of the central processing unit (CPU) (i.e., total CPU time not including queuing for CPU) of transactions of class s at any server of AE_i and $D_{i,s}^{IO}$ be the input-output (IO) service demand (i.e., total IO time not including queuing time for IO) of transactions of class s at any server of AE_i. The service demand at a given device for class s transactions can be measured using the Service Demand Law [10], which says that the service demand at a device is the ratio between the utilization of the device due to class s and the throughput of that class. Then, the response time, $R_{i,s}(n_i)$, of class s transactions at AE_i can be computed as:

$$R_{i,s}(n_i) = \frac{D_{i,s}^{CPU}}{1 - \sum_{t=1}^{S_i} \frac{\lambda_{i,t}}{n_i} \times D_{i,t}^{CPU}} + \frac{D_{i,s}^{IO}}{1 - \sum_{t=1}^{S_i} \frac{\lambda_{i,t}}{n_i} \times D_{i,t}^{IO}}. \tag{17.2}$$

Note that the response time $R_{i,s}$ is a function of the number of servers n_i allocated to AE_i. Eq. (17.2) assumes perfect load balancing among the servers of an AE. This is reflected in the fact that the overall arrival rate $\lambda_{i,s}$ in Eq. (17.2) is divided by n_i. Relaxing this assumption is straightforward.

Each AE i has a utility function U_i that depends on the response times for the classes of that AE. So,

$$U_i = f(\vec{R}_i) = f(\mathcal{M}_i(\vec{\lambda}_i, n_i)). \tag{17.3}$$

The global utility function U_g is a function of the utility functions of each AE. Thus,

$$U_g = h(U_1, \cdots, U_M). \tag{17.4}$$

We use in this chapter the same utility function of our previous work [2]. Other utility functions, such as the ones considered in [14], could be used. This function indicates a decreasing utility as the response time increases. The decrease in utility should be sharper as the response time approaches a desired SLA, $\beta_{i,s}$, for class s at AE_i. Thus, the utility function for class s at online AE_i is defined as

$$U_{i,s} = \frac{K_{i,s} \cdot e^{-R_{i,s} + \beta_{i,s}}}{1 + e^{-R_{i,s} + \beta_{i,s}}} \tag{17.5}$$

where $K_{i,s}$ is a scaling factor. This function has an inflection point at $\beta_{i,s}$ and decreases fast after the response time exceeds this value as shown in Figure 17.2 for $K_{i,s} = 100$ and $\beta_{i,s} = 4$.

The total utility function, U_i, is a weighted sum of the class utility functions. So,

$$U_i = \sum_{s=1}^{S_i} a_{i,s} \times U_{i,s} \tag{17.6}$$

where $0 < a_{i,s} < 1$ and $\sum_{s=1}^{S_i} a_{i,s} = 1$.

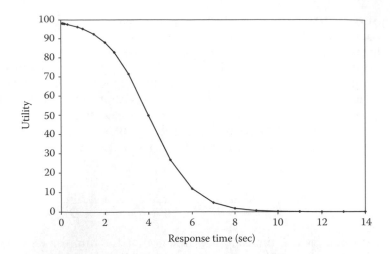

FIGURE 17.2
Utility as a function of response time.

The controller used in the experiments described in the following section is the same we discussed in [2]. We briefly summarize the idea here. The service center can be in any of the states $(n_1, \cdots, n_M)$ such that $\sum_{i=1}^{M} n_i = N^o$ where $N^o \leq N$ is the number of servers that are operational. A utility function value is computed for each state using the response time predictions derived from the performance model. The number of possible states is

$$\binom{N^o + M - 1}{M - 1}. \tag{17.7}$$

An exhaustive search would not be feasible for most real service centers that typically have hundreds of servers and tens of application environments. Thus, we use Beam Search, a heuristic combinatorial search technique [11], to limit the portion of the state space to be searched. A controller runs the Beam Search algorithm combined with the performance model at regular intervals, called control intervals, to find the optimal or close to optimal allocation of servers to AEs.

17.4 Description of Experiments

Two online AEs were considered in all experiments, which were conducted through discrete event simulation using CSim (www.mesquite.com). AE 1 has three classes of transactions and AE 2 has two transaction classes. The total

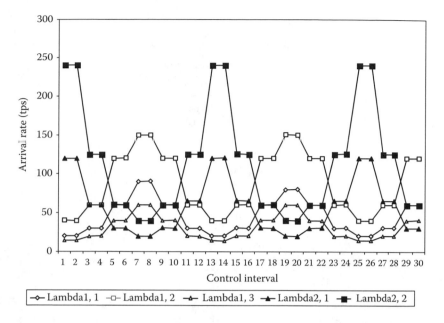

FIGURE 17.3
Variation of the arrival rate for all classes of AEs 1 and 2.

number of servers in the service center is 30 and each server is assumed to have one CPU and one disk.

Three scenarios were considered:

- *No failures with variable workload intensity.* The purpose of this experiment is to illustrate how the controller is capable of dynamically moving servers between the two AEs to cope with a variation in the workload intensity. Figure 17.3 shows the variation of the workload intensity for all five classes during the experiment. On purpose, the peaks in workload intensity for AE 1 coincide with the valleys for AE 2 and vice versa. All 30 servers are assumed to remain operational throughout the entire experiment.

- *Server failure with fixed workload intensity.* In this scenario, one of the servers fails at time unit 5 and recovers at time unit 15. The average workload intensity is maintained constant for all classes as $\vec{\lambda}_1 = (20 \text{ tps}, 40 \text{ tps}, 14 \text{ tps})$ and $\vec{\lambda}_2 = (120 \text{ tps}, 240 \text{ tps})$. The purpose of this experiment is to show how the controller is capable of moving servers between the AEs in response to failure and recovery events.

- *Server failure with variable workload intensity.* This scenario combines the two previous ones. The workload intensity varies as in the first case and one server fails as in the second case.

TABLE 17.1

Input Parameters for the Experiments

Application Environment 1			
s	1	2	3
$D_{1,s}^{CPU}$	0.030	0.015	0.045
$D_{1,s}^{IO}$	0.024	0.010	0.030
$\beta_{1,s}$	0.060	0.040	0.080
$a_{1,s}$	0.350	0.350	0.300
Application Environment 2			
s	1	2	
$D_{2,s}^{CPU}$	0.030	0.015	
$D_{2,s}^{IO}$	0.024	0.010	
$\beta_{2,s}$	0.100	0.050	
$a_{2,s}$	0.450	0.550	

Table 17.1 shows the service demands (in sec) for the CPU and disk, the SLA per class (in sec), as well as the weight of each class for the computation of the utility function of the AE. The global utility function used in all experiments is $U_g = 0.5 \times (U_1 + U_2)$.

17.5 Analysis of Results

The following subsections show the numerical results obtained for the various scenarios described earlier. Each point is an average over 10 runs with 95% confidence intervals shown. In many cases, the confidence intervals are so small that the bars are not visible. The x-axis on all graphs shown in the following subsections indicate the progression of time during the experiment measured in control intervals, which are 2 minutes each.

17.5.1 Variable Workload Intensity and No Failures

Figure 17.4 shows the variation of the global utility function U_g when the workload intensity varies according to Figure 17.3. The utility function shows drops at around 7 and 19 time units when the controller is not used. These drops coincide with the points at which the workloads for AE 1 have their peaks in intensity. Since the workloads of AE 1 have a stricter SLA than AE 2, they have a higher impact on the utility function, as will be seen in Figures. 17.6 and 17.7. When the controller is used, servers are dynamically redeployed and the utility function remains pretty much constant.

Figure 17.5 shows the variation of the number of servers during the same experiment. AE 1 starts with 14 servers and AE 2 with 16 servers. When the controller is used, the number of servers allocated to each AE varies as

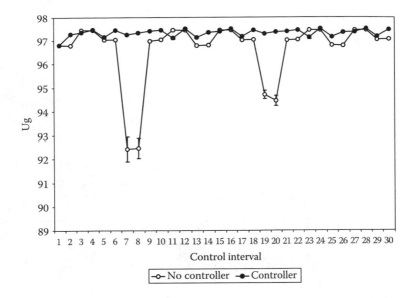

FIGURE 17.4
U_g for variable arrival rate and no failures.

the workload intensity varies. For example, at times 7 and 19, AE 1 experiences a peak in workload intensity at the same time that AE 2 experiences a decline. During these periods, AE 1 gets 23 servers and AE 2 only gets 7. The situation is reversed when AE 2 sees a surge in workload intensity and AE 1 a decline.

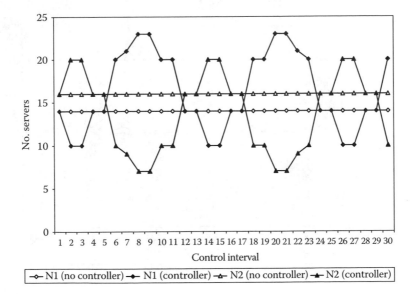

FIGURE 17.5
Variation of number of servers for variable arrival rate and no failures.

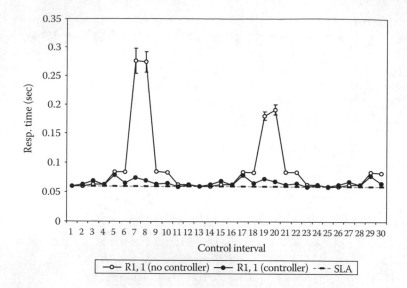

FIGURE 17.6
Variation of the response time for class 1 of AE 1 for variable arrival rate and no failures.

Figure 17.6 shows the variation of the response time for class 1 of AE 1. The figure shows that if the controller is not used, the response time for this class reaches peaks that are much higher than the SLA for that class (i.e., 0.06 sec) at time units 7 and 9. These instants correspond to the peak arrival rates for

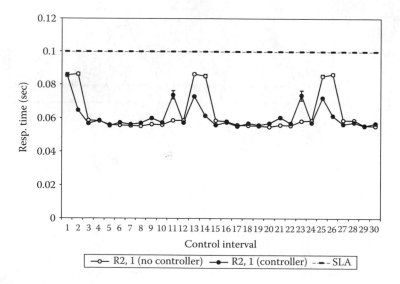

FIGURE 17.7
Variation of the response time for class 1 of AE 2 for variable arrival rate and no failures.

this class. However, when the controller is used, the response time stays close to the SLA even when the workload intensity surges for these classes.

The response time for class 1 of AE 2 is shown in Figure 17.7. In this case, the difference between the controller and no-controller situations is not as marked as in Figure 17.6 because the SLA is not as strict for this class as is the case with the classes of AE 1. As can be seen in Figure 17.7, the response time is always below the SLA of 0.10 sec. There are even cases in which the response time with the controller is worse than that without the controller. This happens because the controller is giving more servers to AE 1 because of its more stringent response time requirements. But, when the workload intensity surges for class 1 of AE 2 (see time units 12 and 24), the controller is able to keep the response time below that which is observed for the noncontroller case.

17.5.2 Fixed Workload Intensity with Failures

In the curves shown in this section, the average arrival rates are fixed for each class. However, since a server allocated to AE 2 fails at time equal to 5 units, the global utility function of the noncontrolled case shows a clear drop at that time, as shown in Figure 17.8. When the server recovers at time 15, U_g increases to its initial level. Note that the controller is able to maintain U_g pretty much constant despite the server failure. It should be emphasized that the small variations of U_g are due to the stochastic nature of the arrival process. What is fixed in these experiments is the average arrival rate.

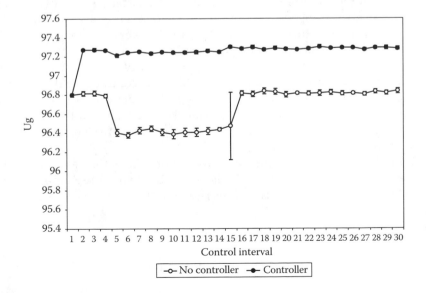

FIGURE 17.8
U_g for fixed arrival rate and failures.

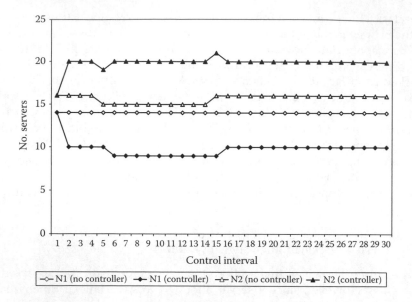

FIGURE 17.9
Variation of number of servers for fixed arrival rate and failures.

The variation of number of servers is illustrated in Figure 17.9. Without the controller, n_1 starts at 14 and n_2 at 16. At time 5, n_2 goes to 15 and remains at that value until time 14. At time 15, n_2 returns to 16. When the controller is used, n_1 starts at 14 but changes immediately to 10 at time 2. The value of n_2 starts at 16 but moves immediately to 20 at time 2. When a server of AE 2 fails at time 5, n_2 goes to 19, but the controller takes one server away from AE 1 at time 6 so that n_2 can return to 20. When the failed server recovers at time 15, AE 1 regains the server it had lost due to the failure of a server allocated to AE 2.

Figure 17.10 shows the response time for class 1 of AE 1 for the controller and no-controller cases. The figure shows that the controller makes the SLA for this class to be slightly violated by 6.7% of its target value of 0.06 sec. This small loss in response time is counterbalanced by a significant gain of 66.7% in the response time of class 1 of AE 2. This gain is reflected as well in the total utility value, as shown in Figure 17.8.

Figure 17.11 shows the variation of the response time for class 1 of AE 2. It can be seen that when the controller is not used, the response time increases at the moment the server fails and decreases only when the server recovers. The controller manages to keep the response time virtually constant due to the dynamic server reallocation.

17.5.3　Variable Workload Intensity with Failures

Figure 17.12 depicts the total utility function U_g. As it can be seen, the controller is able to cope with the simultaneous variability in the workload intensity and server failure and maintain U_g pretty much constant. The curves

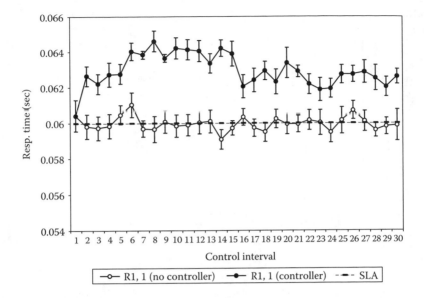

FIGURE 17.10
Variation of the response time for class 1 of AE 1 for fixed arrival rate and failures.

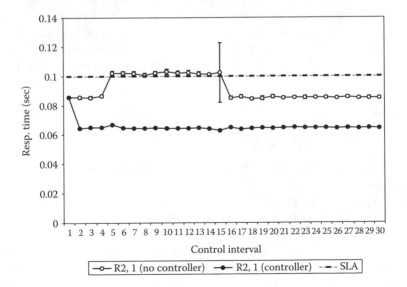

FIGURE 17.11
Variation of the response time for class 1 of AE 2 for fixed arrival rate and failures.

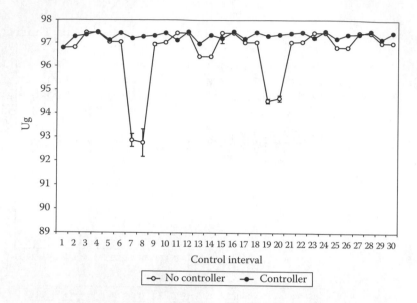

FIGURE 17.12
U_g for variable arrival rate and failures.

for number of servers, not shown here, illustrate how servers are switched among AEs to give each of them the number of servers needed to bring the utility function to its highest possible level for each circumstance. The response time curves, not presented here, are very similar to the ones shown in the case of variable workload and no failures. The reason is that the workload is the same in both scenarios and the controller is able to remove a server that is not needed from one AE and give it to another in the case of a failure.

17.6   Concluding Remarks

This chapter addressed the influence of server failures in the design of autonomic service centers. In this context, autonomic service centers host customers' applications in logically separated AEs. AEs have their own workloads (service demands and workload intensities) and SLA requirements. AEs share the limited physical resources (servers) available in the service center in an exclusive manner. Thus, at any given time, a server is servicing one and only one AE. In this chapter, we considered all service center resources to be homogeneous. Therefore, a server could be initially assigned to and then switched later to any other AE. The service center assigns to each AE a local utility function that reflects the degree of adherence of the AE to its SLAs. A higher local utility implies a better performance for the AE. Local utility functions are combined into a global utility function that is maintained by the service center. The combined global utility function takes into account the relative importance of the individual hosted AEs. The service center goal becomes, then, to keep this global utility function at its maximum value. Our proposed control approach shows how to achieve this goal in an efficient manner. As a matter of fact, we present and evaluate a controller algorithm that is used to drive server allocation decisions. The controller algorithm makes use of a combination of online analytic performance models for the AEs and a combinatorial search technique. The online analytic performance model guides the heuristic search technique in exploring the state of possible configuration vectors. A configuration vector is a vector in which the components denote the number of physical servers assigned to each AE. Online analytic performance models are helpful to the heuristic search technique in the sense that they can predict the performance measurements that correspond to any configuration vector. The heuristic search technique uses that knowledge to avoid exploring configuration vectors that are not very promising performance-wise. Therefore, the heuristic search technique reduces its exploration time and returns the best configuration vector at the end. The controller algorithm instructs, then, the local AEs to implement the new allocation decisions. In this chapter, we show that this approach is not only efficient at maintaining an overall higher global utility value for the service center, but that the technique is robust enough to cope with situations where servers may fail. Our experimental results demonstrate that even in the presence of resource failures, the controller can make the right decisions of replacing the failed servers by operational ones from other AEs that are not that much in need of extra computing capacity. We are currently expanding our research work into the following directions. First, we are investigating the use of different failure models for the resources and making the controller algorithm aware of these models. Second, we are considering the more general case of heterogeneous servers in the service centers where AEs are also constrained by the type of servers they could use.

Acknowledgments

This work is partially supported by grant NMA501-03-1-2022 from the US National Geospatial Intelligence Agency.

References

1. O. Babaoglu, M. Jelasity, and A. Montresor. Grassroots approach to self-management in large-scale distributed systems. In *Proc. EU-NSF Strategic Research Workshop on Unconventional Programming Paradigms*, Mont Saint-Michel, France, 15–17 September 2004.

2. M. N. Bennani and D. A. Menascé. Resource allocation for autonomic data centers using analytic performance models. In *Proceedings 2005 IEEE International Conference on Autonomic Computing (ICAC'05)*, Seattle, Washington, June 13–16, 2005.

3. M. N. Bennani and D. A. Menascé. Assessing the robustness of self-managing computer systems under variable workloads. In *Proceedings IEEE International Conference Autonomic Computing (ICAC'04)*, New York, May 17–18, 2004.

4. J. Chase, M. Goldszmidt, and J. Kephart, eds. In *Proceedings First ACM Workshop on Algorithms and Architectures for Self-Managing Systems*, San Diego, CA, June 11, 2003.

5. Y. Diao, N. Gandhi, J. L. Hellerstein, S. Parekh, and D. M. Tilbury. Using MIMO feedback control to enforce policies for interrelated metrics with application to the Apache Web server. In *Proceedings IEEE/IFIP Network Operations and Management Symposium*, Florence, Italy, April 15–19, 2002.

6. D. A. Menascé, M. N. Bennani, and H. Ruan. On the use of online analytic performance models in self-managing and self-organizing computer systems. In *Self-Star Properties in Complex Information Systems*, O. Babaoglu, M. Jelasity, A. Montresor, C. Fetzer, S. Leonardi, A. van Moorsel, and M. van Steen, eds., Lecture Notes in *Computer Science*, vol. 3460, Springer Verlag, Berlin, 2005.

7. D. A. Menascé and M. N. Bennani. On the use of performance models to design self-managing computer systems. In *Proceedings 2003 Computer Measurement Group Conference*, Dallas, TX, Dec. 7–12, 2003.

8. D. A. Menascé. Automatic QoS control. *IEEE Internet Computing*, Jan./Feb. 2003, vol. 7, no. 1.

9. D. A. Menascé, R. Dodge, and D. Barbará. Preserving QoS of e-commerce sites through self-tuning: A performance model approach. In *Proceedings 2001 ACM Conference E-Commerce*, Tampa, FL, Oct. 14–17, 2001.

10. D. A. Menascé, V. A. F. Almeida, and L. W. Dowdy. *Performance by Design: Computer Capacity Planning by Example*. Prentice Hall, Upper Saddle River, NJ, 2004.

11. V. J. Rayward-Smith, I. H. Osman, and C. R. Reeves, eds., *Modern Heuristic Search Methods*, John Wiley & Sons, Dec. 1996.

12. M. Sopitkamol and D. A. Menascé. A method for evaluating the impact of software configuration parameters on e-commerce sites. In *Proceedings 2005 ACM Workshop on Software and Performance*, Palma de Mallorca, Spain, July 11–14, 2005.

13. G. Tesauro, R. Das, W. Walsh, and J. O. Kephart. Utility-function-driven resource allocation in autonomic systems. In *Proceedings 2005 IEEE International Conference on Autonomic Computing (ICAC'05)*, Seattle, WA, June 13–16, 2005, 342–343.

14. W. E. Walsh, G. Tesauro, J. O. Kephart, and R. Das. Utility functions in autonomic computing. In *Proceedings IEEE International Conference Autonomic Computing (ICAC'04)*, New York, May 17–18, 2004.

18

Effecting Runtime Reconfiguration in Managed Execution Environments

Rean Griffith, Giuseppe Valetto, and Gail Kaiser

CONTENTS

Managed execution environments such as Microsoft's Common Language Runtime (CLR) and Sun Microsystems' Java Virtual Machine (JVM) provide a number of services — including but not limited to application isolation, security sandboxing, and structured exception handling — that are aimed primarily at enhancing the robustness of managed applications. However, none of these services directly enables performing autonomic diagnostics, reconfigurations, or repairs on the managed applications and their constituent subsystems and components.

In this paper we examine how the facilities of a managed execution environment can be leveraged to support autonomic system adaptations, particularly runtime reconfigurations and repairs. We describe a framework we have developed, **Kheiron**, which uses these facilities to dynamically attach/detach an engine capable of performing reconfigurations and repairs on a target system while it continues executing. Kheiron is lightweight and is transparent to the application as well as the managed execution environment: It does not require recompilation of the application nor specially compiled versions of the managed execution runtime. Our initial prototype was implemented for the CLR. To evaluate the prototype beyond toy examples, we searched on SourceForge for potential target systems already implemented on the CLR that might benefit from runtime adaptation. We report on our experience using Kheiron to facilitate runtime reconfigurations in a system that was developed and is in use by others: the Alchemi Enterprise Grid Computing System developed at the University of Melbourne, Australia [1].

18.1 Introduction

A self-healing system "automatically detects, diagnoses and repairs localized hardware and software problems" [2]. Thus we expect a self-healing system to perform runtime reconfigurations and/or repairs of its components as part of a proactive, preventative, and/or reactive response to conditions arising within its operating environment. This runtime response contrasts with the traditional approach to performing system adaptations — stop the system, fix it, then restart — which requires scheduled or unscheduled downtime and incurs costs that cannot always be expressed strictly in terms of money [3, 4]. Keeping the system running while adaptations are being carried out (even if it means operating in a degraded mode [5, 6]) is in many cases more desirable, since it maintains some degree of availability.

One software engineering challenge in implementing a self-healing system is managing the degree of coupling between the components that effect system adaptation (collectively referred to as *the adaptation engine*) and the components that realize the system's functional requirements (collectively referred to as *the target system*). For new systems being built from scratch, designers can either hardwire adaptation logic into the target system or separate the concerns of adaptation and target system functionality, by means of specialized middleware like IQ-Services [7] and ACT [8] or externalized architectures that include a reconfiguration/repair engine, as in Kinesthetics eXtreme (KX) [9] or Rainbow [10]. For legacy systems — which we define as any system for which the source code is not available, or for which it is undesirable to engage in substantial redesign and development — one is limited to using an externalized adaptation engine.

Externalized adaptation architectures may be preferred for a number of software engineering reasons. Hardwiring the adaptation logic inside target system components limits its generalization and reuse [11]. The mixing of code that realizes functional requirements and code that meets nonfunctional requirements ("code tangling" [12]) complicates the analysis and reasoning about the correctness of the adaptations being performed. Moreover, it becomes difficult to evolve the adaptation facilities without affecting the execution and deployment of the target system. Externalized architectures allow the adaptation engine and the target system to evolve independently, rather than requiring that they be developed and deployed in tandem.

We are concerned with identifying and addressing the interactions between the adaptation engine and the target system, while still seeking to minimize their coupling. Examples of interaction issues include, but are not limited to:

1. How does the adaptation engine attach to the target system such that it can effect (i.e., conduct) a reconfiguration or repair?

2. What is the scope of the adaptation actions that can be applied, e.g., can we perform reconfigurations at the granularity of entire programs, subsystems, or components? Can we repair whole programs, subsystems, individual components, classes, methods, or statements? Further, can we add, remove, update, replace, or verify the consistency of elements at the same granularity?

3. What is the impact on the performance of the target system when adaptations are/are not being performed?

4. How do we control and coordinate the adaptation engine and the target application with respect to the timing of adaptation actions given that application consistency must be preserved?

In [13] we presented a framework to partially address questions 1, 2, and 3, in the context of target systems that run in a managed execution environment. Our main focus there was on evaluating performance overhead, using a set of computationally intensive scientific applications written in C#. In this paper we present a case study geared toward exploring some of the issues associated with effecting consistency-preserving reconfigurations (question 4) in a "real-life" system, also augmented using our framework. We chose Alchemi because it meets our technical criteria and is publicly available and apparently actively used.

Our Kheiron prototype uses the profiling facility (accessible via the profiler automatic processing interface (API) of Microsoft's managed execution environment — the CLR — to track the application's execution, and effects changes via bytecode rewriting and creating/augmenting the metadata associated with modules, types, and methods. Conceptually, our approach could be applied to other managed execution environments, e.g., JVM. We chose CLR for our first prototype due to certain technical limitations of most JVM implementations, which we elaborate in [14]; we are currently developing a version of Kheiron that targets the JVM and will attempt to work around those limitations.

Kheiron facilitates attaching an externalized adaptation engine to a running application. The adaptation engine can then perform target-specific consistency checks, reconfigurations, and/or repairs over individual components and subsystems before detaching. Kheiron remains transparent to the application: it is not necessary to modify the target system's source code to facilitate attaching/detaching the adaptation engine or to enable adaptation actions. Further, the adaptations may be fine-grained, e.g., replacing individual method bodies as well as entire components. When no adaptations are being performed, Kheiron's impact on the target system is small, ~5% or less runtime overhead (see [13] for details). Finally — the main point of this paper — it allows adaptations to be enacted at well-understood control points during target system execution, necessary to maintain semantic consistency.

The remainder of this chapter is organized as follows: Section 18.2 covers some background on .NET and the CLR's execution model. Section 18.3 explains how Kheiron works. Section 18.4 describes the target system we selected for our case study, the Alchemi Enterprise Grid Computing System, and outlines the steps involved in reconfiguring that system at runtime. Section 18.5 provides detailed performance measurements and evaluates the impact of Kheiron on the target system. Section 18.6 briefly discusses related work. Finally, Section 18.7 presents our conclusions and directions for future work.

18.2   Background

18.2.1   Common Language Runtime Basics

The CLR is the managed runtime environment in which .NET applications execute. It provides an operating layer between .NET applications and the underlying operating system [15]. The CLR takes on the responsibility of providing services such as application isolation, security sandboxing, and garbage collection. Managed .NET applications are called *assemblies*, and managed executables are called *modules*. Within the CLR, assemblies execute in *application domains*, which are logical constructs used by the runtime to provide isolation from other managed applications.

.NET applications, as generated by the various compilers that target the CLR, are represented in an abstract intermediate form. This representation comprises two main elements, *metadata* and *managed code*. Metadata is "a system of descriptors of all structural items of the application — classes, their members and attributes, global items... and their relationships" [15]. *Tokens* are handles to metadata entries, which can refer to types, methods, members, etc. Tokens are used instead of pointers so that the abstract intermediate representation is memory-model independent. Managed code "represents the functionality of the application's methods... encoded in an abstract binary format known as Microsoft Intermediate Language (MSIL)" [15]. MSIL,

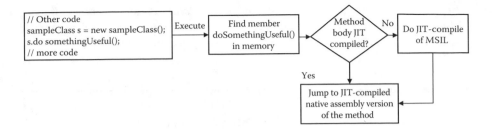

FIGURE 18.1
Overview of the CLR execution cycle.

also referred to as bytecode, is a set of abstract instructions targeted at the CLR. .NET applications written in different languages can interoperate closely, calling each other's functions and leveraging *cross-language inheritance*, since they share the same abstract intermediate representation.

18.2.2   Common Language Runtime Execution Model

Two major components of the CLR interact with metadata and bytecode during execution, the *loader* and the *just-in-time (JIT) compiler*. The loader reads the assembly metadata and creates an in-memory representation and layout of the various classes, members, and methods on demand as each class is referenced. The JIT compiler uses the results of the loader and compiles the bytecode for each method into native assembly instructions for the target platform. JIT compilation normally occurs only the first time the method is called in the managed application. Compiled methods remain cached in memory, and subsequent method calls jump directly into the native (compiled) version of the method, skipping the JIT compilation step (see Figure 18.1).

18.2.3   The CLR Profiler and Unmanaged Metadata APIs

The CLR Profiler APIs allow an interested party (a profiler) to collect information about the execution and memory usage of a running application. There are two relevant interfaces: *ICorProfilerCallback*, which a profiler must implement, and *ICorProfilerInfo*, which is implemented by the CLR. Implementors of ICorProfilerCallback (also referred to as the *notifications API* [16]) can receive notifications about assembly loads and unloads, module loads and unloads, class loads and unloads, function entry and exit, and JIT compilations of method bodies. The ICorProfilerInfo interface is used by the profiler to obtain details about particular events; for example, when a module has finished loading, the CLR will call the **ICorProfilerCallback::ModuleLoadFinished** implementation provided by the profiler, passing the **moduleID**. The profiler can then use **ICorProfilerInfo::GetModuleInfo** to get the module's name, path, and base load address.

The unmanaged metadata APIs are low-level interfaces that provide fast access to metadata, allowing users to emit/import data to/from the CLR [17].

There are two such interfaces of interest, *IMetaDataEmit* and *IMetaDataImport*. IMetaDataEmit generates new metadata tokens as metadata is written, while IMetaDataImport resolves the details of a supplied metadata token.

18.3 Adaptation Framework Prototype Overview

Our Kheiron prototype for CLR is implemented as a single dynamic linked library (DLL), which includes a profiler that implements ICorProfilerCallback. It consists of 3157 lines of C++ code and is divided into four main components:

- The **Execution Monitor** receives "module load," "module unload," and "module attached to assembly" events, JIT compilation events, and function entry and exit events from the CLR.
- The **Metadata Helper** wraps the IMetaDataImport interface and is used by the Execution Monitor to resolve metadata tokens to less cryptic method names and attributes.
- **Internal bookkeeping structures** store the results of metadata resolutions and method invocations, as well as JIT compilation times.
- The **Bytecode and Metadata Transformer** wraps the IMetaDataEmit interface to write new metadata, e.g., adding new methods to a type and adding references to external assemblies, types, and methods. It also generates, inserts, and replaces bytecode in existing methods as directed by the Execution Monitor. Bytecode changes are committed by causing the CLR to JIT-compile the modified methods *again* (referred to hereafter as **re-JIT**).

18.3.1 Model of Operation

Kheiron performs operations on types and methods at various stages in the method invocation cycle, shown in Figure 18.2, to make them capable of interacting with an adaptation engine. In particular, to enable an adaptation engine to interact with a class instance, Kheiron augments the type definition to add the necessary "hooks." Augmenting the type definition is a two-step operation.

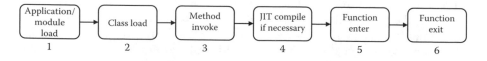

FIGURE 18.2
First method invocation in a managed application.

Step 1 occurs at the end of stage 1, module load time, in Figure 18.2. When the loader loads a module, the bytecode for the method bodies of the module's types is laid out in memory. The starting address of the first bytecode instruction in a method body is referred to as the *Relative Virtual Address* (RVA) of the method. Once the method bodies have been laid out in memory, Kheiron adds what we call *shadow methods*, using **IMetaDataEmit::DefineMethod**, for each of the original public and/or private methods of the type. A shadow method shares all the properties (attributes, signature, implementation flags, and RVA) of the corresponding original method — except the name. By sharing (borrowing) the RVA of the original method, the shadow method thus points at the method body of the original method. Figure 18.3, transition A to B, shows an example of adding a shadow method, **_SampleMethod**, for an original method **SampleMethod**.

It should be noted that extending the metadata of a type by adding new methods must be done before the type definition is installed in the CLR — signaled by a ClassLoadFinished event. Once a type definition is installed, its list of methods and members becomes read-only: Further requests to define new methods or members are silently ignored, even though the call to the API apparently "succeeds."

Step 2 of type augmentation occurs the first time an original method is JIT-compiled, as shown in stage 4 in Figure 18.2. Kheiron uses bytecode rewriting to convert the original method body into a thin *wrapper* that calls the shadow method, as shown in Figure 18.3, transition B to C. Kheiron allocates space for a new method body, uses the Bytecode and Metadata Transformer to generate the sequence of bytecode instructions to call the shadow method, and sets the new RVA for the original method to point at its new method body.

Kheiron's wrappers and shadow methods facilitate the adaptation of class instances. In particular, the regular structure and single return statement of the wrapper method (see Figure 18.4) enables Kheiron to easily inject adaptation instructions into the wrapper as prologues and/or epilogues to shadow method calls.

Adding a prologue to a method requires that new bytecode instructions prefix the existing bytecode instructions. The level of difficulty is the same

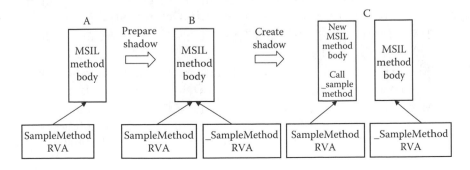

FIGURE 18.3
Preparing and creating a shadow method.

```
SampleMethod( args)
    <room for prolog>
    push args
    call _SampleMethod( args)
    <room for epilog>
    return value/void
```

FIGURE 18.4
Conceptual diagram of a wrapper.

whether we augment the bytecode of the wrapper or the original method. Adding epilogues, however, presents more challenges. Intuitively, we wish to insert new instructions before control leaves a method. In the simple case, a method has a single return statement, and the epilogue can be inserted just before that point. However, for methods with multiple return statements, and/or exception handling routines, finding every possible return point can be an arduous task [18]. Further, the layout and packing of the bytecode for methods that contain exception handling routines is considered a special case that can be complicated to augment correctly [18]. Using wrappers thus delivers a cleaner approach, since we can ignore all of the complexity in the original method.

18.3.2   Performing an Adaptation

To initiate an adaptation, Kheiron augments the wrapper to insert a jump into an adaptation engine at the *control point(s)* before and/or after a shadow method call. Effecting the jump into the adaptation engine is a four-step process:

1. Extend the metadata of the assembly currently executing in the CLR such that a reference to the assembly containing the adaptation engine is added using **IMetaDataEmit::DefineAssemblyRef**.

2. Use **IMetaDataEmit::DefineTypeRef** to add references to the adaptation engine type (class).

3. Add references to the subset of the adaptation engine's methods that we wish to insert calls to, using **IMetaDataEmit::DefineMemberRef**.

4. Augment the bytecode and metadata of the wrapper function to insert bytecode instructions to make calls into the adaptation engine before and/or after the existing bytecode that calls the shadow method.

To persist the bytecode changes made to the method bodies of the wrappers, the Execution Monitor causes the CLR to re-JIT the wrapper method the next time the method is called (i.e., JIT-compile *again*). See [14] for details on CLR re-JITs.

To transfer control to an adaptation engine, Kheiron leverages the control points before and/or after calls to shadow methods. The adaptation engine can then perform any number of operations, such as consistency checks over class instances and components, or reconfigurations and diagnostics of components.

18.4 Dynamic Reconfiguration Case Study

We selected the Alchemi Enterprise Grid Computing System [19], from the University of Melbourne, Australia. Alchemi has several appealing characteristics, relevant for our case study purposes: It was developed and is currently maintained by others, whom we do not know and have not contacted, hence we regard it as a legacy system upon which runtime adaptations can be carried out only via an externalized engine. It is publicly available on SourceForge [20], which makes it possible for other autonomic computing researchers to "repeat" our experiment employing their own technology for comparison purposes. Alchemi is also well documented, which makes it feasible to construct plausible scenarios, where performing runtime reconfigurations and/or repairs on the system could result in real benefits for its real-world users. Alchemi is apparently being used in a number of scientific and commercial grid applications, including an application for distributed, parallel environmental simulations at Commonwealth Scientific and Industrial Research Organisation (CSIRO) Land and Water, Australia, and a micro-array data processing application for early detection of breast cancer developed by Satyam Computers Applied Research Laboratory in India.[1] Finally, Alchemi is implemented as a .NET application on top of the CLR, which is a prerequisite for our current prototype. Alchemi is written in C# and leverages a number of technologies provided by the .NET Framework, including .NET Remoting [21], multithreading, and asynchronous programming.

18.4.1 Alchemi Architecture

The Alchemi Grid follows a master-worker parallel programming paradigm, where a central component (the Manager) dispatches independent units of parallel execution (grid threads) to be executed on grid nodes (Executors) (see Figure 18.5). The Manager is responsible for providing the services associated with the execution of grid applications and their constituent grid threads. It monitors the status of the Executors registered with it, and schedules grid threads to run on them. Executors accept grid threads from the Manager,

---

[1] A list of projects using Alchemi can be found at http://www.alchemi.net/projects.html.

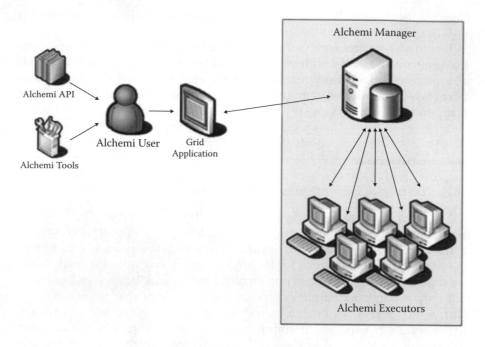

FIGURE 18.5
Alchemi architecture. Source: User guide for Alchemi 1.0 [22].

execute them, and return the completed threads to the Manager. An Executor can be configured as either *dedicated*, i.e., managed centrally, where the Manager "pushes" a computation to an idle, dedicated Executor whenever its scheduling requires, or *nondedicated*, where the Executor instead polls the Manager and hence "pulls" some computational work only during idle periods, e.g., when a screen saver is active.

18.4.2 Motivation Behind Reconfiguring Alchemi

The Alchemi Manager is clearly a key subsystem; and within the Manager, the scheduler — which makes all the grid work allocation decisions — is a key component. As in any resource allocation scenario, the scheduling strategy is critical to the overall efficacy of the system. Further, the efficacy of any particular scheduling algorithm may depend on factors that can vary quite dynamically within the grid, such as the arrival times and rate of jobs submitted for execution, the computational weight of individual work units, the set of currently available Executors, and the overall workload placed on Executors at any point in time. The current version of Alchemi (v1.0 beta) provides a default scheduler, embodied in its *DefaultScheduler* class, that schedules grid threads on a Priority and First Come First Served (FCFS) basis, in that order. This scheduling algorithm is fixed at compile-time and is used throughout the execution lifetime. However, Alchemi conveniently provides a scheduling API that allows custom schedulers to be written.

We do not address whether a one-size-fits-all scheduling algorithm could be implemented to take into account all operating conditions and all kinds of submitted application mixes, but instead intend to enable the Alchemi Manager to switch autonomically among different scheduling algorithms, each potentially tuned for specific conditions and workloads, as the state of the system changes. The same scheduler-swapping provisions could also be used to avert or alleviate situations in which (a subset of) Executors misbehave — for reasons varying from misconfiguration, to the occasional bug in the code of grid threads for some applications, to malicious interference by rogue Executor nodes — in ways that cannot be immediately detected by the monitoring capabilities of the Manager. (In Alchemi, only Executor liveness is currently considered.)

In the next section we describe a proof-of-concept experimental case study that demonstrates how Kheiron can be used to facilitate runtime reconfiguration, specifically replacement of the Alchemi scheduler, without any modifications to the source code of the target system or the underlying CLR-managed execution environment. We show how our adaptation engine attached via Kheiron is able to transparently swap scheduler implementations on the fly, which would enable existing Alchemi installations to take advantage of multiple alternative scheduling algorithms without having to recompile and reinstall any system components. We also discuss how the reconfigurations are carried out in a way that preserves the consistency of the running grid application, as well as the overall distributed grid system.

We should stress that our case study focuses on the feasibility of effecting such *consistency-preserving* reconfigurations of a legacy software system like Alchemi running in a managed execution environment. We do not at all address the optimization issues implied by the concept of dynamic scheduler replacement. We claim only that Kheiron facilitates the development of specific remedies such as optimization: For instance, our approach could enable an adaptive scheduler-swapping scheme that could ensure the grid's performance across a vast range of applications and conditions, which remains an open and interesting research issue. We also do not address here other plausible applications of runtime adaptation, such as patching potential security vulnerabilities as in [38], although we anticipate that the same basic framework should work.

18.4.3 Reconfiguring Alchemi

To swap the grid scheduler in a running instance of the Alchemi grid, we need to implement the reconfiguration engine that interacts with Alchemi's Manager component. Using Kheiron, our CLR profiler described in Section 18.3, we can dynamically attach/detach such an adaptation engine implemented as a separate assembly to/from a running managed application in a fairly mechanical way. However, a first important step is to carefully plan the interactions between the running application, the reconfiguration engine, and the CLR, in such a way that they do not compromise the integrity of either the managed application or the CLR.

Consequently, we — as the developers of the adaptation engine to be attached by Kheiron — must gather some knowledge about the system. Specifically, we need details about how the Alchemi Manager component works, particularly the execution flow in the Manager from startup to shutdown. That enables us to identify potential "safe" control points where reconfiguration actions can take place. We also need to identify those classes the adaptation engine must interact with to effect the scheduler swap. The final step is to implement the special-purpose reconfiguration engine based on what we have learned about the system.

In particular, we learned that when the Alchemi Manager is started (by running the **Alchemi.Manager.exe** assembly), an instance of the *Manager-Container* class, from the **Alchemi.Core.dll** assembly, is created. The instance of the ManagerContainer class represents the Manager proper. On startup, the **ManagerContainer::Start()** routine performs a set of initialization tasks:

1. An object is registered with the .NET Remoting services, allowing Executors to interact with the Manager instance.

2. A singleton instance of the *InternalShared* class is created, holding a reference to the scheduler implementation being used (among other things). The concrete scheduler implementation is referenced as an implementation of the **Alchemi.Core.Manager.IScheduler** interface, which standardizes the scheduler API [19].

3. Two threads, the scheduler thread and the watchdog thread, are started. The scheduler thread runs the **ManagerContainer:: ScheduleDedicated()** method, which loops "forever" on a flag member variable, **_stopScheduler**. It periodically retrieves the scheduler implementation from the InternalShared singleton instance and queries it for a *DedicatedSchedule*. A DedicatedSchedule is a <Grid Thread ID, Executor ID> tuple specifying where the selected grid thread should be scheduled to run. The watchdog thread runs the **ManagerContainer::Watchdog()** method, which loops "forever" on the **_stopWatchdog** flag member variable, periodically checking the status of dedicated Executors.

Based on this Manager startup sequence, we outline below the tasks involved in performing a scheduler swap:

1. Use Kheiron to insert a prologue into the **ManagerContainer::Start()** method such that it jumps into the reconfiguration engine assembly where the instance of the ManagerContainer can be cached.

2. Use Kheiron to insert a prologue into the constructor for the InternalShared class such that it jumps into the reconfiguration engine assembly where the instance can be cached.

3. Once instances of the ManagerContainer and InternalShared classes have been cached, the reconfiguration engine can cause the scheduler thread to exit normally by setting the **_stopScheduler** flag to true, allowing the thread to exit when it next tests the while loop condition.

4. The **Alchemi.Core.Manager.IScheduler** reference stored in the InternalShared singleton can then be replaced by another IScheduler implementation.

5. The **_stopScheduler** flag is set to false and the scheduler thread is restarted.

18.4.4   The Reconfiguration Engine and Replacement Scheduler

Our adaptation engine implementation, found in the **PSL.Alchemi.Reconfig Engine.dll** assembly, consists of two C# classes, *PSLScheduler* and *Reconfig-Engine*. The implementation was done without contacting the Alchemi developers and took about half a day to complete. The total implementation is 465 LOC – 95 LOC for *PSLScheduler.cs* and 370 LOC for *ReconfigEngine.cs*.

PSLScheduler implements the **Alchemi.Core.Manager.IScheduler** interface, and is functionally equivalent to the DefaultScheduler implementation that ships with Alchemi, except for some extra debugging and logging facilities. As noted previously, the goal of PSLScheduler is solely to demonstrate a successful reconfiguration — the scheduler swap — and to exemplify how Kheiron facilitates the development of such a reconfiguration, not to actually improve scheduling.

ReconfigEngine is responsible for caching instances of the Manager classes of interest, ManagerContainer and InternalShared, as well as effecting the scheduler swap. It is implemented according to the singleton design pattern. To effect changes on the ManagerContainer and InternalShared instances, the ReconfigEngine relies on the *Reflection API*, since many of the key variables are private and in some cases read-only. The ReconfigEngine sets up a communication channel after it has attached to the Manager, which allows a Reconfiguration Console to send commands to the ReconfigEngine to trigger reconfigurations (our case study did *not* include sensor monitoring for those conditions under which a different scheduler would be warranted). Table 18.1 shows the method signatures of the ReconfigEngine API.

TABLE 18.1

Reconfiguration Engine API

Method
public static ReconfigEngine GetInstance()
public static void CacheManagerContainer(object o)
public static void CacheInternalShared(object o)
public void SwapScheduler()

18.5   Empirical Evaluation

18.5.1   Experimental Setup

Our experimental testbed was an Alchemi cluster consisting of two Executors (Pentium-4 3GHz desktop machines each with 1GB RAM running Windows XP SP2 and the .NET Framework v1.1.4322), and a Manager (Pentium-III 1.2GHz laptop with 1GB RAM running Windows XP SP2 and the same .NET Framework version).

We ran the PiCalculator sample grid application, which ships with Alchemi, multiple times while requesting that the scheduler implementation be changed during the application's execution. The PiCalculator application computes the value of pi to n decimal digits. In our tests we used the default $n = 100$.

We swapped between the DefaultScheduler and the PSLScheduler. The two schedulers are algorithmically equivalent, except that the PSLScheduler outputs extra logging information to the Alchemi Manager graphical user interface (GUI) so that we could confirm that a scheduler swap actually occurred.

18.5.2   Results

One thing we measured was the time taken to swap the scheduler. We requested scheduler swaps between runs of the the PiCalculator application. The time taken to replace the scheduler instance was about 500 ms, on average; however, that time was dominated by the time spent waiting for the scheduler thread to exit. In the worst case, a scheduler-swap request arrived while the scheduler thread was sleeping (as it is programmed to do for up to 1000 ms on every loop iteration), causing the request to wait until the thread resumes and exits before it is honored. As a result we consider the time taken to actually effect the scheduler swap (modulo the time spent waiting for the scheduler thread to exit) to be negligible.

Table 18.2 compares the job completion times when no scheduler-swap requests are submitted during execution of the PiCalculator grid application, with job completion times when one or more scheduler-swap requests are submitted. As expected, the difference in job completion times is negligible, ~1%, since the scheduler implementations are functionally equivalent. Further, swapping the scheduler had no impact on ongoing execution of the Executors, as an Executor is not assigned an additional work unit (grid thread) until it is finished executing its current work unit.

Thus we were able to demonstrate that Kheiron can be used to facilitate a consistency-preserving reconfiguration of the Alchemi Grid Manager without compromising the integrity of the CLR or the Alchemi Grid Manager, and by extension the Alchemi Grid and jobs actively executing in the grid. The combination of ensuring that the augmentations made by Kheiron to insert hooks for the adaptation engine respect the CLR's verification rules for type and method definitions (see [14] for details on how we guarantee this)

TABLE 18.2

PiCalculator.exe Job Completion Times

Run#	Job Completion Time (ms) w/o Swap	Job Completion Time (ms) w/Swap	#Swaps
1	18.3063232	17.2748400	2
2	18.3163376	18.4665536	1
3	18.3363664	17.3148976	4
4	18.3463808	17.3148976	2
5	18.3063232	17.4150416	2
6	17.1250560	18.2662656	2
7	18.3463808	18.3163376	4
8	17.5352144	18.5266400	1
9	17.5252000	18.4965968	2
10	18.3363664	18.3463808	2
Avg	18.07799488	17.97384512	2.2

and relying on human analysis to determine what transformations Kheiron should perform, and when they should be performed, can guarantee that the operation of the target system is not compromised. Human analysis leverages the consistency guarantees of Kheiron with respect to the CLR, allowing the designers of adaptations to focus on preserving the consistency of the target system (at the application level) based on knowledge of its operation.

18.6 Related Work

The techniques — bytecode rewriting, metadata augmentation, and method call interposition — used by Kheiron to attach/detach an adaptation engine to/from an application running in a managed execution environment are similar to techniques used by dynamic Aspect Oriented Programming (AOP) engines. In general, AOP [12] is an approach to designing software that allows developers to modularize cross-cutting concerns that manifest themselves as nonfunctional system requirements. Modularized cross-cutting concerns, "aspects," allow developers to cleanly separate the code that meets system requirements from the code that meets the nonfunctional system requirements. In the context of adaptive systems, AOP is an approach to designing a system such that the nonfunctional requirement of having adaptation mechanisms available is cleanly separated from the system's functional logic. An AOP engine is still necessary to realize the final system. AOP engines weave together the code that meets the functional requirements of the system with the aspects that encapsulate the nonfunctional system requirements — in our case inserting hooks where reconfiguration and repair actions can be performed.

There are three kinds of AOP engines: those that perform weaving at compile time (static weaving), e.g., AspectJ [23] and Aspect C# [24]; those that perform weaving after compile time but before load time, e.g., Weave .NET [25] and Aspect.NET [26], which pre-process .NET assemblies, operating directly on type and assembly metadata; and those that perform weaving at runtime (dynamic weaving) at the bytecode level, e.g., A Dynamic AOP-Engine for .NET [27] and CLAW [28]. Our adaptation framework prototype exhibits analogous dynamic weaving functionality.

A Dynamic AOP-Engine for .NET exhibits the basic behavior necessary to enable method call interposition before, after, and around a given method. Injection and removal of aspects is done at runtime using the CLR Profiler API for method re-JITs and Unmanaged Metadata APIs. However, their system requires that applications run with the debugger enabled — which incurs as much as a 3X performance slowdown. CLAW uses dynamically generated proxies to intercept method calls before passing them on to the "real" callee. CLAW uses the CLR Profiler interface and the Unmanaged Metadata APIs to generate dynamic proxies and insert aspects. An implementation of CLAW was never released, and development seems to have tapered off, so we were unable to investigate its capabilities and implementation details.

Effecting runtime reconfigurations in software systems falls under the topic of *change management* [29]. Change management is a principled aspect of runtime system evolution that helps identify what must be changed; provides a context for reasoning about, specifying, and implementing change; and controls change to preserve system integrity as well as meeting extrafunctional requirements such as availability, reliability, etc.

A number of existing systems support runtime reconfiguration at various granularities. The Dynamically Alterable System (DAS) operating system [30] provides support for reconfiguring applications by letting a module be replaced by another module with the same interface. DAS's replugging mechanism requires special memory addressing hardware and a complex virtual-memory architecture to work. The DMERT operating system [31] supports the reconfiguration of the C functions that make up the switching software running on AT&T's 3B20D processor. Entire procedures can be interchanged, provided that the function signature remains constant. DMERT uses a level of indirection between a function call and the actual target of a function in memory. It is, however, very specific to the telecommunications application domain. K42 [32] is an example of an operating system that supports reconfiguration of its constituent components by virtue of its design. Explicit component boundaries, a notion of quiescent states (for consistency preservation), support for state transfers between functionally compatible components, and indirection mechanisms for accessing system components all play a role in supporting reconfigurations such as component swaps and object interposition.

Argus [33] supports coarse-grained reconfigurations in distributed systems. Argus is a language based on Clu [34] and an underlying operating system. Argus' unit of reconfiguration is a "guardian," a server that implements a set of functions via a set of handlers. The approaches and techniques for reconfiguring a system are tightly tied to the Argus system and language. Conic [29, 35] provides a powerful environment for reconfiguring distributed systems following the change management model. However, it also restrains the language and runtime system.

18.7   Conclusions and Future Work

We have described a dynamic runtime framework, Kheiron, which uses the facilities of a managed execution environment to transparently attach/detach an adaptation engine to/from a target system executing in that managed environment. We also present an example of using Kheiron in tandem with a reconfiguration engine implementation, to effect consistency preserving reconfigurations in the Alchemi Enterprise Grid Computing System. We leverage knowledge of the Alchemi system obtained from its public documentation to identify "safe" control points during program execution where reconfiguration actions can be performed. This approach to change management [29] is in part motivated by the results of Gupta et al. [36], who present a proof of the undecidability of automatically finding all the control points in an application where a consistency-preserving adaptation can be performed.

Our proof-of-concept case study shows the feasibility of using managed execution environment facilities to effect runtime reconfiguration on a legacy target system. In future work we seek to apply our approach to other managed execution environments, e.g., the Jikes Research Virtual Machine (RVM) [37] or Sun Microsystems JVM. Further, we are interested in investigating how our adaptation framework could be used to effect fine-grained reconfigurations or repairs coordinated by an existing externalized adaptation architecture such as Rainbow [10] or KX [9]. Finally, we are investigating whether we can develop similar techniques for effecting adaptations in applications running in a nonmanaged execution environment, e.g., conventional legacy C applications.

Acknowledgments

The Programming Systems Laboratory is funded in part by National Science Foundation grants CNS-0426623, CCR-0203876, and EIA-0202063, and in part by Microsoft Research.

References

1. The University of Melbourne, Alchemi — Plug and Play Grid Computing. Available at http://www.alchemi.net
2. Jeffrey O. Kephart and David M. Chess. The vision of autonomic computing. *IEEE Computer*, 41–52, January 2003.
3. George Candea et al. Improving availability with recursive micro-reboots: A soft-state case study. In *Dependable Systems and Networks — Performance and Dependability Symposium (DNS-PDS)*, 213–248, June 2002.

4. Mark E. Segal and Ophir Frieder. On-the-fly program modification systems for dynamic updating. In *IEEE Software magazine*, 53–65, March 1993.
5. Charles Shelton and Philip Koopman. Using architectural properties to model and measure system-wide graceful degradation. In *Proceedings of ICSE Workshop on Architecting Dependable Systems (WADS)*, 267–289, May 2002.
6. Philip Koopman. Elements of the self-healing system problem space. In *Proceedings of ICSE Workshop on Architecting Dependable Systems (WADS)*, 31–36, May 2003.
7. G. Eisenhauer and K. Schwan. An object-based infrastructure for program monitoring and steering. In *Proceedings of the 2nd SIGMETRICS Symposium on Parallel and Distributed Tools (SPDT98)*, 10–20, August 1998.
8. S. M. Sadjadi and P. K.McKinley. Transparent self-optimization in existing CORBA applications. In *Proceedings of the 1st IEEE International Conference on Autonomic Computing*, 88–95, May 2004.
9. Gail Kaiser et al. Kinesthetics eXtreme: An external infrastructure for monitoring distributed legacy systems. In *Proceedings of The Autonomic Computing Workshop 5th Workshop on Active Middleware Services (AMS)*, 22–30, June 2003.
10. Shang-Wen Cheng et al. Rainbow: Architecture-based self-adaptation with reusable infrastructure. In *IEEE Computer*, 46–54, October 2004.
11. Bradley Schmerl and David Garlan. Exploiting architectural design knowledge to support self-repairing systems. In *Proceedings of the 14th International Conference of Software Engineering and Knowledge Engineering*, 241–248, July 2002.
12. Gregor Kiczales et al. Aspect-oriented programming. In *Proceedings of European Conference on Object-Oriented Programming*, 220–242, June 1997.
13. Rean Griffith and Gail Kaiser. Manipulating managed execution runtimes to support self-healing systems. In *Proceedings of the 2005 Workshop on Design and Evolution of Autonomic Application Software*, 1–7, May 2005.
14. Rean Griffith and Gail Kaiser. *Adding Self-healing Capabilities to the Common Language Runtime*. Technical Report CUCS-005-05, Department of Computer Science, Columbia University in the City of New York, February 2005. Available at http://www.cs.columbia.edu/techreports/cucs-005-05.pdf.
15. Serge Lidin. Inside Microsoft .NET IL Assembler. *Microsoft Press*, 2002.
16. Microsoft. Common Language Runtime Profiling, 2002.
17. Microsoft. Common Language Runtime Metadata Unmanaged API, 2002.
18. Aleksandr Mikunov. Rewrite MSIL code on the fly with the .NET Framework Profiling API. *MSDN Magazine*, September 2003. Available at http://msdn.microsoft.com/msdnmag/issues/03/09/NETProfilingAPI
19. Akshay Luther et al. Alchemi: A .NET-Based Enterprise Grid Computing System. In *Proceedings of the 6th International Conference on Internet Computing (ICOMP'05)*, June 2005.
20. Project: Alchemi [.NET Grid Computing Framework]: Summary. Available at http://sourceforge.net/projects/alchemi
21. Ingo Rammer. Advanced .NET Remoting (C# Edition) (Paperback), Apress, Berkeley, CA, April 2002.
22. Akshay Luther et al. Alchemi: *A .NET-based Enterprise Grid System and Framework, User Guide for Alchemi 1.0*. July 2005. Available at http://www.alchemi.net/files/1.0.beta/docs/AlchemiManualv.1.0.htm
23. G. Kiczales et al. An overview of AspectJ. In *Proceedings of European Conference on Object-Object Programming*, 327–353, June 2001.

24. Howard Kim. AspectC#: An AOSD implementation for C#, Masters thesis, Department of Computer Science, Trinity College, Dublin, September 2002. Available at https://www.cs.tcd.ie/publications/tech-reports/reports.02/TCD-CS-2002-55.pdf

25. Donal Lafferty et al. Language independent aspect-oriented programming. In *Proceedings of the 18th ACM SIGPLAN Conference on Object-Oriented Programming, Systems, Languages and Applications*, 1–12, October 2003.

26. Bjorn Rasmussen et al. *Aspect.NET — A Cross-Language Aspect Weaver*. Department of Computer Science, Trinity College, Dublin, 2002.

27. Andreas Frei et al. *A Dynamic AOP-Engine for .NET*. Technical Report 445, Department of Computer Science, ETH Zurich, May 2004. Available at http://www.iks.inf.ethz.ch/publications/files/daopnet.pdf

28. John Lam. CLAW: Cross-Language Load-Time Aspect Weaving on Microsoft's Common Language Runtime. Demonstration at the 1st International Conference on Aspect-Oriented Software Development, April 2002. Available at http://trese.cs.utwente.nl/aosd2002/index.php?content=clawclr

29. J. Kramer and J. Magee. The evolving philosophers problem: Dynamic change management. In *IEEE Transactions on Software Engineering*, 1293–1306, November 1990.

30. Hannes Goullon et al. Dynamic restructuring in an experimental operating systems. In *IEEE Transactions on Software Engineering*, 298–307, July 1978.

31. R. Yacobellis et al. The 3B20D Processor and DMERT operating system: Field administration sub-system. *Bell Systems Technical Journal*, 323–339, January 1983.

32. C. Soules et al. System support for online reconfiguration. In *Proceedings of USENIX Annual Technical Conference*, 141–154, June 2003.

33. Toby Bloom. *Dynamic Module Replacement in a Distributed Programming System*, PhD thesis, Technical Report MIT/LCS/TR-303, MIT Laboratory for Computer Science, Cambridge MA, March 1983. Available at http://www.lcs.mit.edu/publications/pubs/pdf/MIT-LCS-TR-303.pdf

34. Barbara Liskov. *A History of CLU*. Technical Report MIT/LCS/TR-561, MIT Laboratory for Computer Science, Cambridge MA, April 1992. Available at http://www.lcs.mit.edu/publications/pubs/pdf/MIT-LCS-TR-561.pdf

35. Jeff Magee et al. Constructing distributed systems in Conic. In *IEEE Transactions on Software Engineering*, 663–675, June 1989.

36. Deepak Gupta et al. A formal framework for on-line software version change. In *IEEE Transactions on Software Engineering*, 120–131, February 1996.

37. Jikes. Available at http://jikes.sourceforge.net

38. Stelios Sidiroglou et al. Building a reactive immune system for software services. In *USENIX Annual Technical Conference*, 149–161, April 2005.

19

Self-Organizing Scheduling on the Organic Grid

Arjav Chakravarti, Gerald Baumgartner, and Mario Lauria

CONTENTS

The Organic Grid is a biologically inspired and fully decentralized approach to the organization of computation that is based on the autonomous scheduling of strongly mobile agents on a peer-to-peer network. Through the careful design of agent behavior, the emerging organization of the computation can be customized for different classes of applications.

We report our experience in adapting the general framework to run two representative applications on our Organic Grid prototype: the National Center for Biotechnology Information's Basic Local Alignment Search Tool

(NCBIBLAST) code for sequence alignment and Cannon's algorithm for matrix multiplication. The first is an example of an independent task application, a type of application commonly used for grid scheduling research because of its easily decomposable nature and absence of intranode communication. The second is a popular block algorithm for parallel matrix multiplication and represents a challenging application for grid platforms because of its highly structured and synchronous communication pattern.

Agent behavior completely determines the way computation is organized on the Organic Grid. We intentionally chose two applications at opposite ends of the distributed computing spectrum having very different requirements in terms of communication topology, resource use, and response to faults. We detail the design of the agent behavior and show how the different requirements can be satisfied. By encapsulating application code and scheduling functionality into mobile agents, we decouple both computation and scheduling from the underlying grid infrastructure. In the resulting system every node can inject a computation onto the grid; the computation naturally organizes itself around available resources.

19.1   Introduction

Many scientific fields, such as genomics, phylogenetics, astrophysics, geophysics, computational neuroscience, and bioinformatics, require massive computational power and resources, which might exceed those available on a single supercomputer. There are two drastically different approaches for harnessing the combined resources of a distributed collection of machines: traditional grid computing schemes and centralized master-worker schemes.

Research on Grid scheduling has focused on algorithms to determine an optimal computation schedule based on the assumption that sufficiently detailed and up-to-date knowledge of the system state is available to a single entity (the metascheduler) [1, 3, 20, 41]. While this approach results in a very efficient utilization of the resources, it does not scale to large numbers of machines. Maintaining a global view of the system becomes prohibitively expensive, and unreliable networks might even make it impossible.

A number of large-scale systems are based on variants of the master-worker model [2, 6, 13, 15, 16, 21, 24, 25, 30, 31, 39, 46]. The fact that some of these systems have resulted in commercial enterprises shows the level of technical maturity reached by the technology. However, the obtainable computing power is constrained by the performance of the single master (especially for data-intensive applications) and by the difficulty of deploying the supporting software on a large number of workers.

At a very large scale, much of the conventional wisdom we have relied upon in the past is no longer valid, and new design principles must be developed. First, very few assumptions (if any) can be made about the systems, in particular about the amount of knowledge available about the system.

Second, since the system is constantly changing (in terms of operating parameters, and resource availability), self-adaption is the normal mode of operation and must be built in from the start. Third, the deployment of the components of an infrastructure is a nontrivial issue, and should be one of the fundamental aspects of the design. Fourth, any dependence on specialized entities such as schedulers, masters nodes, etc., needs to be avoided unless such entities can be easily replicated in a way that scales with the size of the system.

We propose a completely new approach to large-scale computations that addresses all these points simultaneously with a unified design methodology. While known methods of organizing computation on large systems can be traced to techniques that were first developed in the context of parallel computing on traditional supercomputers, our approach is inspired by the organization of complex systems. Nature provides numerous examples of the emergence of complex patterns derived from the interactions of millions of organisms that organize themselves in an autonomous, adaptive way by following relatively simple behavioral rules. In order to apply this approach to the organization of computation over large complex systems, a computation must be broken into small, self-contained chunks, each capable of expressing autonomous behavior in its interaction with other chunks.

The notion that complex systems can be organized according to local rules is not new. Montresor et al. [33] showed how an ant algorithm could be used to solve the problem of dispersing tasks uniformly over a network. Similarly, the Routing Information Protocol (RIP) routing table update protocol uses simple local rules that result in good overall routing behavior. Other examples include autonomous grid scheduling protocols [26] and peer-to-peer file sharing networks [19, 40].

Our approach is to encapsulate computation and behavior into mobile agents, which deliver the computation to available machines. These mobile agents then communicate with one another and organize themselves in order to use the resources effectively. We envision a system where every node is capable of contributing resources for ongoing computations and starting its own arbitrarily large computation. Once an application is started at a node, e.g., the user's laptop, other nodes are called in to contribute resources. New mobile agents are created that, under their autonomous control, readily colonize the available resources and start computing.

Only minimal support software is required on each node, since most of the scheduling infrastructure is encapsulated along with the application code inside an agent. In our experiments we deployed only a Java Virtual Machine (JVM) and a mobile agent environment on each node. The scheduling framework described in this chapter is being implemented as a library that a developer will be able to adapt for his or her purposes.

Computation organizes itself on the available nodes according to a pattern that emerges from agent-to-agent interaction. In the simplest case, this pattern is an overlay tree rooted at the starting node; in the case of a data-intensive application, the tree can be rooted at one or more separate, presumably well-connected machines at a supercomputer center. More complex patterns can be developed as required by the applications requirements, by using different topologies than the tree and/or by having multiple overlay networks, each specialized for a different task.

In our system, the only knowledge each agent relies upon is what it can derive from its interaction with its neighbor and with the environment, plus an initial *friends list* needed to bootstrap the system. The nature of the information required for successful operation is application dependent and can be customized. E.g., for our first (data-intensive) application, both neighbor computing rate and communication bandwidth of the intervening link were important; this information was obtained using feedback from the ongoing computation.

Agent behavior completely determines the way computation is organized. In order to demonstrate the feasibility and generality of this approach, we report our experience in designing agent behavior for running two representative applications on an Organic Grid.

The first, the NCBI BLAST code for sequence alignment, is an example of an independent task application. This type of application is commonly used for grid scheduling research because of its easily decomposable nature and absence of intranode communication. The second, Cannon's algorithm for matrix multiplication, is a block algorithm for parallel matrix multiplication that interleaves communication with computation. Because of its highly structured and synchronous communication pattern, it is a challenging application for grid platforms.

The most important contribution of the experiments described here is to demonstrate how the very different requirements — in terms of communication topology, resource use, and response to faults — of each of these two applications at the opposite ends of the distributed computing spectrum can be satisfied by the careful design of agent behavior in an Organic Grid context.

19.2   Background and Related Work

This section contains a brief introduction to the critical concepts and technologies used in our work on autonomic scheduling, as well as the related work in these areas. These include peer-to-peer and Internet computing, self-organizing systems and the concept of emergence, and strongly mobile agents.

19.2.1   Peer-to-Peer and Internet Computing

The goal of utilizing the central processing unit (CPU) cycles of idle machines was first realized by the Worm project [23] at Xerox PARC. Further progress was made by academic projects such as Condor [30]. The growth of the Internet made large-scale efforts like GIMPS [46], SETI@home [39], and folding@home [15] feasible. Recently, commercial solutions such as Entropia [13] and United Devices [44] have also been developed.

The idea of combining Internet and peer-to-peer computing is attractive because of the potential for almost unlimited computational power, low cost, ease, and universality of access — the dream of a true Computational Grid. Among the technical challenges posed by such an architecture, scheduling is one of the most formidable — how to organize computation on a highly dynamic system at a planetary scale while relying on a negligible amount of knowledge about its state.

19.2.2 Decentralized Scheduling

Decentralized scheduling has recently attracted considerable attention. Two-level scheduling schemes have been considered [22, 38], but these are not scalable enough for the Internet. In the scheduling heuristic described by Leangsuksun et al. [29], every machine attempts to map tasks onto itself as well as its K best neighbors. This appears to require that each machine have an estimate of the execution time of subtasks on each of its neighbors, as well as of the bandwidth of the links to these other machines. It is not clear that their scheme is practical in large-scale and dynamic environments.

G-Commerce was a study of dynamic resource allocation on the Grid in terms of computational market economies in which applications must buy resources at a market price influenced by demand [45]. While conceptually decentralized, this scheme, if implemented, would require the equivalent of centralized commodity markets (or banks, auction houses, etc.) where offer and demand meet and commodity prices can be determined.

Recently, a new autonomous and decentralized approach to scheduling has been proposed to address the needs of large grid and peer-to-peer platforms. In this bandwidth-centric protocol, the computation is organized around a tree-structured overlay network with the origin of the tasks at the root [26]. Each node sends tasks to and receives results from its K best neighbors, according to bandwidth constraints. One shortcoming of this scheme is that the structure of the tree, and consequently the performance of the system, depends completely on the initial structure of the overlay network. This lack of dynamism is bound to affect the performance of the scheme and might also limit the number of machines that can participate in a computation.

19.2.3 Self-Organization of Complex Systems

The organization of many complex biological and social systems has been explained in terms of the aggregations of a large number of autonomous entities that behave according to simple rules. According to this theory, complicated patterns can emerge from the interplay of many agents — despite the simplicity of the rules [43, 18]. The existence of this mechanism, often referred to as *emergence*, has been proposed to explain patterns such as shell motifs, animal coats, neural structures, and social behavior. In particular, complex behaviors of colonial organisms such as social insects (e.g., ants, bees) have been studied in detail, and their applications to the solution of classic computer science problems such as task scheduling and TSP (Task Scheduler Pro) has been proposed [4, 33].

The dynamic nature and complexity of mobile ad-hoc networks (MANETs) have motivated some research in self-organization as an approach to reducing the complexity of systems installation, maintenance, and management. Self-organizing algorithms for several network functions of MANETs have been proposed, including topology control and broadcast [47, 8]. Recently, the network research community has even tried to formalize the concept of self-organization; the four design paradigms proposed by Prehofer and Bettstetter represent a first attempt to provide guidelines for developing a self-organized network function [35].

In a departure from the methodological approach followed in previous projects, we did not try to accurately reproduce a naturally occurring behavior. Rather, we started with a problem and then designed a completely artificial behavior that would result in a satisfactory solution to it.

Our work is somewhat closer to the self-organizing computation concept explored in the Co-Fields project [32]. The idea behind Co-Fields is to drive the organization of autonomous agents through artificial potential fields.

Our work was inspired by a particular version of the emergence principle called Local Activation, Long-range Inhibition (LALI) [42]. The LALI rule is based on two types of interactions: a positive, reinforcing one that works over a short range, and a negative, destructive one that works over longer distances. We retain the LALI principle but in a different form: We use a definition of distance which is based on a performance-based metric. Nodes are initially recruited using a friends list (a list of some other peers on the network) in a way that is completely oblivious of distance, therefore propagating computation on distant nodes with the same probability as close ones. During the course of the computation, the agent behavior encourages the propagation of computation among well-connected nodes while discouraging the inclusion of distant (i.e., less responsive) agents.

19.2.4   Strongly Mobile Agents

To make progress in the presence of frequent reclamations of desktop machines, current systems rely on different forms of checkpointing: automatic, e.g., SETI@home, or voluntary, e.g., Legion. The storage and computational overheads of checkpointing put constraints on the design of a system. To avoid this drawback, desktop grids need to support the asynchronous and transparent migration of processes across machine boundaries.

Mobile agents [28] have relocation autonomy. These agents offer a flexible means of distributing data and code around a network, of dynamically moving between hosts as resource availability varies, and of carrying multiple threads of execution to simultaneously perform computation, decentralized scheduling, and communication with other agents. There have been some previous attempts to use mobile agents for grid computing or distributed computing [5, 17, 34, 36].

The majority of the mobile agent systems that have been developed until now are Java-based. However, the execution model of the JVM does not permit an agent to access its execution state, which is why Java-based mobility libraries can provide only *weak mobility* [14]. Weakly mobile agent systems, such as IBM's Aglets framework [27], do not migrate the execution state of methods. The go() method, used to move an agent from one virtual machine to another, simply does not return. When an agent moves to a new location, the threads currently executing in it are killed without saving their state. The lifeless agent is then shipped to its destination and restarted there. Weak mobility forces programmers to use a difficult programming style, i.e., the use of callback methods, to account for the absence of migration transparency.

By contrast, agent systems with *strong mobility* provide the abstraction that the execution of the agent is uninterrupted, even as its location changes. Applications where agents migrate from host to host while communicating with one another are severely restricted by the absence of strong mobility. Strong mobility also allows programmers to use a far more natural programming style.

The ability of a system to support the migration of an agent at any time by an external thread is termed *forced mobility*. This is essential in desktop grid systems, because owners need to be able to reclaim their resources. Forced mobility is difficult to implement without strong mobility.

We provide strong and forced mobility for the full Java programming language by using a preprocessor that translates an extension of Java with strong mobility into weakly mobile Java code that explicitly maintains the execution state for all threads as a mobile data structure [11, 12]. For the target weakly mobile code we currently use IBM's Aglets framework [27]. The generated weakly mobile code maintains a movable execution state for each thread at all times.

19.3 Design of an Organic Grid

The purpose of this section is to describe the architecture of the proof-of-concept, small-scale prototypes of the Organic Grid we have built so far.

19.3.1 Agent Behavior

In designing the behavior of the mobile agents, we faced the classic issues of performing a distributed computation in a dynamic environment: distribution of the data, discovery of new nodes, load balancing, collection of the results, tolerance to faults, detection of task completion. The solutions for each of these issues had to be cast in terms of one-to-one interactions between pairs of agents and embedded in the agent behavior. Using an empirical approach, we developed some initial design decisions and we then refined them through an iterative process of implementation, testing on our experimental testbed, performance analysis, redesign, and new implementation. To facilitate the process, we adopted a modular design in which different aspects of the behavior were implemented as separate and well-identified routines.

As a starting point in our design process, we decided to organize the computation around a tree-based overlay network that would simplify load balancing and the collection of results. Since such a network does not exist at the beginning of the computation, it has to be built on the fly as part of the agents' takeover of the system.

In our system, a computational task represented by an agent is initially submitted to an arbitrary node in the overlay network. If the task is too large to be executed by a single agent in a reasonable amount of time, agents will clone themselves and migrate to other nodes; the clones will be assigned a small section of the task by the initiating agent. The new agents will complete the subtasks that they were assigned and return the results to their parent. They will also, in turn, clone and send agents to available nodes and distribute subtasks to them. The overlay network is constituted by the connections that are created between agents as the computation spreads out.

For our preliminary work we used the Java-based Aglets weak mobility library, on top of which we added our own strong mobility library. An Aglets environment is set up when a machine becomes available (for example, when the machine has been idle for some time; in our experiments we assumed the machines to be available at all times).

Every machine has a list of the URLs of other machines that it could ask for work. This list is known as the *friends list*. It is used for constructing the initial overlay network. The problem of how to generate this initial list was not addressed in our work; one could use one of the mechanisms used to create similar lists in tools such as Gnutella, CAN [37], and Chord [40].

The environment creates a stationary agent, which asks the friends for work by sending them messages. If a request arrives at a machine that has no computation running on it, the request is ignored. Nothing is known about the configurations of the machines on the friends list or of the bandwidths or latencies of the links to them, i.e., the algorithm is zero-knowledge and appropriate for dynamic, large-scale systems.

A large computational task is written as a strongly mobile agent. This task should be divisible into a number of independent and identical subtasks by simply dividing the input data. A user sets up the agent environment on his/her machine and starts up the computation agent. One thread of the agent begins executing subtasks sequentially. This agent is now also prepared to receive requests for work from other machines. On receiving such a request, it checks whether it has any uncomputed subtasks, and if it does, it creates a clone of itself and sends that clone to the requesting machine. The requester is now this machine's *child*.

A clone is ready to do useful work as soon as it reaches a new location. It asks its parent for a certain number of subtasks s to work on. When the parent sends the subtasks, one of this agent's threads begins to compute them. Other threads are created as needed to communicate with the parent or to respond to requests from other machines. When such a request is received, the agent clones itself and dispatches its own clone to the requester. The computation spreads in this manner. The topology of the resulting overlay network is a tree with the originating machine at the root node.

When the subtasks on a machine have been completely executed, the agent on that machine requests more subtasks to work on from its parent. The parent attempts to comply. Even if the parent does not have the requested number of subtasks, it will respond and send its child what it can. The parent keeps a record of the number of subtasks that remain to be sent, and sends a request for those tasks to its own parent.

Every time a node of the tree obtains some r results, either computed by itself or obtained from a child, it needs to send the results to its parent. It also sends along a measurement of the time that has elapsed since the last time it computed r results. The results and the timing measurement are packaged into a single message. At this point, the node also checks whether its own — or any of its children's — requests were not fully satisfied. If that is the case, a request for the remaining number of subtasks is added to the message and the entire message is sent to the node's parent. The parent then uses the timing measurements to compare the performance of its children and to restructure the overlay network. The timing measurement was built into the agent behavior in order to provide some feedback on its own performance (in terms of both computational power and communication bandwidth).

19.3.2 Details of the Agent Behavior

Maintenance of Child Lists A node cannot have an arbitrarily large number of children because this will adversely affect the synchronization delay at that node. Since the data transfer times of the independent subtasks are large, a node might have to wait for a very long time for its request to be satisfied. Therefore, each node has a fixed number of children, c. The number of children also should not be too small so as to avoid deep trees which will lead to long delays in propagating the data from the root to the leaf nodes. These children are ranked by the rate at which they send in results. When a child sends in r results with the time that was required to obtain them, its ranking is updated. This ranking is a reflection of the performance of not just a child node, but of the entire subtree with the child node as its root. This ranking is used in the restructuring of the tree, as described below.

Restructuring of the Overlay Network The topology of a typical overlay network is a tree with the root being the node where the original computation was injected. It is desirable for the best-performing nodes to be close to the root. This minimizes the communication delay between the root and the best nodes, and the time that these nodes need to wait for their requests to be handled by the root. This principle to improve system throughput is applicable down the tree, i.e., a mechanism is required to structure the overlay network such that the nodes with the highest throughput are closer to the root, while those with low throughput are near the leaves.

A node periodically informs its parent about its best-performing child. The parent then checks whether its grandchild is present in its list of former children. If not, it adds the grandchild to its list of potential children and tells this node that it is willing to consider the grandchild. The node then informs the grandchild that it should now contact its grandparent directly. This results in fast nodes percolating toward the root of the tree.

When a node updates its child list and decides to remove its slowest child, sc, it does not simply discard the child. It sends sc a list of its other children, which sc attempts to contact in turn. If sc had earlier been propagated to this node, a check is made as to whether sc's original parent is still a child of this node. In that case, sc's original parent, op, is placed first in the list of nodes being sent for sc to attempt to contact. Since sc was op's fastest child at some point, there is a good chance that it will be accepted by op again.

Fault Tolerance   A node depends on its parent to supply it with new subtasks to work on. However, if the parent were to become inaccessible due to machine or link failures, the node and its own descendents would be unable to do any useful work. A node must be able to change its parent if necessary; every node keeps a list of a of its ancestors in order to accomplish this. A node obtains this list from its parent every time the parent sends it a message. The updates to the ancestor list take into account the possibility of the topology of the overlay network changing frequently.

A child waits a certain user-defined time for a response after sending a message to its parent — the ath node in its ancestor list. If the parent is able to respond, it will, irrespective of whether it has any subtasks to send its child at this moment or not. The child will receive the response, check whether its request was satisfied with any subtasks, and begin waiting again if that is not the case.

If no response is obtained within the timeout period, the child removes the current parent from its ancestor list and sends a message to the $(a - 1)$st node in that list. This goes on until either the size of the list becomes 0 or an ancestor responds to this node's request.

If a node's ancestor list does go down to size 0, the node has no means of obtaining any work to do. The mobile agent that computes subtasks informs the agent environment that no useful work is being done by this machine, and then self-destructs. Just as before, a stationary agent begins to send out requests for work to a list of friends.

However, if an ancestor does respond to a request, it becomes the parent of the current node and sends a new ancestor list of size a to this node. Normal operation resumes with the parent sending subtasks to this node and this node sending requests and results to its parent.

Prefetching   A potential cause of slowdown in the basic scheduling scheme described earlier is the delay at each node due to its waiting for new subtasks. This is because it needs to wait while its requests propagate up the tree to the root and subtasks propagate down the tree to the node.

We found that it is beneficial to use prefetching for reducing the time that a node waits for subtasks. A node determines that it should request t subtasks from its parent. The node then makes an optimistic prediction of how many subtasks it might require in the future and requests $t + i(t)$ subtasks from its parent. When a node finishes computing one set of subtasks, more subtasks are readily available for it to work on, even as a request is submitted to the parent. This interleaving of computation and communication reduces the time for which a node is idle.

While prefetching will reduce the delay in obtaining new subtasks to work on, it also increases the amount of data that needs to be transferred at a time from the root to the current node, thus increasing the synchronization delay and data transfer time. This is why excessively aggressive prefetching will end up performing worse than a scheduling scheme with no prefetching.

19.4   Measurements

We have demonstrated the applicability of our scheduling approach using two very different types of applications, the NCBI BLAST code for sequence alignment [10], and Cannon's algorithm for parallel matrix multiplication [9]. In this section, we summarize the results of these experiments, with emphasis on Cannon's algorithm.

19.4.1   Independent Task Application: BLAST

For our initial experiments, we used BLAST, an application that is representative of a class of applications commonly used in grid scheduling research called an *independent task application* (ITA) [10]. The lack of communication between the tasks of an ITA simplifies scheduling, because there are no constraints on the order of evaluation of the tasks.

The application consisted of 320 tasks, each matching a given 256KB sequence against a 512KB chunk of a data base. When arriving at a node, a mobile agent installs the BLAST executable and then repeatedly requests new tasks from its parent and returns the results to its parent until no more tasks are available. If the agent receives requests for work from an idle machine, it sends a clone of itself to the idle machine. The computation thus spreads out from its source in the form of a tree. The source distributes the data in the form of computational subtasks that flow down the tree; results flow toward the root. This same tree structure was also used as the overlay network for making scheduling decisions. In general, there could be separate overlay networks: for data distribution, for scheduling, and for communication between subtasks. For this application, there is no communication between subtasks while the overlay trees for data distribution and scheduling overlap.

We ran the experiments with an arbitrary initial configuration of the overlay network, as shown in Figure 19.1. To simulate the effect of heterogeneity,

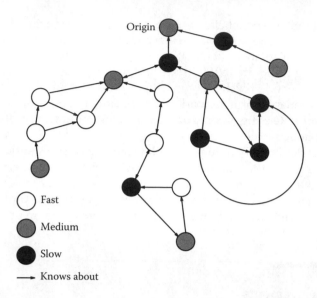

FIGURE 19.1
BLAST: Original configuration of machines.

we introduced delays into the application code, resulting in fast, medium, and slow nodes. We performed a variety of experiments with different parameters of our scheduling algorithm, such as the width of the overlay tree or the number of results over which to average the performance of a node, and measured the running time and the time needed for the computation to reach all nodes. The parameters that resulted in the best performance were a maximum tree width of 5 and a result burst size of 3. Figure 19.2 shows the

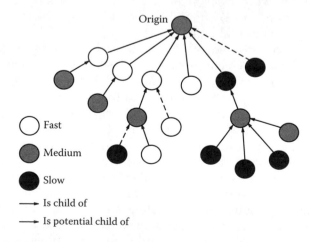

FIGURE 19.2
Final node organization, result-burst size = 3, with child propagation.

resulting overlay tree at the end of the computation, in which most of the fast nodes had been propagated closer to the root.

19.4.2 Communicating Tasks: Cannon's Matrix-Matrix Multiplication

For demonstrating the generality of the self-organizing approach and the flexibility of the Organic Grid scheduling framework, we selected a second application at the opposite end of the spectrum, characterized by a highly regular and synchronous pattern of communication — Cannon's matrix multiplication algorithm [7]. Cannon's algorithm employs a square processor grid of size $k = p \times p$ in which computation is alternated (and can be interleaved) with communication. The initial node waits until k machines are available for the computation. Each processor in the grid then gets one tile of each of the argument matrices. After multiplying these tiles, one of the argument matrices is rotated along the first dimension of the processor grid; the other argument matrix is rotated along the second dimension of the processor grid. Each processor gets new tiles of the argument matrices and adds the result of multiplying these tiles to its tile of the result matrix. The algorithm terminates after p of these tile multiplications. We have implemented this algorithm in Java.

This application employs three different overlay networks: a star topology for data distribution, a torus for the communication between subtasks, and the tree overlay of the scheduling framework. The metric used for restructuring the tree was the time to multiply two matrix tiles. While for the ITA the resource constraint was the communication bandwidth of the root, for Cannon's algorithm it was the number of machines that belong to the torus. Below we report a subset of the results of our experiment; more results are available in [9].

Three aspects of the Organic Grid implementation of Cannon's matrix multiplication were sought to be evaluated: (i) performance and scalability, (ii) fault tolerance, and (iii) decentralized selection of compute nodes. A good evaluation of this application required tight control over the experimental parameters. The experiments were therefore performed on a Beowulf cluster of homogeneous Linux machines, each with dual AMD Athlon MP processors (1.533 GHz) and 2 GB of memory. When necessary, artificial delays were introduced to simulate a heterogeneous environment. The accuracy of the experiments was improved by multiplying the matrices 16 times instead of just once.

19.4.3 Scalability

We performed a scalability evaluation by running the application on various sizes of tori and matrices. The tree adaptation mechanism was temporarily disabled in order to eliminate its effect on the experiments.

Table 19.1 and Figure 19.3 present a comparison of the running times of 16 rounds of matrix multiplications on tori with 1, 2, and 4 agents along each

TABLE 19.1

Running Time on 1 and 16 Machines, 16 Rounds

| Matrix | Single Agent | | | 4 × 4 Agent Grid | | |
Size (MB)	Tile (MB)	Time (sec)	Tile (MB)	Time (sec)	Speedup
1	1	75	0.0625	34	2.2
4	4	846	0.25	43	19.7
16	16	14029	1	454	30.9

dimension. Superlinear speedups are observed with larger numbers of nodes because of the reduction in cache effects with a decrease in the size of the tiles stored at each machine.

19.4.4 Adaptive Tree Mechanism

We then made use of the adaptive tree mechanism to select the best available machines for the torus in a decentralized manner. The feedback sent by each child to its parent was the time taken by the child to complete its two previous tile multiplications.

We experimented with a desktop grid of 20 agents in Figure 19.4. These 20 agents then formed a tree overlay network, of which the first 16 to contact

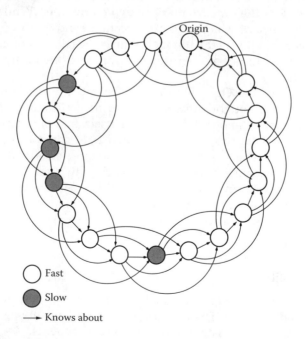

○ Fast

● Slow

→ Knows about

FIGURE 19.3

Cannon: Original configuration of machines.

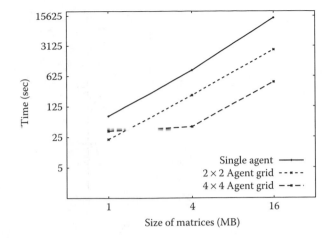

FIGURE 19.4
Running time on 1, 4 and 16 machines, 16 rounds.

the distribution agent were included in a torus with 4 agents along each dimension; the remaining agents acted as extras in case any faults occurred. The initial tree and torus can be seen in Figures 19.5 and 19.6 with 4 slow nodes in the torus and 4 extra, fast nodes.

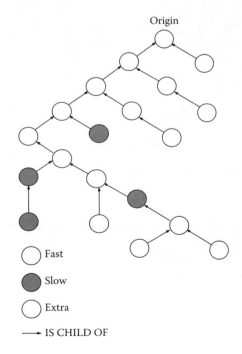

FIGURE 19.5
Original tree overlay.

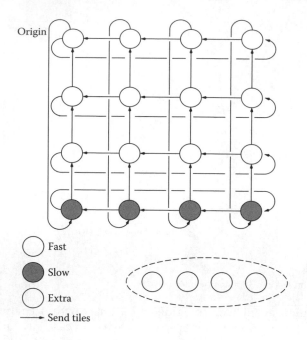

FIGURE 19.6
Original torus overlay.

The structure of the tree continually changed, and the high-performance nodes were pushed up toward the root. When a fast, extra node found that one of its children was slower than itself and part of the torus, it initiated a swap of roles. The topology of the tree and the torus before and after the fourth swap are shown in Figures 19.7 and 19.8.

Each matrix multiplication on the 4 × 4 agent grid had 4 tile multiplication stages; our experiment consisted of 16 rounds — 64 stages. A tile multiplication took 7 sec on a fast node and 14 sec on a slow one. Table 19.2 presents the average execution time of these stages. This began at 10 sec., then increased to 13 sec. before the first swap took place. The fast nodes were

TABLE 19.2

Performance at Different Stages of Experiment,
4 × 4 Agent Grid

Stage	Swap Position on Torus	Avg. Tile Mult. Time (sec)
1–3	—	10
4	12	13
5	—	15
6	13,15	15
7–42	—	14
43	12	14
44–47	12	13
48-64	—	7

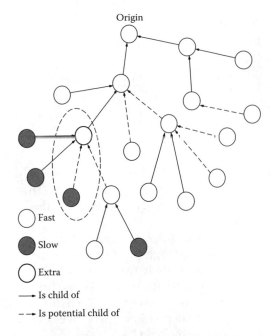

FIGURE 19.7
Tree overlay before fourth swap.

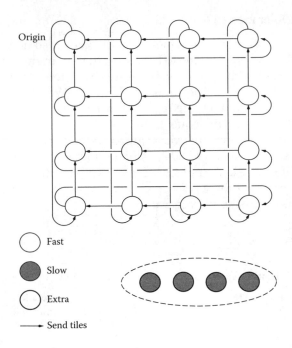

FIGURE 19.8
Torus overlay after fourth swap.

TABLE 19.3

Running Time of 16 Rounds on 4 × 4 Grid,
16MB Matrix, 1MB Tiles, Adaptive Tree

Slow Nodes	Extra Nodes	Time (sec)
4	0	898
0	0	462
4	4	759

inserted into the torus on stages 4, 6 and 43. Once the slow nodes had been swapped out, the system required 4 rounds until all the 16 agents sped up and reached high steady-state performance. The effect of this on overall running time can be seen in Table 19.3.

While the adaptive tree mechanism undoubtedly results in a performance improvement in the presence of high-performance extra nodes, it also introduces some overhead. Nodes provide feedback to their parents, who in turn rank their children and propagate the best ones. We first ran the Cannon application without any extra nodes present, and then disabled the adaptive tree mechanism for a second set of experiments. The overhead of this mechanism was negligible, as can be seen in Table 19.4.

19.4.5   Fault Tolerance

We introduce crash failures by bringing down some machines during application execution. We were interested in observing the amount of time that the system would stall in the presence of failures. Different numbers of failures were introduced at different positions on the torus. When multiple nodes on the same column crash, they are replaced in parallel. The replacements for crashes on a diagonal occur sequentially.

The system recovers rapidly from failures on the same column and diagonal, as can be seen in Table 19.5. For a small number of crashes (one or two), there is little difference in the penalty of crashes on columns or diagonals. This difference increases for three crashes, and we expect it to increase further for larger numbers of crashes on larger tori.

TABLE 19.4

Overhead of Adaptive Tree, 16 Rounds, 4 × 4 Grid, 16MB Matrix, 1MB Tiles

No Adaptation			Adaptation		
Slow Nodes	Extra Nodes	Time (sec)	Slow Nodes	Extra Nodes	Time (sec)
4	0	898	4	0	899
0	0	454	0	0	462

TABLE 19.5

Running Time of 16 Rounds on 4 × 4 Grid, 16MB Matrix, 1MB Tiles

No. of Failures	Failures on Column		Failures on Diagonal	
	Positions	Time (sec)	Positions	Time (sec)
0	—	454	—	454
1	5	466	5	466
2	5, 9	479	6, 9	464
3	5, 9, 13	486	6, 9, 12	540

19.5 Conclusions and Future Work

We have designed a desktop grid in which mobile agents are used to deliver applications to idle machines. The agents also contain a scheduling algorithm that decides which task to run on which machine. Using simple scheduling rules in each agent, a tree-structured overlay network is formed and restructured dynamically, such that well-performing nodes are brought closer to important resources, thus improving the performance of the overall system.

We have demonstrated the applicability of our scheduling scheme with two very different styles of applications: an independent task application, a BLAST executable, and an application in which individual nodes need to communicate, a Cannon-style matrix multiplication application.

Because of the unpredictability of a desktop grid, the scheduler does not have any a priori knowledge of the capabilities of the machines or the network connections. For restructuring the overlay network, the scheduler relies on measurements of the performance of the individual nodes and makes scheduling decisions using application-specific cost functions. In the case of BLAST, where the data were propagated along the same overlay tree, nodes with higher throughput were moved closer to the root to minimize congestion. In the case of Cannon's algorithm, where the data came from a separate data-center, the fastest nodes were moved closer to the root, to prevent individual slow nodes from slowing down the entire application.

The common aspect in scheduling the tasks for these very different applications is that access to a resource needs to be managed. In the case of BLAST, the critical resource is the available communication bandwidth at the root and at intermediate nodes in the tree. If a node has too many children, communication becomes a bottleneck. Conversely, if a node has too few children, the tree becomes too deep and the communication delay between the root and the leaves too long. The goal for BLAST was, therefore, to limit the width of the tree and to propagate high-throughput nodes closer to the root. In the case of Cannon's algorithm, the critical resource is the communication torus. Since any slow node participating in the torus would slow down the entire application, the goal is to propagate the fast nodes closer to the root and to keep the slower nodes further from the root.

By selecting the appropriate parameters to our scheduling algorithm, an application developer can tune the scheduling algorithm to the characteristics of an individual application. This choice of parameters includes constraints on how the overlay tree should be formed, e.g., the maximum width of the tree, and a metric with which the performance of individual nodes can be compared to decide which nodes to propagate up in the tree. Our scheduling scheme is inherently fault tolerant. If a node in the overlay tree fails, the tree will be restructured to allow other nodes to continue participating in the application. If a task is lost because of a failing node, it will eventually be assigned to another node. However, in the case of communication between tasks, such as in Cannon's algorithm, it is necessary for the application developer to write application-specific code to recover from a failed node and to reestablish the communication overlay network.

In the near future we plan to harness the computing power of idle machines by running the agent platform inside a screen saver. Since computing resources can become unavailable (e.g., if a user wiggles the mouse to terminate the screen saver), we are planning to extend our scheduling cost functions appropriately to allow agents to migrate a running computation, while continuing the communication with other agents.

We are also planning to investigate combinations of distributed, zero-knowledge scheduling with more centralized scheduling schemes to improve the performance for parts of the grid with known machine characteristics. Similar as in networking, where decentralized routing table update protocols such as RIP coexist with more centralized protocols such as OSPF (Open Shortest Path First), we envision a grid in which a decentralized scheduler would be used for unpredictable desktop machines, while a centralized scheduler would be used for, say, a Globus host.

The system described here is a reduced-scale proof-of-concept implementation. Clearly, our results need to be validated on a large-scale system. In addition to a screen saver–based implementation, we are planning the construction of a simulator. Some important aspects of the Organic Grid approach that remain to be investigated are more advanced forms of fault detection and recovery, the dynamic behavior of the system in relation to changes in the underlying system, and the management of the friends lists.

Acknowledgments

This research was done when all authors were at The Ohio State University. It was partially supported by the Ohio Supercomputer Center grants PAS0036-1 and PAS0121-1.

References

1. D. Abramson, J. Giddy, and L. Kotler. High performance parametric modeling with Nimrod/G: Killer application for the global grid? in *Proceedings of International Parallel and Distributed Processing Symp.*, May 2000, 520–528.
2. Berkeley Open Infrastructure for Network Computing (BOINC). [Online]. Available: http://boinc.berkeley.edu/
3. F. Berman, R. Wolski, H. Casanova, W. Cirne, H. Dail, M. Faerman, S. Figueira, J. Hayes, G. Obertelli, J. Schopf, G. Shao, S. Smallen, N. Spring, A. Su, and D. Zagorodnov, Adaptive computing on the grid using AppLeS. *IEEE Transactions on Parallel and Distributed Systems*, 14(4):369–382, 2003.
4. E. Bonabeau, M. Dorigo, and G. Theraulaz. *Swarm Intelligence: From Natural to Artificial Systems*. Oxford University Press, Santa Fe Institute Studies in the Sciences of Complexity, 1999.
5. J. Bradshaw, N. Suri, A. J. Cañas, R. Davis, K. M. Ford, R. R. Hoffman, R. Jeffers, and T. Reichherzer, Terraforming cyberspace, *Computer*, 34(7):48–56, July 2001.
6. D. Buaklee, G. Tracy, M. K. Vernon, and S. Wright. Near-optimal adaptive control of a large grid application, in *Proceedings of the International Conference on Supercomputing*, June 2002, 315–326.
7. L. Cannon. *A Cellular Computer to Implement the Kalman Filter Algorithm*, Ph.D. thesis, Montana State University, 1969.
8. A. Cerpa and D. Estrin. ASCENT: Adaptive self-configuring sEnsor networks topologies, *IEEE Transactions on Mobile Computing*, 3(3): 272–285, 2004.
9. A. J. Chakravarti, G. Baumgartner, and M. Lauria. Application-specific scheduling for the Organic Grid, in *Proceedings of the 5th IEEE/ACM International Workshop on Grid Computing (GRID 2004)*, Pittsburgh, November 2004, 146–155.
10. A. J. Chakravarti, G. Baumgartner, and M. Lauria. The Organic Grid: Self-organizing computation on a peer-to-peer network, in *Proceedings of the International Conference on Autonomic Computing*. IEEE Computer Society, May 2004, 96–103.
11. A. J. Chakravarti, X. Wang, J. O. Hallstrom, and G. Baumgartner. Implementation of strong mobility for multi-threaded agents in Java, in *Proceedings of the International Conference on Parallel Processing*, IEEE Computer Society, October 2003, 321–330.
12. A. J. Chakravarti, X. Wang, J. O. Hallstrom, and G. Baumgartner. Implementation of strong mobility for multi-threaded agents in Java, Dept. of Computer and Information Science, The Ohio State University, Tech. Rep. OSU-CISRC-2/03-TR06, Feb. 2003.
13. A. A. Chien, B. Calder, S. Elbert, and K. Bhatia. Entropia: architecture and performance of an enterprise desktop grid system, *Journal Parallel and Distributed Computing*, vol. 63, no. 5, 597–610, 2003.
14. G. Cugola, C. Ghezzi, G. P. Picco, and G. Vigna. Analyzing mobile code languages, in *Mobile Object Systems: Towards the Programmable Internet*, ser. Lecture Notes in Computer Science, J. Vitek, Ed., no. 1222, Springer-Verlag, 1996, 93–110. [Online]. Available: http://www.polito.it/~picco/papers/ecoop96.ps.gz

15. folding@home. [Online]. Available: http://folding.stanford.edu
16. J. Frey, T. Tannenbaum, I. Foster, M. Livny, and S. Tuecke. Condor-G: A computation management agent for multi-institutional grids, in *Proceedings IEEE Symp. on High Performance Distributed Computing (HPDC)*, San Francisco, CA, August 2001, 7–9.
17. R. Ghanea-Hercock, J. Collis, and D. Ndumu. Co-operating mobile agents for distributed parallel processing, in *Third International Conference on Autonomous Agents AA99*, Minneapolis, MN, ACM Press, May 1999.
18. A. Gierer and H. Meinhardt. A theory of biological pattern formation. *Kybernetik*, vol. 12, 30–39, 1972.
19. Gnutella. [Online]. Available: http://www.gnutella.com
20. A. S. Grimshaw and W. A. Wulf. The Legion vision of a worldwide virtual computer, *Communications of the ACM*, 40, no. 1, 39–45, January 1997.
21. E. Heymann, M. A. Senar, E. Luque, and M. Livny. Adaptive scheduling for master-worker applications on the computational grid, in *Proceedings of the First International Workshop on Grid Computing*, 2000, 214–227.
22. H. James, K. Hawick, and P. Coddington. Scheduling independent tasks on metacomputing systems, in *Proceedings of Parallel and Distributed Computing Systems*, August 1999.
23. J.A.H. John F. Shoch, The "Worm" programs — early experience with a distributed computation, *Communications of the ACM*, 25(3): 172–180, March 1982.
24. N. T. Karonis, B. Toonen, and I. Foster. MPICH-G2: A grid-enabled implementation of the message passing interface, *Journal of Parallel and Distributed Computing*, 63(5): 551–563, 2003.
25. T. Kindberg, A. Sahiner, and Y. Paker. Adaptive Parallelism under Equus, in *Proceedings of the 2nd International Workshop on Configurable Distributed Systems*, March 1994, 172–184.
26. B. Kreaseck, L. Carter, H. Casanova, and J. Ferrante, Autonomous protocols for bandwidth-centric scheduling of independent-task applications, in *Proceedings of the International Parallel and Distributed Processing Symposium*, April 2003, 23–25.
27. D. B. Lange and M. Oshima. *Programming and Deploying Mobile Agents with Java Aglets*. Addison-Wesley, 1998.
28. Danny B. Lange and Mitsuru Oshima. Seven good reasons for mobile agents, *Communications of the ACM*, 42(3): 88–89, March 1999.
29. C. Leangsuksun, J. Potter, and S. Scott. Dynamic task mapping algorithms for a distributed heterogeneous computing environment. in *Proceedings of Heterogeneous Computing Workshop*, April 1995, 30–34.
30. M. Litzkow, M. Livny, and M. Mutka, Condor — a hunter of idle workstations, in *Proceedings of the 8th International Conference of Distributed Computing Systems*, June 1988, 104–111.
31. M. Maheswaran, S. Ali, H. J. Siegel, D. A. Hensgen, and Richard F. Freund. Dynamic matching and scheduling of a class of independent tasks onto heterogeneous computing systems, in *Proceedings of the 8th Heterogeneous Computing Workshop*, April 1999, 30–44.
32. M. Mamei and F. Zambonelli. Co-Fields: a Physically Inspired Approach to Distributed Motion Coordination, *IEEE Pervasive Computing*, 3(2): April 2004.
33. A. Montresor, H. Meling, and O. Babaoglu. Messor: Load-balancing through a swarm of autonomous agents, in *Proceedings of 1st Workshop on Agent and*

Peer-to-Peer Systems, ser. Lecture Notes in Artificial Intelligence, no. 2530. Springer-Verlag, Berlin, July 2002, 125–137.

34. B. Overeinder, N. Wijngaards, M. van Steen, and F. Brazier. Multi-agent support for Internet-scale Grid management, in *AISB'02 Symposium on AI and Grid Computing*, O. Rana and M. Schroeder, eds., April 2002, 18–22.

35. C. Prehofer and C. Bettstetter, Self-Organization in Communication Networks: Principles and Design Paradigms, *IEEE Communications Magazine*, 43(7): 78–85, July 2005.

36. O. Rana and D. Walker, The Agent Grid: Agent based resource integration in PSEs, in *16th IMACS World Congress on Scientific Computation, Applied Mathematics and Simulation*, Lausanne, Switzerland, August 2000.

37. S. Ratnasamy, P. Francis, M. Handley, R. Karp, and S. Shenker, A scalable content addressable network, in *Proceedings of ACM SIGCOMM'01*, 161–172, 2001.

38. J. Santoso, G. D. van Albada, B. A. A. Nazief, and P. M. A. Sloot. Hierarchical job scheduling for clusters of workstations, in *Proceedings of Conference Advanced School for Computing and Imaging*, 99–105, June 2000.

39. SETI@home. [Online]. Available: http://setiahomes.ssl.berkeley.edu

40. I. Stoica, R. Morris, D. Karger, M. F. Kaashoek, and H. Balakrishnan, Chord: A scalable peer-to-peer lookup service for internet applications, in *Conference on Applications, Technologies, Architectures, and Protocols for Computer Communications*, San Diego, CA, 149–160, 2001.

41. I. Taylor, M. Shields, and I. Wang. *Grid Resource Management*, Kluwer, June 2003, ch. 1 — Resource Management of Triana P2P Services.

42. G. Theraulaz, E. Bonabeau, S. C. Nicolis, R. V. Sol, V. Fourcassi, S. Blanco, R. Fournier, J.-L. Joly, P. Fernandez, A. Grimal, P. Dalle, and J.-L. Deneubourg, Spatial patterns in ant colonies, *PNAS*, vol. 99, no. 15, 9645–9649, 2002.

43. A. Turing, The chemical basis of morphogenesis, *Philosophical Transcripts of Royal Society of London*, vol. 237, no. B, 37–72, 1952.

44. United Devices. Grid computing solutions. [Online]. Available: http://www.ud.com

45. R. Wolski, J. Plank, J. Brevik, and T. Bryan, Analyzing market-based resource allocation strategies for the computational grid, *International Journal of High-Performance Computing Applications*, vol. 15, no. 3, 258–281, 2001.

46. G. Woltman. GIMPS: The great internet mersenne prime search. [Online]. Available: http://www.mersenne.org/prime.htm

47. J. Wu and I. Stojmenovic, Ad Hoc Networks, *IEEE Computer*, 37(2): 29–13, February 2004.

20

Autonomic Data Streaming for High-Performance Scientific Applications

Viraj Bhat, Manish Parashar, and Nagarajan Kandasamy

CONTENTS

Efficient and robust data-streaming services are a critical requirement of emerging Grid applications, which are based on seamless interactions and coupling between geographically distributed application components. Furthermore, the dynamism of Grid environments and applications requires that these services be able to continually manage and optimize their operation based on system state and application requirements. This chapter presents a design and implementation of such a self-managing data-streaming service that is based on online control and optimization strategies. A Grid-based fusion workflow scenario is used to evaluate the service and demonstrate its feasibility and performance.

20.1   Introduction

Grid computing has established itself as the dominant paradigm for wide-area high-performance distributed computing. As Grid technologies and testbeds mature, they are enabling a new generation of scientific and engineering application formulations which are based on seamless interactions and couplings between geographically distributed computational, data, and information services. A key requirement of these applications is the support for high-throughput low-latency robust data streaming between the different distributed components of the applications.

For example, a typical Grid-based fusion simulation workflow consists of coupled simulation codes, which run simultaneously on separate high-performance computing (HPC) resources at supercomputing centers and must interact at runtime with services for interactive data monitoring, online data analysis and visualization, data archiving, and collaboration.

The fusion codes generate a huge amount of data, which must be streamed efficiently and effectively between these distributed components. Furthermore, the data-streaming services themselves must have minimal impact on the execution of the simulations, should satisfy stringent application/user space and time constraints, and should guarantee that no data are lost.

Satisfying these requirements in large-scale, heterogeneous, and highly dynamic Grid environments, where computational and communication resources are shared and their behaviors and performances are highly variable, is a significant challenge. It typically involves multiple functional and performance-related aspects that must be dynamically adapted to match current application requirements and existing operating conditions. As systems and applications grow in scale and complexity, and with many of these applications running in batch mode with limited or no runtime access, maintaining desired quality of service (QoS) using current approaches based on ad hoc manual tuning and heuristics is not just tedious and error prone but is infeasible. The applications requirements and the characteristics of the Grid executions environments warrant that the data-streaming service itself be able to dynamically detect and respond, quickly and correctly, to changes in the application's behaviors and the state of the Grid; that is, the service should be largely autonomic.

This chapter presents the design, implementation, and experimental evaluation of such a self-managing data-streaming service for wide-area Grid environments. The service is constructed using an infrastructure for self-managing Grid services that provides a programming system for specifying self-managing behaviors, and models and mechanisms for enforcing these behaviors at runtime [22]. A key contribution of this chapter is the combination of typical rule-based self-management approaches with formal model-based online control strategies. While the former are more flexible and relatively simple and easy to implement, they require a great deal of expert knowledge and are very tightly coupled to specific applications, and their performance is difficult to analyze in terms of optimality, feasibility, and stability properties. Advanced control formulations offer a theoretical basis for self-managing adaptations in distributed applications. Specifically, this chapter combines model-based limited lookahead controllers (LLCs) with rule-based managers to dynamically enforce adaptation behaviors under various operating conditions [6].

The operation and performance of the self-managing data-streaming service are demonstrated using a Grid-based fusion simulation workflow consisting of long-running coupled simulations, executing on remote supercomputing sites at the National Energy Research Scientific Computing Center (NERSC) in California and the Oak Ridge National Laboratory (ORNL) in Tennessee, and generating several terabytes of data, which must then be streamed over the network for live analysis and visualization at the Princeton Plasma Physics Laboratory (PPPL) in New Jersey and for archiving at ORNL. The goal of the data-streaming service is to enable the required data streaming while minimizing overheads on the simulation, adapting to network conditions, and preventing loss of data.

The rest of this chapter is organized as follows. Section 20.2 describes the driving Grid-based fusion simulation project and highlights its data-streaming requirements and challenges. Section 20.3 describes the models and mechanisms for enabling self-managing Grid services and applications. Section 20.4 presents the design, implementation, operation, and evaluation of the self-managing data-streaming service. Section 20.5 presents related work. Section 20.6 concludes the chapter.

20.2 Wide-Area Data Streaming in the Fusion Simulation Project

20.2.1 Fusion Simulation Workflow

The Center of Plasma Edge Simulation (CPES) fusion simulation project [17] of the U.S. Department of Energy's (DoE) Scientific Discovery through Advanced Computing (SciDAC) program is developing a new integrated

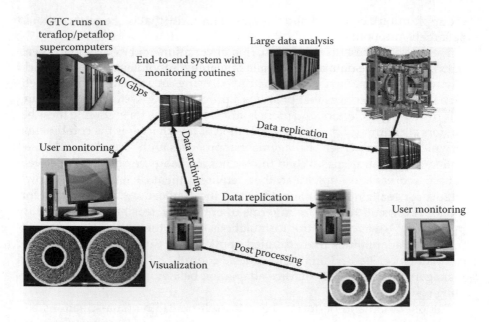

FIGURE 20.1
Workflow for the fusion simulation project.

Grid-based predictive plasma edge simulation capability to support next-generation burning plasma experiments, such as the International Thermonuclear Experimental Reactor (ITER). Effective online management and transfer of the simulation data is a critical part for this project and is essential to the scientific discovery process. A typical workflow for the project is illustrated in Figure 20.1. It consists of coupled simulation codes, i.e., the edge turbulence particle-in-cell (PIC) code (Gyrokinetic Toroidal Code [GTC]) [20] and the microscopic Magnetohydrodynamic (MHD) code (M3D) [10], which run simultaneously on thousands of processors on separate HPC resources at supercomputing centers. The data produced by these simulations must be streamed live between the coupled simulations and to remote sites, for online simulation monitoring and control, data analysis and visualization, online validation, and archiving.

20.2.2 Requirements for a Wide-Area Data-Streaming Service

The fundamental requirement of the wide-area data-streaming service is to efficiently and robustly stream data from live simulations to remote services while satisfying the following constraints: (1) Enable high-throughput, low-latency data transfer to support near real-time access to the data; (2) Minimize overheads on the executing simulation. The simulation executes in batch for days, and it is desired that the overhead of the streaming on the simulation be less than 10% of the simulation execution time; (3) Adapt to network conditions to maintain desired QoS. The network is a shared resource, and the usage patterns vary constantly; (4) Handle network failures without loss of data. Network failures usually lead to buffer overflows, and data have to be written to local disks to avoid loss. This increases the overhead in the simulation. Further, the data are no longer available for remote analysis.

20.3 Model and Mechanisms for Self-Management

The data-streaming service presented in this chapter is constructed using the Accord infrastructure [6], [22], which provides the core models and mechanisms for realizing self-managing Grid services. Its key components are shown in Figure 20.2, and are described in the following sections.

20.3.1 A Programming System for Self-Managing Services

The programming system extends the service-based Grid programming paradigm to relax assumptions of static (defined at the time of instantiation) application requirements and system/application behaviors and allows them to be dynamically specified using high-level rules. Further, it enables the behaviors of services and applications to be sensitive to the dynamic state of the system and the changing requirements of the application and to adapt to these changes at runtime. This is achieved by extending Grid services to include the specifications of policies (in the form of rules) and mechanisms for self-management, and providing a decentralized runtime infrastructure for consistently and efficiently enforcing these policies to enable self-managing functional, interaction, and composition behaviors based on current requirements, state, and execution context.

A self-managing service extends a Grid service with a control port for external monitoring and steering. An element manager monitors and controls the runtime behaviors of the managed service/element according to changing requirements and state of applications as well as their execution environment.

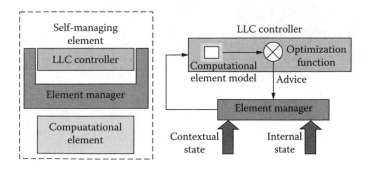

FIGURE 20.2
A self-managing element and interactions between the element manager and local controller.

The control port consists of sensors and actuators, which may be parameters, variables, or functions, and enable the state of the service to be queried and the behaviors of the service to be modified. The control port and service port are used by the service manager to control the functions, performance, and interactions of the managed service. The control port is described using WSDL (Web Service Definition Language) [12] and may be a part of the general service description, or may be a separate document with access control. Policies are in the form of simple if-condition then-action rules described using XML and include service adaptation and service interaction rules. Examples of control ports and policy specifications can be found in [22].

20.3.2 Online Control Concepts

Figure 20.3 shows the overall LLC framework [1, 15], where the management problem is posed as a sequential optimization under uncertainty. Relevant parameters of the operating environment (such as data generation patterns and effective network bandwidth) are estimated and used by a mathematical model to forecast future application behavior over a prediction horizon N. The controller optimizes the forecast behavior as per the specified QoS requirements by selecting the best control inputs to apply to the system. At each time step k, the controller finds a feasible sequence $u^*(i)|i \in [k+1, k+N]$ of inputs (or decisions) within the prediction horizon. Then, only the first move is applied to the system, and the whole optimization procedure is repeated at time $k+1$ when the new system state is available.

The LLC approach allows for multiple QoS goals and operating constraints to be represented in the optimization problem and solved for each control step. It can be used as a management scheme for systems and applications where control or tuning inputs must be chosen from a finite set, and those exhibiting both simple and nonlinear dynamics. In addition, it can accommodate

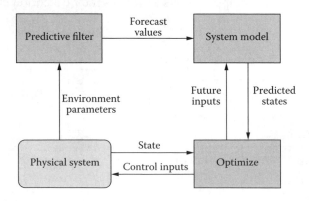

FIGURE 20.3
The LLC control structure.

runtime modifications to the system model itself caused by resource failures, dynamic data injection, and time-varying parameter changes. The following discrete-time state-space equation describes the system dynamics:

$$x(k + 1) = f(x(k), u(k), \omega(k)),$$

where $x(k) \in \Re^k$ is the system state at time step k, and $u(k) \in U \subset \Re^m$ and $\omega(k) \in \Re^r$ denote the control inputs and environment parameters at time k, respectively. The system dynamics model f captures the relationship between the observed system parameters, particularly those relevant to the QoS specifications, and the control inputs that adjust these parameters.

Though environment parameters such as workload patterns in Grid environments are typically uncontrollable, they can be estimated online with some bounded error using appropriate forecasting techniques; for example, a Kalman filter [16]. Since the current values of the environment inputs cannot be measured until the next sampling instant, the corresponding system state can only be estimated as:

$$\hat{x}(k + 1) = f(x(k), u(k), \hat{\omega}(k)),$$

where $\hat{x}(k+1)$ is the estimated system state and $\hat{\omega}(k)$ denotes the environment parameters estimated by the forecasting model(s).

A self-managing application must achieve specific QoS objectives while satisfying its operating constraints. These objectives may be expressed as a *set-point specification*, where the controller aims to operate the system close to the desired state $x^* \in X$, where X is the set of valid system states. The application must also operate within strict constraints on both the system variables and control inputs. A general form is used to describe the operating constraints of interest as $H(x(k)) \leq 0$, while $u(x(k)) \subseteq U$ denotes the control-input set $u(x(k))$ permitted in state $x(k)$. It is also possible to consider *transient* or *control costs* as part of the system operating requirements, indicating that certain trajectories toward the desired state are preferable to others in terms of their cost to the system. The overall performance specification will then require that the system reach its setpoint while minimizing the corresponding control costs. This specification is captured by the following norm-based function J that defines the overall operating cost at time k:

$$J(x(k), u(k)) = \| x(k) - x^* \|_P + \| u(k) \|_Q + \| \Delta u(k) \|_R,$$

where $\Delta u(k) = u(k) - u(k - 1)$ is the change in control inputs and P, Q, and R are user-defined weights denoting the relative importance of the variables in the cost function. The optimization problem of interest is then posed in Figure 20.4, and solved using the LLC structure introduced in Figure 20.3.

20.3.3 Operation

The element (service) managers provided by the programming system are augmented with controllers, allowing them to use model-based control and

$$\text{Minimize } \sum_{i=k+N}^{k+N} J(x(i), u(i))$$

Subject to:

$$\hat{x}(i + 1) = f(x(i), u(i), \hat{w}(i)),$$

$$H(x(i)) \leq 0,$$

$$u(i) \subset U(x(i))$$

FIGURE 20.4
The lookahead optimization problem.

optimization strategies [6]. A manager monitors the state of its underlying elements and their execution context, collects and reports runtime information, and enforces adaptation actions determined by its controller. The enhanced managers thus augment human-defined rules, which may be error prone and incomplete, with mathematically sound models, optimization techniques, and runtime information. Specifically, the controller decides when and how to adapt the application behavior, and the managers focus on enforcing these adaptations in a consistent and efficient manner.

20.4 The Self-Managing Data-Streaming Service

This section describes a self-managing data-streaming service to support a Grid-based fusion simulation, based on the models and mechanisms presented in the previous section. A specific driving simulation workflow is shown in Figure 20.5, consisting of a long-running GTC fusion simulation

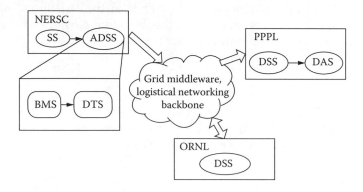

FIGURE 20.5
The self managing data streaming application.

executing on a parallel supercomputer at NERSC and generating terabytes of data over its lifetime. This data must be analyzed and visualized in real time, while the simulation is still running, at a remote site at PPPL, and also archived either at PPPL or ORNL. Data-streaming techniques from a large number of processors have been shown to be more beneficial for such a runtime analysis than writing data to the disk [18].

The data-streaming service in Figure 20.5 is composed of four core services:

1. A *Simulation Service (SS)* executing on an IBM SP machine at NERSC and generating data at regular intervals that have to be transferred at runtime for analysis and visualization at PPPL and archived at data stores at PPPL or ORNL.

2. A *Data Analysis Service (DAS)* executing on a computer cluster located at PPPL to analyze the data streamed from NERSC.

3. A *Data Storage Service (DSS)* to archive the streamed data using the Logistical Networking backbone [36], which builds a Data Grid of storage services located at ORNL and PPPL.

4. An *Autonomic Data Streaming Service (ADSS)* that manages the data transfer from SS (at NERSC) to DAS (at PPPL) and DSS (at PPPL/ORNL). It is a composite composed of two services:

    (a) The *Buffer Manager Service (BMS)* manages the buffers allocated by the service based on the rate and volume of data generated by the simulation and determines the granularity of blocks used for data transfer.

    (b) The *Data Transfer Service (DTS)* manages the transfer of blocks of data from the buffers to remote services for analysis and visualization at PPPL and archiving at PPPL or ORNL. The data transfer service uses the Internet BackPlane Protocol (IBP) [35] to transfer data.

The objectives of the self-managing ADSS are the following:

1. *Prevent any loss of simulation data*: Since data are continuously generated and the buffer sizes are limited, the local buffer at each data transfer node must be eventually emptied. Therefore, if the network link to the analysis cluster is congested, then data from the transfer nodes must be written to a local hard disk at NERSC itself.

2. *Minimize overhead on the simulation*: In addition to transferring the generated data, the transfer nodes must also perform useful computations related to the simulation. Therefore, the ADSS must minimize the computational and resource requirements of the data transfer process on these nodes.

3. *Maximize the utility of the transferred data*: It is desirable to transfer as much of the generated data as possible to the remote cluster for analysis and visualization. Storage on the local hard disk is an

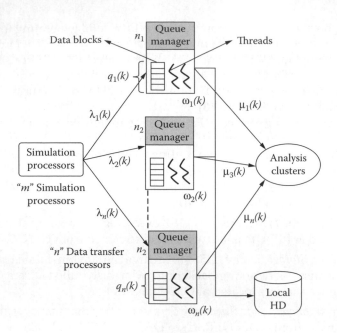

FIGURE 20.6
LLC model for the ADSS controller.

option only if the available network bandwidth is insufficient to accommodate the data generation rate and there is a danger of losing simulation data.

20.4.1   Design of the ADSS Controller

The ADSS controller is designed using the LLC concepts discussed in Section 20.3. Figure 20.6 shows the queuing model for the streaming service, where the key operating parameters for a data transfer node n_i at time step k are as follows: (1) State variable: The current average queue size at n_i denoted as $q_i(k)$; (2) Environment variables: $\lambda_i(k)$ denotes the data generation rate into the queue q_i and $B(k)$ the effective bandwidth of the network link; (3) Control or decision variables: Given the state and environment variables at time k, the controller decides $\mu_i(k)$ and $\omega_i(k)$, the data transfer rate over the network link and to the hard disk, respectively. The system dynamics at each node n_i evolves as per the following equations:

$$\hat{q}_i(k+1) = \hat{q}_i(k) + (\hat{\lambda}_i(k) \cdot (1 - \mu_i(k) - \omega_i(k))) \cdot T$$
$$\lambda_i(k) = \phi(\lambda_i(k-1), k).$$

The queue size at time $k + 1$ is determined by the current queue size, the estimated data generation rate $\lambda_i(k)$, and the data transfer rates, as decided by the controller, to the network link and the local hard disk. The data generation rate is estimated using a forecasting model ϕ, implemented here by an exponentially weighted moving-average (EWMA) filter. The sampling duration for the controller is denoted as T. Both $0 \le \mu_i(k) \le 1$ and $0 \le \omega_i(k) \le 1$ are chosen by the controller from a finite set of appropriately quantized values. Note that in practice, the data transfer rate is a function of the effective network bandwidth $B(k)$ at time k, the number of sending threads, and the size of each data block transmitted from the queue. These parameters are decided by appropriate components within the data-streaming service (discussed in Section 20.4.2).

The LLC problem is now formulated as a set-point specification where the controller aims to maintain each node n_i's queue q_i around a desired value q^* while maximizing the utility of the transferred data, that is, by minimizing the amount of data transferred to the hard disk/local depots [35].

$$Minimize : \sum_{j=k}^{k+N} \sum_{i=1}^{n} \alpha_i (q^* - q_i(j))^2 + \beta_i \omega_i(j)^2$$

$$Subject\ to : \sum_{i=1}^{n} \mu_i(j) \le B(j)\ and\ q_i(j) \le q_{max}\ \forall i$$

Here, N denotes the prediction horizon, q_{max} the maximum queue size, and α_i and β_i denote user-specified weights in the cost function.

When control inputs must be chosen from a set of discrete values, the LLC formulation, as posed above, will show an exponential increase in worst-case complexity with an increasing number of control options and longer prediction horizons — the so-called "curse of dimensionality." Since the execution time available for the controller is often limited by hard application bounds, it is necessary to consider the possibility that it may have to deal with suboptimal solutions. For adaptation purposes, however, it is not critical to find the global optimum to ensure system stability; a feasible suboptimal solution will suffice. Taking advantage of the fact that the operating environment does not change drastically over a short period of time, suboptimal solutions are obtained using *local search methods*, where given the current values of $\mu_i(k)$ and $\omega_i(k)$, the controller searches a limited neighborhood of these values for a feasible solution for the next step.

20.4.2   Implementation and Deployment of ADSS

ADSS (refer to Figure 20.7) is implemented as a composite service comprising a BMS that manages the buffers allocated by the ADSS, and a DTS that manages the transfer of blocks of data from the buffers. The BMS supports two buffer management schemes, Uniform and Aggregate buffering. *Uniform buffering* divides the data into blocks of fixed sizes and is more suitable when the simulation can transfer all its data items to a remote storage. *Aggregate buffering*, on the other hand, aggregates blocks across multiple time steps for

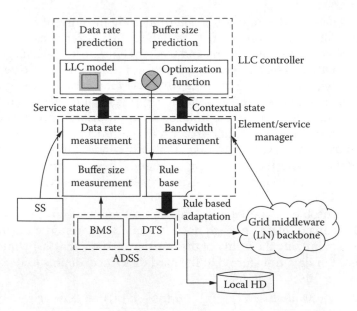

FIGURE 20.7
Implementation overview of the ADSS.

network transfer, and can be used when the network is congested. The control ports for these services are described in [22].

The *ADSS Online Controller* consists of the system model, the set-point specification, and the LLC scheme. The system model obtains inputs from the data generation rate prediction and buffer size prediction submodule, which provides it with future values of the data rates (sizes) and future buffer capacities, respectively. The prediction of the data generation rate uses a EWMA filter with a smoothing constant of 0.95. A single-step LLC scheme ($N = 1$) is implemented on each node/data transfer processor n_i with a desired queue size of $q^* = 0$. The weights in the multiobjective cost function are set to $\alpha_i = 1$ and $\beta_i = 10^8$, to penalize the controller very heavily for writing data to the hard disk. The decision variables μ_i and ω_i are quantized in intervals of 0.1. The controller sampling time T is set to 80 sec in the implementation.

The *ADSS Element Manager* supplies the controller with internal state of the ADSS and SS services, including the observed buffer size on each node, n_i, the simulation-data generation rate, and the network bandwidth. The effective network bandwidth of the link between NERSC and PPPL is measured using Iperf [32], which reports the transmission control protocol (TCP) bandwidth available, delay jitter, and datagram loss.

The element manager also stores a set of rules, which are triggered based on controller decisions and enforce adaptations within the DTS/BMS. For example, the controller decides the amount of data to be sent over the network or to local storage, and the element manager decides the corresponding buffer management scheme to be used within the BMS to achieve this. The element

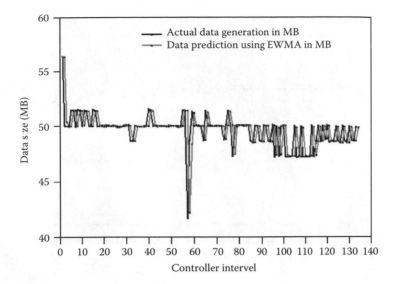

FIGURE 20.8
Accuracy of predictions of data generation using EWMA for GTC.

manager also adapts the DTS service to send data to local/low-latency storage — for example, NERSC/ORNL — when the network is congested.

20.4.3 Experiments

The setup for the experiments presented in this section consisted of the GTC fusion simulation running on 32 to 256 processors at NERSC, and streaming data for analysis to PPPL. A 155 Mbps ESNET [19] connection between PPPL and NERSC was used. A single controller was used, and the controller and managers were implemented using threading. Up to four simulations processors were used for data streaming.

20.4.3.1 Accuracy of Prediction of Data Generation Using EWMA

Figure 20.8 plots a trace of the predicted and actual data generated by the simulation, plotted for a controller interval (sampling time) of 80 sec. The simulation ran for 3 hours at NERSC on 64 processors, and used four data-streaming processors. The incoming data rate into each transfer processor was estimated with good accuracy by EWMA filter as follows: $\hat{\lambda}_i(k) = \gamma \cdot \lambda_i(k) + (1 - \gamma) \cdot \hat{\lambda}_i(k-1)$, where $\gamma = 0.95$ is the smoothing factor. It follows from the plot that the EWMA can accurately predict the data generation for GTC simulations.

20.4.3.2 Controller Behavior for Long-Running Simulations

Figure 20.9 plots a representative snapshot of the streaming behavior for a long-running GTC simulation. During the plotted period, DTS always transfers data to remote storage, and no data are transferred to local storage, as the

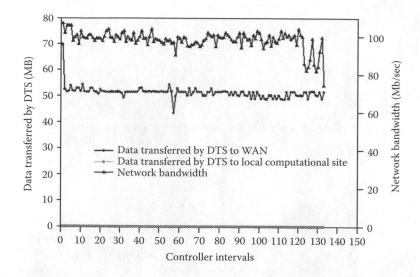

FIGURE 20.9
Controller and DTS operation for the GTC simulation.

effective network bandwidth remains steady and no congestions are detected. Since the network bandwidth is sufficient to stream all of the generated data, there are no adaptations triggered by the controller during the controller interval and the controller operates in a stable mode.

20.4.3.3 DTS Adaptations Based on Control Strategies

To observe adaptation in the DTS, the network was congested between NERSC and PPPL during the controller intervals 13 to 20 (recall that each controller interval is 80 sec), as observed in Figure 20.10. During controller intervals 1 to 13 there is no congestion in the network and all the data are transferred by DTS over the network to PPPL. During the intervals of network congestion (from 13 to 20), the controller observes the environment and state variables and advises the element manager to adapt DTS behavior. The adaptation causes the DTS to send data to a local storage/hard disk in addition to sending data to the remote location, to prevent data loss due to buffer overflows. In Figure 20.10 it is observed that this adaptation is triggered multiple times until the network is no longer congested at around the 21st controller interval. The data sent to the local storage fall to zero at this point.

20.4.3.4 Adaptations in the BMS

This scenario demonstrates the adaptation in the BMS service. A uniform BMS scheme is triggered in cases where data generation is constant and in cases where the congestion increases, and thus an aggregate buffer management is triggered. The triggering of the appropriate buffering scheme in the BMS

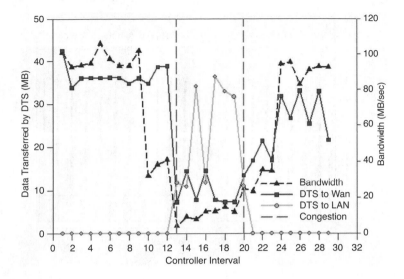

FIGURE 20.10
Adaptations in the DTS due to network congestion.

is prescribed by the controller to overcome network congestion. The adaptations are shown in Figure 20.11. As seen in the figure, during intervals 0 to 7, the uniform blocking scheme is used, and during the intervals 7 to 16, the aggregate blocking scheme is used to compensate for network congestion.

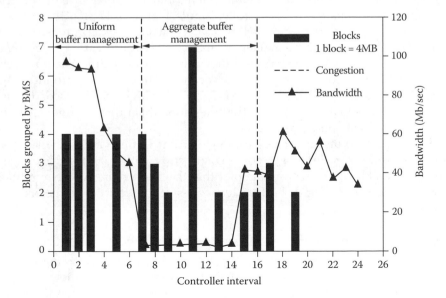

FIGURE 20.11
BMS adaptations due to varying network conditions.

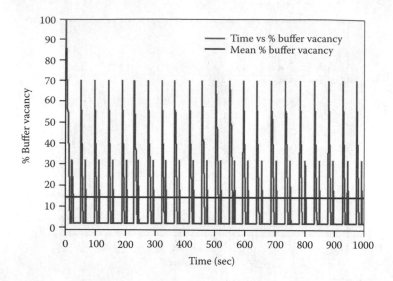

FIGURE 20.12
%Buffer Vacancy using heuristically based rules.

20.4.3.5   Heuristic vs. Control-Based Adaptation in ADSS

This evaluation illustrates how the percentage buffer vacancy (that is, the empty space in the buffer) varies over time for two scenarios; one in which only rules are used for buffer management, and the other in which rules are used in combination with controller inputs. Figure 20.12 plots the percent buffer vacancy for the first case. In this case, management was purely reactive and based on heuristics, and the element manager was not aware of the current and future data generation rate and the network bandwidth. It can be seen from the figure that the average buffer vacancy in this case was around 16%, i.e., in most cases 84% of the buffer was full. Such a high occupancy of data in the buffer leads to a slowdown in the simulation [5] and results in increased loss of data due to buffer overflows.

Figure 20.13 plots the corresponding percent buffer vacancy when the model-based controller was used in conjunction with rule-based management. It can be observed from the plot that the mean buffer vacancy in this case is around 75%. Higher buffer vacancy results in reduced overheads and data loss.

20.4.3.6   Overhead of the Self-Managing Data Streaming

Overheads on the simulation due to the self-managing data-streaming service are due primarily to two factors. The first are the activities of the controller during a controller interval. This includes the controller decision time, the

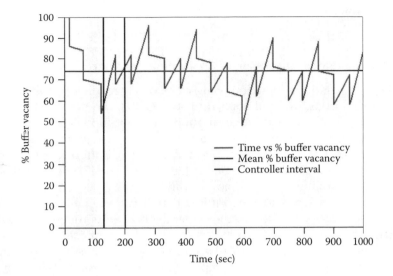

FIGURE 20.13
%Buffer Vacancy using control-based self-management.

cost of adaptations triggered by rule executions, and the operation of BMS and DTS. The second is the cost of the data streaming itself. These overheads are presented below.

Overheads due to controller activities: For a controller interval of 80 sec, the average controller decision time was ≈2.1 sec (2.5%) at the start of the controller operation. This reduced to ≈0.12 sec (0.15%) as the simulation progressed due to local search methods used. The network measurement cost was 18.8 sec (23.5%). The operating cost of the BMS and DTS was 0.2 sec (0.25%) and 18.8 sec (23.5%), respectively. Rule execution for triggering adaptations required less than 0.01 sec. The controller was idle for the rest of the control interval. Note that the controller was implemented as a separate thread (using *pthread* [31]) and its execution overlapped with the simulation.

Overhead of data streaming: A key requirement of the self-managing data streaming was that its overhead on the simulation be less than 10% of the simulation execution time. Percent overhead of the data streaming is defined as: $(\hat{T}_s - T_s)/T_s$, where $\hat{T}_s$ and T_s denote the simulation execution time with and without data streaming, respectively. The percent overhead of data streaming on the GTC simulation was less than 9% for 16–64 processors and reduced to about 5% for 128–256 processors. The reduction was due to the fact that as the number of simulation processors increased, the data generated per processor decreased.

20.5   Related Work

20.5.1   Rule-Based Adaptation of Application Behavior

Application/service adaptation using rule-based techniques was systematically studied in Accord and applied to objects [21], components [23, 33] and Grid services [22] for scientific applications and workflows. Active buffering, a buffering scheme for collective input/output (I/O), in which processors actively organize their idle memory into a hierarchy of buffers for periodic data output using heuristics, was studied in [28, 29]. RESAS [7] supports dynamic rule-based adaptation of real-time software and provides tools for programmers. Specifically, it provides algorithms to modify the reliability and/or timeliness of software without affecting other aspects of its functionality. A key challenge in rule-based adaptations is the generation of rules, which is typically manual. Correctness of rule-based management has been investigated for business applications using complex mechanisms based on databases [4] or business models [11], and in the security domain using types [27] as part of the policy specification process and using auctions [8] at runtime.

20.5.2   Control-Based Adaptation and Performance Management

Recent research efforts [13, 14] have investigated using feedback (or reactive) control for resource and performance management for single-processor computing applications. These techniques observe the current application state and take corrective action to achieve specified QoS, and have been successfully applied to problems such as task scheduling [9, 25], bandwidth allocation, and QoS adaptation in Web servers [3]; load balancing in e-mail and file servers [13, 24, 34]; network flow control [30, 38]; and processor power management [26, 37]. Feedback control theory was similarly applied to data streams and log processing for controlling the queue length and for load balancing [40]. Classical feedback control, however, has some inherent limitations. It usually assumes a linear and discrete-time model for system dynamics with an unconstrained state space, and a continuous I/O domain. The objective of the research presented in this chapter is to address this limitation and manage the performance of distributed applications that exhibit hybrid behaviors comprising both discrete-event and time-based dynamics [2] and execute under explicit operating constraints using the proposed LLC method. Predictive and change-point detection algorithms have been proposed for managing application performance, primarily to estimate key performance parameters such as achieved response time, throughput, etc., and predict corresponding threshold violations [39].

20.6 Conclusion

This chapter presented the design and implementation of a self-managing data streaming service that enables efficient data transport to support emerging Grid-based scientific workflows. The presented design combines rule-based heuristic adaptations with more formal model-based online control strategies to provide a self-managing service framework that is robust and flexible and can address the dynamism in application requirements and system state. A fusion simulation workflow was used to evaluate the data-streaming service and its self-managing behaviors. The results demonstrate the ability of the service to meet Grid-based data-streaming requirements, as well as its efficiency and performance.

Acknowledgment

The research presented in this work is supported in part by the National Science Foundation via grants numbers ACI 9984357, EIA 0103674, EIA 0120934, ANI 0335244, CNS 0305495, CNS 0426354, and IIS 0430826, and by the Department of Energy via the grant number DE-FG02-06ER54857.

References

1. S. Abdelwahed, N. Kandasamy, and S. Neema. A control-based framework for self-managing distributed computing systems. In *Workshop on Self-Managed Systems (WOSS'04)*, Newport Beach, CA 2004.
2. S. Abdelwahed, N. Kandasamy, and S. Neema. Online control for self-management in computing systems. In *10th IEEE Real-Time and Embedded Technology and Applications Symposium*, 368–376, Le Royal Meridien, King Edward, Toronto, Canada, 2004.
3. T. F. Abdelzaher, K. G. Shin, and N. Bhatti. Performance guarantees for Web server end-systems: A control theoretic approach. *IEEE Transactions on Parallel and Distributed Systems*, 13(1):80–96, 2002.
4. A. Abrahams, D. Eyers, and J. Bacon. An asynchronous rule-based approach for business process automation using obligations. In *Third ACM SIGPLAN Workshop on Rule-Based Programming (RULE'02)*, 323–345, Pittsburgh, PA, 2002. ACM Press.

5. V. Bhat, S. Klasky, S. Atchley, M. Beck, D. McCune, and M. Parashar. High performance threaded data streaming for large scale simulations. In *5th IEEE/ACM International Workshop on Grid Computing (Grid 2004)*, 243–250, Pittsburgh, PA, 2004.

6. V. Bhat, M. Parashar, H. Liu, M. Khandekar, N. Kandasamy, and S. Abdelwahed. Enabling self-managing applications using model-based online control strategies. In *3rd IEEE International Conference on Autonomic Computing*, Dublin, Ireland, 2006.

7. T. E. Bihari and K. Schwan. Dynamic adaptation of real-time software. *ACM Transactions on Computer Systems*, 9(2):143–174, 1991.

8. L. Capra, W. Emmerich, and C. Mascolo. A micro-economic approach to conflict resolution in mobile computing. In *Workshop on Self-Healing Systems (SIGSOFT'02)*, 31–40, Charleston, SC, 2002.

9. A. Cervin, J. Eker, B. Bernhardsson, and K. Arzen. Feedback-feedforward scheduling of control tasks. *Real-Time Systems*, 23(1-2):25–53, 2002.

10. J. Chen. *M3D Home*. http://w3.pppl.gov/%7ejchen, 2005.

11. J. J. Cheng, D. Flaxer, and S. Kapoor. RuleBAM: A rule-based framework for business activity management. In *IEEE International Conference on Services Computing(SCC'04)*, 262–270, Shanghai, China, 2004.

12. E. Christensen, F. Curbera, G. Meredith, and S. Weerawarana. *Web Services Description Language (WSDL) 1.1*. http://www.w3.org/TR/wsdl, 15 March 2001.

13. J. L. Hellerstein, Y. Diao, S. Parekh, and D. M. Tilbury. *Feedback Control of Computing Systems*. Wiley-IEEE Press, Hoboken, NJ, 2004.

14. J. L. Hellerstein, Y. Diao, and S. S. Parekh. *Applying Control Theory to Computing Systems*. Technical report, December 7, 2004.

15. N. Kandasamy, S. Abdelwahed, and J. P. Hayes. Self-optimization in computer systems via online control: Application to power management. In *1st IEEE International Conference on Autonomic Computing (ICAC'04)*, 54–61, New York, 2004.

16. M. Khandekar, N. Kandasamy, S. Abdelwahed, and G. Sharp. A control-based framework for self-managing computing systems. *Multiagent and Grid Systems, an International Journal*, 1(2):63–72, 2005.

17. S. Klasky, M. Beck, V. Bhat, E. Feibush, B. Ludscher, M. Parashar, A. Shoshani, D. Silver, and M. Vouk. Data management on the fusion computational pipeline. *Journal of Physics: Conference Series*, 16(2005):510–520, 2005.

18. S. Klasky, S. Ethier, Z. Lin, K. Martins, D. McCune, and R. Samtaney. Grid-Based Parallel Data Streaming implemented for the Gyrokinetic Toroidal Code. In *Supercomputing Conference (SC 2003)*, vol. 24, Phoenix, AZ, 2003.

19. Lawrence-Berkeley-National-Laboratory. *Energy Sciences Network*. http://www.es.net/, 2004.

20. Z. Lin, T. S. Hahm, W. W. Lee, W. M. Tang, and R. B. White. Turbulent transport reduction by zonal flows: Massively parallel simulations. *Science*, 281(5384):1835–1837, 1998.

21. H. Liu. *Accord: A Programming System for Autonomic Self-Managing Applications*. PhD thesis, Rutgers University, 2005.

22. H. Liu, V. Bhat, M. Parashar, and S. Klasky. An autonomic service architecture for self-managing grid applications. In *6th International Workshop on Grid Computing (Grid 2005)*, 132–139, Seattle, WA, 2005.

23. H. Liu, M. Parashar, and S. Hariri. A component-based programming framework for autonomic applications. In *1st IEEE International Conference on Autonomic Computing (ICAC-04)*, 10–17, New York, 2004.

24. C. Lu, G. A. Alvarez, and J. Wilkes. Aqueduct: Online data migration with performance guarantees. In *USENIX Conference on File Storage Technologies (FAST'02)*, 219–230, Monterey, CA, 2002.

25. C. Lu, J. A. Stankovic, S. H. Son, and G. Tao. Feedback control real-time scheduling: Framework, modeling, and algorithms. *Real-Time Systems*, 23(1-2):85–126, 2002.

26. Z. Lu, J. Hein, M. Humphrey, M. Stan, J. Lach, and K. Skadron. Control-theoretic dynamic frequency and voltage scaling for multimedia workloads. In *International Conference on Compilers, Architectures, & Synthesis Embedded Systems (CASES)*, 156–163, Grenoble, France, 2002. ACM Press.

27. E. C. Lupu and M. Sloman. Conflicts in policy-based distributed systems management. *IEEE Transactions on Software Engineering*, 25(6):852–869, 1999.

28. X. Ma. *Hiding Periodic I/O Costs in Parallel Applications*. PhD thesis, University of Illinois at Urbana-Champaign, 2003.

29. X. Ma, J. Lee, and M. Winslett. High-level buffering for hiding periodic output cost in scientific simulations. *IEEE Transactions Parallel Distributed Systems*, 17(3):193–204, 2006.

30. S. Mascolo. Classical control theory for congestion avoidance in high-speed internet. In *38th IEEE Conference on Decision and Control*, volume 3, 2709–2714, Phoenix, Arizona, 1999.

31. B. Nichols, D. Buttlar, and J. P. Farrell. *PThreads Programming: A POSIX Standard for Better Multiprocessing*. O'Reilly, Sebastopol, CA, 1996.

32. NLANR/DAST. *Iperf 1.7.0 : The TCP/UDP Bandwidth Measurement Tool*. http://dast.nlanr.net/Projects/Iperf/, 2005.

33. M. Parashar and J.C. Browne. Conceptual and implementation models for the Grid. *IEEE, Special Issue on Grid Computing*, 93(2005):653–668, 2005.

34. S. Parekh, N. Gandhi, J. Hellerstein, D. Tilbury, T. Jayram, and J. Bigus. Using control theory to achieve service level objectives in performance management. *Real Time Systems*, 23(1-2):127–141, 2002.

35. J. S. Plank, M. Beck, W. R. Elwasif, T. Moore, M. Swany, and R. Wolski. The Internet backplane protocol: Storage in the network. In *NetStore99: The Network Storage Symposium*, Seattle, WA, 1999.

36. J. S. Plank and M. Beck. The Logistical Computing Stack: A Design for wide-area, scalable, uninterruptible computing. In *Dependable Systems and Networks, Workshop on Scalable, Uninterruptible Computing (DNS 2002)*, Bethesda, Maryland, 2002.

37. V. Sharma, A. Thomas, T. Abdelzaher, K. Skadron, and Z. Lu. Power-aware QoS management in Web servers. In *Real-Time Systems Symposium*, 63–72, Cancun, Mexico, 2003.

38. R. Srikant. Control of communication networks. In T. Samad, ed., *Perspectives in Control Engineering: Technologies, Applications, New Directions*, 462–488. Wiley-IEEE Press, 2000.

39. R. Vilalta, C. Apte, J. L. Hellerstein, S. Ma, and S. M. Weiss. Predictive algorithms in the management of computer systems. *IBM Systems Journal*, 41(3):461–474, 2002.

40. X. Wu and J. L. Hellerstein. Control theory in log processing systems. In *Summer 2005 RADS (Reliable Adaptive Distributed systems Laboratory) Retreat*, 2005.

21

Autonomic Power and Performance Management of Internet Data

Bithika Khargharia and Salim Hariri

CONTENTS

21.1 Motivations and Introduction

As portable and hand-held devices shrink in size while adding more and more features, the biggest impediment to progress in this technology seems to be the power and energy consumption. Advanced features and improved performance comes at the cost of consuming increased power that requires increased battery capacity as well as the higher cooling needs of the device components. However, large battery capacity increases the device's size. Naturally, there has been extensive research dedicated towards low-power and low-energy design and conservation to further the mobile and hand-held technology [1–9]. The battery-size and hence the available battery capacity places an obvious upper limit on the available power budget for portable and hand-held devices.

However, the same bound has not appeared for bigger systems such as servers, switches, and interconnects, the main reason being that power is derived from the electricity grid. However, the rapid growth in Internet infrastructure and services has significantly increased the demand for power. Driven by the Internet economy, recent years have seen an almost insatiable demand for power by data centers that house thousands of dense servers within a relatively small real-estate. These data centers host different types of applications/services where each customer has a different Service Level Agreements (SLA) contract. It is a rather stark disparity between the distributed nature of the Internet economy supported by the nation's fiber optic backbone and the concentration of electricity demands at specific locations along the same fiber optic backbone [10].

Data centers also have the advantage of scaling out with standardized, commercial off-the shelf (COTS) dense, inexpensive resources [11]. This makes it a very appealing commodity for both the supplier and the consumer. While the consumer can use the data center services and applications without bothering about the cost of installing, tuning, and managing computer systems, the supplier or the data center owners make a good return on investment (ROI) by provisioning this infrastructure that can scale in an inexpensive manner to meet high demands, and thus enable large and mixed enterprise computing by supporting a myriad of services and applications on this infrastructure.

However, the large scale data centers have severely stressed the power grid networks and are posing a technical, financial, and environmental threat to the economy and the society at large. For example data centers host thousands of servers, storage, and switches densely packed to maximize floor space utilization [12]. With the current density levels of 200 watts per square foot for these servers, storages, switches, and data-center support for 24 × 7 could require 80 TWh per year, costing $8B per year at $100 per MWh and releasing 50 tons of CO_2 annually [13]. These figures hint at a number of issues related to data center growth and management.

i. First, high energy consumption leads to a high Total Cost of Ownership (TCO) for data center owners. For example, a white paper published by American Power Conversion (APC) [14] computes the TCO for a typical data center rack with the following characteristics:
Power rating: 100KW
Power density: 50W/sq ft
Life Cycle: 10 years
Average rock power: 1500W
Redundancy: 2N

ii. As can be seen from Figure 21.1, data centers require a significant capital investment in equipment for cooling and backup power generation and this cost scales with high demand for power [13]. Moreover, given a fixed budget, one dollar spent in cooling is one dollar not spent in a hardware that can actually produce a better service [15].

iii. The tightly packed server racks in data centers create high power density zones within the data center, making cooling both expensive

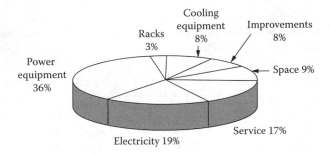

FIGURE 21.1
The pie chart plots [14] the electrical requirements for a typical data center.

and complicated. Insufficient cooling leads to intermittent failures of computing nodes. Furthermore, constraints on the amount of power that can be delivered to server racks makes energy conservation critical for fully utilizing the available space on these racks [13].

iv. Data centers typically bundle energy costs into the cost they charge consumers for hosting. With high energy costs, rack energy usage alone could account for up to 23% to 50% of the collocation revenue. Similar arguments apply for managed hosting as well, where the data center controls and owns the IT equipment and provides the service [12].

v. Excessive power demand leaves Internet services unnecessarily vulnerable to energy supply disruptions. For example, wholesale electricity spiked to $800 per MWh on the California spot market [13].

vi. High demand for power by data centers has caused a number of data center projects to be delayed or cancelled because most often the state is unable to meet the power demand. For example, the New York Times reported that the planned 46 data centers for the New York City metropolitan area alone have asked for a minimum of 500 megawatts of power, which is enough to power 500,000 households [16].

vii. Finally, electricity production also harms the environment. For example, excessive production of CO_2 from coal-burning power plants can significantly contribute to global warming [13].

American Power Conversion (APC) points out in one of their white paper publications [14] that the ideal data center would be "rightsized" and only incur the infrastructure costs that are actually required at a given time. To achieve the theoretically available cost savings, the ideal data center would only have the power and cooling infrastructure needed at the moment; it would only take the space that it needed at the moment, and it would only incur service costs on capital infrastructure capacity that was actually being used. It would be perfectly scalable. The fact that today's data centers are far from ideal is corroborated by the statistics referenced in [11] that say that there is about 3 watts of cooling and backup for every 1 watt of usage or average utilization of 20% with the remaining 80% over-provisioned for increased availability and rare peak loads.

The question then boils down to determine if it is possible to take traditional statically provisioned, power hungry data centers to a more dynamic infrastructure with the objective of minimizing power consumption but providing equally good quality of service. This would mean building a dynamic data center that can scale well to incoming traffic such that it handles peak traffic efficiently and is power efficient during off-peak hours.

Researchers have started exploring the field of power management related to servers, server-clusters, and data centers starting around 2000. However it heavily draws upon prior research on power and energy management in portable devices. As an example, IBM first made the case for power management in web servers indicating dynamic voltage scaling (DVS) as a means to save energy not only in embedded devices but also in high performance web servers [17]. Much of the earlier work in this area used server turn off/on mechanisms for power management using heuristic based algorithms to decide when and how many servers to turn on/off [13–15]. Other researchers have attempted to combine dynamic voltage scaling with server on/off for power management [18]. In references [19–20] QoS and performance management is considered in addition to power. Most of the techniques proposed to manage power were ad-hoc except a few recent techniques that have adopted mathematically rigorous optimization techniques [21–22].

The rest of this chapter is organized as follows. Section 2 provides a brief overview of existing power management techniques for both mobile devices and server systems. This section also features a discussion of the unique features of our research approach as compared to existing approaches. Section 3 introduces the Autonomic Computing Paradigm and the general notion of an *autonomic component*. Section 4 casts the data center power and performance management problem into the Autonomic Computing Paradigm. It discusses how we can build an Autonomic Data Center by using Autonomic Clusters, Autonomic Servers, and Autonomic device components such as Autonomic Processor, Autonomic Memory, and Autonomic IO. Section 5 discusses a case study of autonomic power and performance management for a three-tier data center. It presents a detailed discussion of the models and their formalisms. It also discusses the formulation of the optimization problem for power management while maintaining performance at the cluster and the server level. Section 6 presents the evaluation of the optimization algorithms as applied to a memory system case-study. Section 7 concludes the chapter with a summary of our research approach towards the data center power and performance management problem.

21.2   Overview of Power Management Techniques

We can categorize power management techniques into three broad classes as discussed in [13]: 1) Hardware based power management, 2) Turning-off unused device/system capacity, and 3) QoS and energy tradeoff.

21.2.1   Hardware Based Power Management

In this section we present power management techniques that can be applied to specific hardware technology. For example, processors with DVS (Dynamic Voltage Scaling) capability such as Transmeta's Crusoe processor [27] and Intel's Pentiums with SpeedStep [28] allow varying the processor voltage in proportion with its frequency. By varying the processor voltage and frequency, it is possible to obtain a quadratic reduction in power consumption [17].

This is a type of Dynamic Power Management (DPM) technique that reduces power dissipation by slowing or shutting down components that are idle or underutilized. Frequency-scaling, clock throttling, and DVS are three DPM techniques available on existing processors. Another example is the *Direct Rambus DRAM (RDRAM)* [29] technology that delivers high bandwidth (1.6 GB per sec per device) using a narrow bus topology operating at a high clock-rate. As a result each *RDRAM* chip in a memory system can be set to an appropriate power state independently. [30] uses multiple power modes of RDRAM and dynamically turns off memory chips with power-aware page allocation in operating systems. These studies suggest that future memory should support power management to turn on/off part of the memory.

21.2.2   Turning off Unused Device/System Capacity

A significant amount of research has been dedicated towards the goal of reducing the energy cost of maintaining surplus capacity. This is equally true for power management in the realm of portable devices as well as servers and server clusters. DPM (Dynamic Power Management) is one such very popular technique that reduces power dissipation by slowing or shuting down components that are idle or underutilized. As mentioned in the previous section, some DPM techniques require hardware support such as a DVS enabled processor. However, there are other DPM techniques that can work with any kind of hardware. In the battery-operated domain, researchers have targeted the power management of an entire range of devices from the processor, to cache, memory, disk, NIC, etc., based on DPM techniques. DPM schemes in the storage area cover three levels of storage cache, memory, and disk. The relationship between memory and disk is established by the fact that the smaller the memory size, the higher the page misses and this results in a higher disk access. This relationship can be used to achieve power savings by proactively changing disk IO by expanding or contracting the size of the memory depending on the workload [31]. However, lower miss rate does not necessarily save disk energy as shown by the work of [32]. Similarly, there has been work that addresses NIC energy consumption while achieving high throughput [33]. In addition there are a myriad of heuristic-based approaches that determine the idle time that a device should be in before it is switched to a low power mode [3–5]. The DPM technique has been widely and successfully applied to the server domain for power management. The fact that maximum energy savings can be gained from CPU power management has been established by an earlier work [17]. This has led to a number of techniques for power management that use DVS for the processor power management in conjunction with heuristics based software power management techniques.

Another technique that falls into this category of power management is to turn whole devices on/off dynamically in order to reduce the surplus capacity and hence the power consumption. This technique has been applied both in the battery-operated domain as well as in the server domain. Some of the policies for cluster-wide power management in server farms employ various combinations of DVS and node Vary-On/Vary-Off (VOVO) to reduce the aggregate power consumption of a server cluster during periods of reduced workload [18]. Another policy focuses on processor power-management and presents a request-batching scheme where jobs are forwarded to the processor in batches such that the response time constraint is met and the processor can remain in a lower power state for a longer period of time [12]. Yet another policy concentrates the workload on a limited number of servers in the cluster such that the rest of the servers can remain switched off. This leads to a load concentration to reduce the idle power consumed [15].

21.2.3   QoS and Energy Trade-Offs

Research that has focused on QoS trade-offs for energy savings has specifically tried to study the impact of power savings on performance. They have investigated if additional power savings can be obtained at the cost of QoS trade-off within acceptable limits. This has given rise to the application of proactive online optimization techniques as well as reactive control theoretic techniques being applied for power management while maintaining performance both in battery-operated and server domains. For example [6–9] have developed a myriad of stochastic optimization techniques for portable devices. In the server domain, [21] has presented three online approaches for server provisioning and DVS control for multiple applications — namely a predictive stochastic queuing technique, reactive feedback control technique, and another hybrid technique where they use predictive information for server provisioning and feedback control for DVS. [19] has studied the impact of reducing power consumption of large server installations subject to QoS constraints. They developed algorithms for DVS in QoS-enabled web servers to minimize energy consumption subject to service delay constraints. They use a utilization bound for schedulability of a-periodic tasks [20] to maintain the timeliness of processed jobs while conserving power. [22] investigates autonomic power control policies for Internet servers and data centers. They use both the system load and thermal status to vary the utilized processing resources to achieve acceptable delay and power performance. They use Dynamic Programming to solve their optimization problem. [34] addresses base power consumption for web servers by using a *power-shifting* technique that dynamically distributes power among components using workload sensitive polices.

Most of the work related to power management of servers has either focused on the processor or used heuristics to address base power consumption in server clusters. This motivates us to adopt a holistic approach for system level power management where we exploit the interactions and dependencies between different devices that constitute a whole computing system. We apply the technique at different hierarchies of a high-performance computing system similar to data centers where systems assume different definition at each hierarchy — going from device, to server, to server-cluster. This makes our approach more comprehensive as compared to [35]. They solve the problem of hierarchical power management for an Energy Managed Computer (EMC) System with self-power managed components (specifically IO) while exploiting application level scheduling. We adopt a mathematically rigorous optimization approach for determining optimal power states that a system can be in under the performance constraints. The closest to combining device power models to build a whole system has been presented in [36]. We use modeling and simulation initially for development and testing of our framework. This enables us to develop power management techniques reasonably unlimited by current hardware technologies lending us with important insights in terms of technology enhancements in existing hardware to exploit greater savings in power.

21.3　Autonomic Computing Paradigm

The increase in complexity, dynamism, and heterogeneity of resources in network centric systems in general and data centers in particular has made the control and management of such systems and their services a challenging research problem. Current state of the art control and management techniques are incapable of managing such systems and services. And therefore new approaches must be developed. One approach (Autonomic Computing) that is analogous to the biological nervous systems is emerging as a promising paradigm to efficiently address the control and management challenges of these systems and their services.

In previous work we have laid the foundation for an Autonomic Computing System [23]. An autonomic computing system is a system that has the capability to self-configure, self-protect, self-heal, and self-optimize its operations automatically at runtime to meet its overall system and service requirements. In this section, we will review the underlying principles of Autonomic Computing and our approach to use this paradigm to manage both power and performance in large scale data centers.

21.3.1　Notion of an Autonomic Component

An *autonomic component* (AC) is the smallest unit of an autonomic application or system. It is a self-contained software module or system with specified input/output interfaces and explicit context dependencies. It also has embedded mechanisms for self-management responsible for providing functionalities, exporting constraints, managing its own behavior in accordance with context and policies, and interacting with other autonomic components [23]. In summary, it is a system or application augmented with intelligence to manage and maintain itself under changing circumstances impacted by the internal or external environment. Figure 21.2 gives the structure of an *autonomic component*. As seen in Figure 21.2, the *managed element* is the traditional application or system to be managed. It has been extended by a *component*

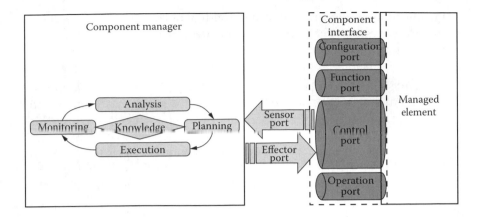

FIGURE 21.2
Any autonomic component consists of the managed element, the component interface, and the component manager.

interface to enable autonomic management by the *component manager*. Essentially, the *autonomic component* consists of two parts: i) *component interface* that defines the procedures to measure the operating states of the *managed element* as well as procedures to change the *managed element*'s configurations or operations at runtime, and ii) *component manager* that continuously monitors the *managed element*'s states, analyzes them, and executes the appropriate actions to maintain the component within its operational requirements. In the following sections we describe each element of an *autonomic component* in detail.

21.3.1.1   Managed Element

The *managed element* is an existing hardware or software resource that is modified to manage itself autonomously. Any *managed element* can be converted into an *autonomic component* by adopting the *component interface* through which the *component manager* manages the system or the software.

21.3.1.2   Component Interface

The *component interface* consists of four ports referred to as the *configuration port, function port, control port*, and *operation port*. The nature and type of information furnished by these ports depend on the self-management objective (self-optimization, self-healing, self-protection, or self-configuration). As shown in Figure 21.2, the *control port* consists of two parts — *sensor* and *effecter* ports. The *component manager* uses the *sensor* port of the *component interface* in order to characterize and quantify its current operational state by observing and analyzing the appropriate operational attributes specific to the self-management objective. The *component manager* also uses the *effecter port* to enforce the self-management decision in order to maintain the *managed element* within its operational requirements.

In what follows, we describe each individual port. In Section 4 we describe each port with specific examples related to the power and performance management objective.

i. **Operation Port**   This port defines the policies and rules that must be enforced to govern the operations of the *managed element* as well as its interactions with other *managed elements*. Consequently, this port defines two types of rules/policies — *behavior rules* and *interaction rules*. The *behavior rules* define the normal operational region for the *managed element*. *Interaction rules* define the rules that govern the interactions of the *managed element* with other *managed elements*.

ii. **Function Port**   The function port defines the control and management functions provided by the *managed element* that can be used to enforce the management decisions.

iii. **Configuration Port**   This port maintains configuration information related to the *managed element* such as its current configuration, its current activity level, etc. This information is used by the *control port sensor* to monitor the state of the *managed element*.

iv. **Control Port**   This is the "external port" that defines all the information that can be monitored and the control functions that can be invoked on the *managed element*. This *port* uses the policies and rules specified in the other ports (*function, operation*, and *configuration* port) to achieve the desired autonomic management behavior for the *autonomic component*. This port consists of the *sensor* port and *effecter* port. The *sensor* port defines the necessary monitoring parameters and measurable attributes that can accurately characterize the state of the *managed element* at any instant of time. The *effecter* port defines all the control actions that can be invoked at runtime to manage the behavior of the *managed element*.

21.3.1.3   *Component Manager*

The *component manager* is the augmented intelligence of the *autonomic component* to maintain itself under changing circumstances impacted by the internal or external environment [23]. In order to achieve the autonomic management objectives for the *autonomic component*, the *component manager* continuously cycles through four phases of operations — *monitoring, analysis, planning*, and *execution* as shown in Figure 21.2. In each phase, the *component manager* may use previously stored knowledge to aid this decision making process. The *component manager* controls and manages the *autonomic component* using the *component interface* ports. Note that different performance parameters and their acceptable values can be set a priori by the user or can be changed dynamically by the *component manager* of the managed resource or software module. Furthermore, we can also dynamically add new methods in the *sensor* and *effecter* ports.

21.4   Autonomic Power and Performance Management

In Section 3 we studied the general structure and mechanisms of an *autonomic component*. In this section we devise and instrument an autonomic component to autonomously manage its power and performance. We discuss how an *autonomic component* can be coupled together with other *autonomic components to build a complete autonomic system*. We use this as an archetypal system to build *autonomic data centers* from *autonomic clusters, autonomic servers*, and *autonomic devices* within a system such as *autonomic processor, autonomic memory*, and *autonomic IO*. In essence, we use these *autonomic components* as building blocks to build a fully-autonomic *data center*. In what follows, we describe each component of an *autonomic component* for power and performance management.

21.4.1   Power and Performance Managed Autonomic Component

An *autonomic component* that manages its power and performance under changing workload conditions has the same structure as that of a general *autonomic component* as shown in Figure 21.2. The *component manager* interacts with the *managed element* (processor, memory, IO, server, server cluster, or whole data center) through the equivalent *component interface sensor port* to continuously *monitor* its power consumption and performance, *analyses* and *plans* alternate strategies as necessary to minimize the power consumption while maintaining performance, and *executes* the strategy on the *managed element* using its *effecter port*.

21.4.1.1   *Managed Element — Processor, Memory, IO, Server, Server Cluster, or Data Center*

In our approach for data center power management we provide for power and performance management at every hierarchy of the data center ranging from the smallest device components, such as processor, memory, etc. to servers, server clusters, and eventually whole data centers. Thus, for autonomic power and performance management, the *managed element* assumes different definitions at different hierarchies of the high-performance data center. At the top-level, the *managed element* is the entire data center, at the next-level the *managed element* is a cluster within the data center, then the *managed element* is a server within the cluster, and at the bottom-most rung the *managed element* is a device within the server such as a processor, memory, disk, network card, etc.

We model the *managed element* at any hierarchy, as a set of states and transitions. Each state is associated with power consumption and performance values. A transition takes the *managed element* from one state to another. It is the task of the *component manager* to enable the *managed element* to transition to a state where power consumption is minimal without violating any performance constraints. These states and transitions are specific to the type and nature of the *managed element*. It exposes this information through the *component interface* that is used by the *component manager* to maintain the *managed element* in an optimal power state where power consumption is minimal while meeting all performance constraints.

21.4.1.2   Component Interface

We have already described the *component interface* for a generic *autonomic component*. In this section we describe each individual port with respect to the power and performance management objective using a front-end web server as an example *managed element*.

  i. **Operation Port**
     Given below is an example of the XML schema for the operation port of a front-end web server in a three-tier data center:

```
<operation_port>
        <component name="WEBSERVER">
        <operation type="behavior">
                <MAXCLIENTS> </MAXCLIENTS>
                <BASE_POWER> </BASE_POWER>
                <MAX_POWER> </MAX_POWER>
                <ACTIVE_POWER> </ACTIVE_POWER>
                <SLEEP_POWER> </SLEEPPOWER_POWER>
                <STANDBY_POWER> </ STANDBY_POWER>
         </operation>

        <operation type="interaction"
                <input type="PROCESS">
                        <job> r1 </job>
                </input>
                <input type="FORWARD">
                <job1> r2 </job1>
                        <to> APPSERVER </to>
                <job2> r3 </job2>
                        <to> DBSERVER </to>
                </input>
                <input type="PROCESS AND FORWARD">
                <job> </job>
                </input>
        </operation>
           ...
</operation_port>
```

     The *'behavior'* rule specifies the maximum number of clients that can be simultaneously supported by the web server as well as the base and maximum power consumption for the web server as specified by the manufacturer. As shown in Figure 21.6a, it also defines the power states and power consumption in each state for the web server. This is also along the lines of ACPI [25] power state definitions. The *'interaction'* rules specify the type of inputs that can be accepted by the web server. For example, job type 'PROCESS' is directly processed by the web server, job type 'FORWARD' is forwarded by the web server to an application server (APPSERVER) or a database server (DBSERVER) depending on the requirements of the request. There is also a third type of input, 'PROCESS AND FORWARD' that is first processed by the web server and then forwarded for further processing.

ii. **Function Port**

Given below is an example of the XML schema for the function port of the web server.

```
<function_port>
    <component name="WEBSERVER">
      <function name="CHANGE POWER STATE">
             <_toACTIVE> </_toACTIVE>
             <_toSTANDBY> </_toSTANDBY>
             <_toSLEEP> </_toSLEEP>
      </function>
      ...
</function_port>
```

The field <_toACTIVE> defines the specific set of actions that need to be taken in hardware in order to bring the web server from any power state to the 'Active' power state. The web server power states are shown in Figure 21.6a. For example, it may involve a call to an IPMI [26] method to implement this power state transition. The other fields can be similarly interpreted.

iii. **Configuration Port**

The configuration port for the web server maintains information about the current power state the web server is in, the total number of jobs (of each type) received and the average number of jobs processed by the web server in a fixed interval of time. This is a measure of the web server's activity. The XML schema for the configuration port is given below.

```
<configuration_port>
             <component name="WEBSERVER">
        <component attr_name="CURRENT POWER STATE" value=""/>
           <component attr_name="NUM JOBS IN"
                 type="r1" value=""
                 type="r2" value=""
                 type="r3" value=""/>
<component attr_name="NUM JOBS PROCESSED"
                 type="r1" value=""
                 type="r2" value=""
                 type="r3" value=""/>
</configuration_port>
```

iv. **Control Port**

The XML schema for the *control* port *sensor* and *effecter* for the web server are given below.

```
<control_port>
       <sensor>
              <function name="getPOWERSTATE">
                     <parameter1> NULL </parameter1>
                     <returnType> STRING </returnType>
               </function>
               <function name="getAVPROCESSEDJOBS">
                     <parameter1> JOBTYPE </parameter1>
                                  <type> STRING </type>
                     <returnType> DOUBLE </returnType>
              </function>
               <function name="getTOTALJOBSIN">
                     <parameter1> JOBTYPE </parameter1>
                                  <type> STRING </type>
                  <returnType> INT </returnType>
              </function>
       </sensor>
       ......
       <effector>
          <function name="setPOWERSTATE">
                     <parameter1> POWERSTATE </parameter1>
                           <type> STRING </type>
                  <returnType> ERRORCODE </returnType>
              </function>
       </effector>
</control_port>
```

The *sensor* port contains functions getPOWERSTATE, getAVPROCESSED-
JOBS, getTOTALJOBSIN to get the current power state, the average number
of jobs processed since the last time inquired, and the total number of jobs that
have arrived, respectively. These function calls internally use the information
provided by the *configuration* port to return results. Similarly, the *effecter* port
function, setPOWERSTATE, lets a certain power state be enforced upon the
web server as determined by the *component manager*. This function implements the desired power state transition by using the *function* port method
'CHANGE POWER STATE'.

21.4.1.3   *Component Manager*

In what follows, we will use the web server discussed in Section 4.1.2 as a
running example to explain the functionality of the *component manager*. In
the *monitoring* phase, the *component manager* invokes the 'getPOWERSTATE',
'getTOTALJOBSIN', and 'getAVPROCESSEDJOBS' methods provided by the
sensor port of the web server. In the *analysis* phase it uses the returned values
of 'getTOTALJOBSIN', and 'getAVPROCESSEDJOBS' method to predict the
number of new jobs and the number of processed jobs in the next observation
window. In the *planning* phase, the *component manager* uses previously stored

knowledge to determine if the predicted number of jobs processed is within the normal operational region for the web server. If the *component manager* determines that at the current power state the web server cannot maintain the desired *throughput* value preset for the web server, it determines an alternate optimal power state such that the web server does not fall below the threshold *throughput* value and yet the total power consumed by the web server is minimal. During the *planning* phase the *component manager* can arrive at this decision using different evaluation techniques such as heuristics, optimization techniques, control theoretic approaches, or game theory. In the *execution* phase, the *component manager* enforces the optimal power state on the web server by using the 'setPOWERSTATE' method provided by the *effecter* port.

21.4.2   Autonomic Data Center: Data Center of Autonomic Building Blocks

Figure 21.3 shows a typical three-tier data center. It consists of three tiers of clusters — *Front End Tier (FE)* with cluster of web servers, the *Mid Tier (MT)* with cluster of application servers, and the *backend Tier (BE)* with cluster of database servers. Each *managed system* in the data center, be it the whole data center, a server cluster within the data center, a single server within the cluster, or devices within a server such as processor, memory, and IO, is an *autonomic component*. Thus, there is a *component manager* associated with the *autonomic data center*, a *component manager* associated with each of the *autonomic FE clusters, the autonomic BE cluster* and *autonomic MT cluster*, a *component manager* associated with each *autonomic server* within an *autonomic cluster*, and *component manager*, associated with devices within an *autonomic server*. For ease of illustration we show the data center level *component manager*, cluster level *component managers*, and server level *component managers* in Figure 21.3. In essence, this introduces a hierarchical management relationship between the different *autonomic components* with *components* higher up in the hierarchy setting the bounds for the search space for the optimal power state. *Components* lower down the hierarchy search for optimal power states within those bounds. Thus, a *cluster-level component manager* prunes the search space based on its knowledge of the entire cluster and the corresponding *server-level component managers* perform an exhaustive search for the optimal state within the pruned space. In that sense, the cluster-level *component manager* comes up with a rough estimate based on the global knowledge of the whole cluster, whereas the server-level *component manager* refines that solution with local knowledge of the specific server. This hierarchical approach reduces the complexity of the search algorithm and makes the solutions more scalable and efficient for online use. In Figure 21.3, the *inner control loop* refers to this local control of the server-level *component manager* while the *outer control loop* refers to the global control performed by the cluster-level *component manager*. This hierarchical relationship exists across all the *autonomic components* at different hierarchies of the *autonomic data center*.

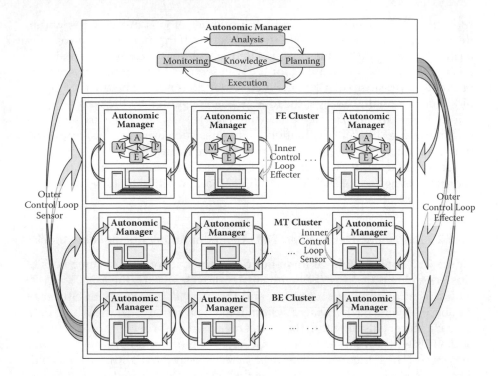

FIGURE 21.3

An autonomic data center is built using autonomic clusters, autonomic servers, and autonomic device components such as autonomic processor, autonomic memory, and autonomic IO. These autonomic components work together in a hierarchical fashion for power and performance management.

21.5   Modeling and Analysis of Hierarchical Autonomic Data Center for Power and Performance Management

In this section, we describe our modeling and simulation approach to analyze and evaluate the effectiveness of power and performance management strategies for an *autonomic data center*.

Figure 21.4 shows the model for the power and performance managed three-tier *autonomic data center*. We model the data center consisting of three distinguishable hierarchies: *i)*cluster level, where the whole data center is a collection of networked clusters, *ii)* server level, where each cluster is a collection of networked servers, and *iii)* device level, where each server is a collection of networked devices. A single unit of each hierarchy is modeled as an *autonomic component*. We use DEVS modeling and simulation [37] environment for executing the model.

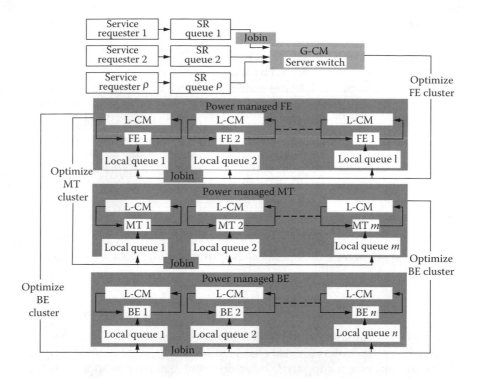

FIGURE 21.4
This is the model for a three-tier autonomic data center.

21.5.1 Discrete Event System (DEVS) Modeling and Simulation Framework

We model and simulate the *autonomic data center* using the DEVS modeling and simulation framework developed at the University of Arizona [37]. DEVS provides a sound modeling and simulation framework and is derived from mathematical dynamical system theory. It supports hierarchical, modular composition and reuse and can express discrete time, continuous, and hybrid models. DEVS allows for the construction of hierarchical simulation models composed of atomic and coupled models. Each atomic model is assigned to an atomic simulator, and atomic models as components within coupled models are assigned to a coupled simulator. Coupled models are assigned to coordinators, while coupled models as components within larger models are assigned to coupled-coordinator simulators. The simulators keep track of the events and execute the simulation model-defined methods based on the events list. Each component of our model shown in Figure 21.4 is formulated as a DEVS atomic model. A DEVS atomic model is a finite state machine and is formally defined as,

$$M_{Atomic} = \langle X, S, Y, \delta_{int}, \delta_{ext}, \lambda, t_a \rangle$$

where,
  X is set of inputs accepted by the model
  S is the set of states of the model
  Y is the output set generated by the model
  $\delta_{int} : S \rightarrow S$, captures internal state transitions for the model
  $\delta_{ext} : Q \times X \rightarrow S$, captures state transitions for the model in response
      to external inputs
  $\lambda : S \rightarrow Y$, is the output function that maps a state to an output from
      the output set.
  t_a : is the time-advance function for remaining in a state before an
      internal state transition occurs
  $Q = \{(s.e)|s \in S, 0 \le e \le ta(s)\}$ is the *total state* set
  e is the elapsed time since the last transition

In what follows, we discuss each atomic model of Figure 21.4 in detail and
how they work together to maintain the power and performance of the entire
data center.

21.5.2   Hierarchical Autonomic Power and Performance Management

To explain our model, we consider a homogenous cluster of four servers at
each tier. A data center can also receive different types of requests requiring
different application support and/or different database support. For example, a request for a specific web page can be serviced directly from the web
server's cache if available. Another request, for example, the number of books
written by a certain author in a book-selling website, requires the web server to
query the database server to retrieve the result and send it back to the client.
The database server that services this request in the back-end may be an
Oracle server or a SQL server. Similarly, a request may require PhP application support to service it. To capture this wide range of request types, our
model takes into consideration different types of applications and databases.
Depending on the type of the request coming in it can be processed in different
ways. This is shown in Figure 21.5.

  i. r_1: Request processed by FE web server.
  ii. r_2: Request forwarded by FE web server to the MT application server
      for processing.
      There are two sub-types of this request type:
      $r_{2.1}$: Request processed by MT application server.
      $r_{2.2}$: Request further forwarded by MT application server to the BE
      database server for processing.
  iii. r_3: Request forwarded by FE web server to the BE database server
      for processing. There is one sub-type of this request type.
      $r_{3.1}$: Request processed by BE database server.

Each server can be individually in one of the three power states as shown
in Figure 21.6a. The *active* state contains five sub-states — *active-A1, active-A2,*

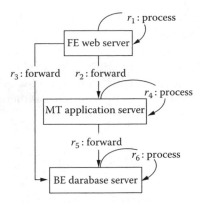

FIGURE 21.5
Different types of requests arriving into a data center.

active-A3, active-A4, active-A5. Each sub-state corresponds to a particular operating voltage and frequency of a DVS enabled processor. The states and transitions for the cluster at each tier are shown in Figure 21.6b.

21.5.2.1 Service Requester

The Service Requester in Figure 21.7 simulates a client sending requests to the data center. Since there can be multiple concurrent clients generating requests, our model contains multiple Service Requesters. Each Service Requester has a different Service Level Agreement (SLA) with the data center. The Service Requester is modeled based on [38]. Each Service Requester opens a session with the data center.

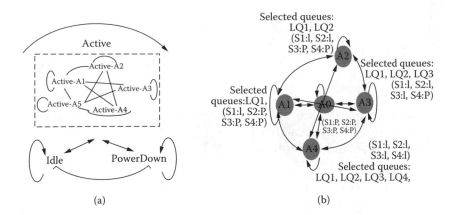

FIGURE 21.6
a) The power states of a single server with multiple sub-states within the power state 'Active'. Each sub-state corresponds to a voltage and frequency value for the processor. b) The global power states for a server cluster consisting of four servers. We only consider the 'Idle' and 'PowerDown' states for a single server within the cluster.

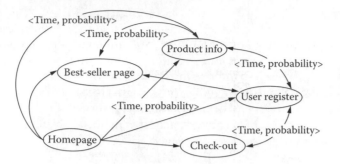

FIGURE 21.7

Figure shows an example of a client modeled as a service requester clicking on links in a book-selling web site.

The session remains alive for a period of time, called *session time*, at the end of which the connection is closed. Each session is a persistent HTTP session. Using this connection the Service Requester repeatedly makes a request, parses the server's response to the request, and makes a new request based on the parsed response. The 'thinking time' of the real client is emulated by the period between two simultaneous requests generated by a Service Requester. An example for a book-selling web page is shown in Figure 21.7. In our model we consider three Service Requesters, each representing a single connection, and the 'thinking time' is simulated as request inter-arrival time for a single Service Requester. The jobs generated by the Service Requester are stored in the respective Service Queue. The DEVS atomic model formulation for the Service Requester is as follows

$$M_{PM} = \langle X, S, Y, \delta_{int}, \delta_{ext}, \lambda, t_a \rangle$$

$S = \{active\}$
$X = \{\}$
$Y = \text{jobId}$
$\delta_{int} : S \rightarrow S$
Transitions from '*active*' to '*active*' after time t_a of 'jobInterArrivalTime'
$t_a = \text{'requestInterArrivalTime'}$
$\lambda : S \rightarrow Y$, is the output function

21.5.2.2 *Service Queue*

A Service Queue stores incoming jobs (requests) if there are free slots in its internal queue. Jobs that could not be stored in the Service Queue are treated as lost and the Service Queue measures the *request loss* as a performance parameter. Also, the amount of time that a job waits in the queue before it is processed is measured by its *wait time*. The Service Queue thus measures the *average wait time* of jobs in the queue. The SLA of the Service Requester determines the threshold values for *request loss rate* and *average wait time* for each Service Requester. For example, a premium level Service Requester has a smaller threshold *average wait time* and smaller threshold *request loss rate* as compared to other less important clients. We could also use other performance parameters here (e.g average connections rate for the data center, etc). Once a job is processed and acknowledged by a server, it is deleted from the respective Service Queue creating space for new incoming jobs. There is one such Service Queue associated with each Service Requester. The DEVS atomic model formulation for the Service Queue is as follows

$$M_{Queue} = \langle X, S, Y, \delta_{int}, \delta_{ext}, \lambda, t_a \rangle$$

$S = \{wait, out\}$,
$X : \{jobIn, jobAck, reqState\}$
$\qquad Y : \{jobId, (requestLoss, avWaitTime)\}$
$\delta_{ext} : S \times X \to S$,
In response to input 'jobId' it stores the jobId in the queue if the queue is not full.
In response to input 'jobAck' it deletes the job from the queue.
In response to input 'reqState' it sends the pair (*requestLoss, avWaitTime*).
$\delta_{int} : S \to S$,
is the transition from *out* to *out* after time t_a which is the time taken to send the output is the immediate transition from *out* to *wait* where time t_a is zero
t_a : is the time taken to send an output while in phase '*out*', or time taken to transition to phase '*wait*' which is zero.
$\lambda : S \to Y$, is the output function

21.5.2.3   Local Queue

The Local Queue stores jobs for a FE Web Server, an MT App Server, or a BE Database Server to be processed. The Local Queue receives jobs from the Service Queues based on the global power state decision of the Global Component Manager. It keeps forwarding jobs to the associated server for processing. It deletes a job from the queue once it receives an acknowledgement for that job. The Local Queue has a finite queue length and cannot receive new jobs if the queue is full. This is treated as a request loss. Just like the Service Queue, the Local Queue measures the *request loss* and *average wait time* for jobs in the queue. These two parameters are used as performance parameters by the Local Component Manager when optimizing the corresponding server for power and performance. The DEVS atomic model formulation for the Local Queue is similar to the Service Requester. There is a Local Queue associated with each such server.

### 21.5.2.4	*FE Web Server*

The FE Web Server processes client requests. It waits in the initial phase of *'pwrDown'* as long as its Local Queue is empty. Once the Local Queue forwards jobs to the FE Web Server for processing, it holds in phase *'transitioning'* for the power state transition time and then holds in *'processing'* for the processing time of the incoming job, and finally passivates in phase *'idle'* waiting for new jobs. The *'idle'* phase is when the FE Web Server is powered on but not doing any work and the *'active'* phase is when it is processing requests. If the FE Web Server is already in the *'idle'* phase and receives incoming jobs, it immediately transitions to *'active-A1'*, holds in *'processing'* for job processing time, and finally passivates in phase *'idle'*. This is because we consider the transition time between *'idle'* and *'active-A1'* as zero. After processing a request, the FE Web Server outputs an acknowledgement for the processed job by sending the job id out to the Local Queue and the Service Queue. If the FE Web Server receives an input command from the Local Component Manager, and it is in any of the *'active-A1'*, *'active-A2'*, *'active-A3'*, *'active-A4'*, *'active-A5'*,*'idle'*, or *'pwrDown'* states it will hold in *'transitioning'* for the transition time and then waits in the target phase *'active-A1'*, *'active-A2'*, *'active-A3'*, *'active-A4'*, *'active-A5'*,*'idle'*, or *'pwrDown'*) depending on the command received from the Local Component Manager. The FE Web Server does not receive any inputs if it is either in phase *'processing'* or phase *'transitioning'*. Occasionally, the Local Component Manager can query the FE Web Server its current power state. The FE Web Server transitions to the *'computeState'* phase, holds there for a certain time, and then outputs its current power state to the Local Component Manager.

The FE Web Server can be formally defined as a DEVS atomic model as follows:

$$M_{FE-WEBSERVER} = \langle X, S, Y, \delta_{int}, \delta_{ext}, \lambda, t_a \rangle$$

S: {*'active-A1'*, *'active-A2'*, *'active-A3'*, *'active-A4'*, *'active-A5'*, *'idle'*, *'pwrDown'*, *computeState, transitioning, processing*}

X :{{$d|d \in d_{PM_{FE-WEBSERVER}}$}, *job_id, reqState$_{FE-WEBSERVER}$*}, where d is a set of decisions output by the Local Component Manager,

where,

$d_{PM_{FE-WEBSERVER}} = \{go\_active\,A1, go\_active\,A2, go\_active\,A3, go\_active\,A4,$
$go\_active\,A5, go\_idle, go\_pwrdown\}$

job_id is the id of the job sent from the job queue
reqState$_{FE-WEBSERVER}$ is a request from the Local Component Manager to get the current state

Y: {sendState$_{FE-WEBSERVER}$, jobAck}
$\delta_{ext} : S \times X \rightarrow S$,
in response to input d it transitions to the target phase *active-A1'*, *' active-A2'*, *' active-A3'*, *' active-A4'*, *' active-A5'*, *' idle'*,*' pwrDown'*, effected by the input $x \in X$

In response to $reqState_{FE-WEBSERVER}$ it transitions to *computeState* and then sends its current power state to the Local Component Manager

In response to job_id it transitions to *transitioning* (if not in the 'active' phase) and then *processing*. At the end of *processing* phase it outputs the job id as an acknowledgement to the Local Queue and Service Queue

$\delta_{int} : S \rightarrow S$,

is the transition from *transitioning* to *processing* after time t_a which is the transition time

is the transition from *processing* to *active* after time t_a which is the job processing time

is the transition from *computeState* to *active* after time t_a which is the time taken to compute state.

t_a :transition time, computeState time or execution time for job currently processed.

21.5.2.5 Cluster-Level Power and Performance Optimization

The cluster-level power and performance optimization is performed by the Global Component Manager.

21.5.2.5.1 Global Component Manager

In a three-tier data center, all client requests are first addressed by the web server cluster in the FE tier. Thus, the task of the Global Component Manager is to determine the optimal power state for the entire FE tier cluster such that the *global request loss rate* RL_M^G and *global average wait time* W_M^G constraints are met. It achieves this objective by solving the optimization problem with respect to the predicted *global request loss rate* $RL_M^G|_{pred}$ and predicted *global average wait time* $W_M^G|_{pred}$ constraints for the next n cycles where n is a configurable parameter. Now, given that there are multiple Service Requesters the *request loss rate* RL_M and the *average wait time* W_M correspond to each Service Requester. That would mean solving the optimization problem for each Service Requester where RL_M^G and W_M^G values are replaced by the RL_M and W_M values for that Service Requester. However, we use the algorithm (shown below) to bring down that number to a single pair of values by using the following logic. If $B_M^G|_{pred}$ and $R_M^G|_{pred}$ are lower than the threshold values B_M and R_M for each Service Requester, then the FE tier cluster as a whole can be brought to a *lower* power state. The global performance constraints B_M^G and R_M^G are dictated by the Service Requester that has the smallest values of R_M and B_M. Similarly, if $B_M^G|_{pred}$ and $R_M^G|_{pred}$ are higher than the threshold values B_M and R_M for each Service Requester, then the FE tier cluster as a whole can be brought to a *higher* power state. The global performance constraints B_M^G and R_M^G are dictated by the Service Requester that has the smallest values of R_M and B_M. Notice that the most premium customer has the lowest values for R_M and B_M. Thus we let the most premium customer dictate the global performance constraints in either case. Now, if the predicted values $B_M^G|_{pred}$

and $R_M^G|_\text{pred}$ are lower than the threshold values for some Service Requesters and higher for others, the FE tier cluster as a whole can be brought either to a *lower* power state or to a *higher* power state. The Global Component Manager chooses to do the latter and optimizes the FE tier cluster to operate at a *higher* global power state. In this case, the global performance constraint values B_M^G and R_M^G are dictated by the Service Requester that has the smallest values of R_M and B_M amongst the ones that have the predicted values $B_M^G|_\text{pred}$ and $R_M^G|_\text{pred}$ higher than the threshold values R_M and B_M. This results in a significant reduction in the number of optimization equations that need to be solved at the beginning of a prediction cycle. Mathematically,

Case I: $\underset{\forall i:1:p}{SR}$, if $(R_{M_i}|_\text{pred}, B_{M_i}|_\text{pred}) \leq (R_{M_i}, B_{M_i})$ then $R_M^G = R_{M_1}, B_M^G = B_{M_1}$

Case II: $\underset{\forall i:1:p}{SR}$ if $(R_{M_i}|_\text{pred}, B_{M_i}|_\text{pred}) \geq (R_{M_i}, B_{M_i})$ then $R_M^G = R_{M_1}, B_M^G = B_{M_1}$

Case III: $\underset{\substack{i \in \{1, 2...p\} \\ j \in \{1, 2..p\}}}{SR}$ if $(R_{M_i}|_\text{pred}, B_{M_i}|_\text{pred}) \geq (R_{M_i}, B_{M_i})$ and

if $(R_{M_i}|_\text{pred}, B_{M_i}|_\text{pred}) \geq (R_{M_j}, B_{M_j})$ and

$R_{M_i} \leq R_{M_j}, B_{M_i} \leq B_{M_j}$

then $R_M^G = R_{M_i}, B_M^G = B_{M_i}$

To give an example, let us assume that the current state of the FE tier cluster is $A1$ as shown in Figure 21.6b, where $S1$ is in state *active (A)* and $S2$, $S3$, $S4$ are in the *sleep (S)* state. The Global Component Manager predicts the R_M and B_M values for the next n cycles for each Service Requester. It solves the optimization equation given below with the constraints determined as described above. Let us presume that the optimal power state is determined to be $A2$ where $S1$ and $S2$ are in state *idle (I)* and $S3$ and $S4$ are in state *powerDown (P)*. The *Server Switch* implements this global state, by forwarding jobs from the Service Requesters to the Local Queues of servers $S1$ and $S2$. It does not forward any jobs to the Local Queues of $S3$ and $S4$. Thus, $S3$ and $S4$ will remain in the *powerDown (P)* state because their Local Queues are empty. There can be many variants of the order of selecting jobs from the Service Queues. It could be fair-share where the *Server Switch* forwards jobs from each Service Queue one after another. It could also give priority to the premium customers by forwarding jobs from their Service Queue first before moving to another Service Requester. We use the second scheme in our approach. The DEVS atomic model formulation for the Global Component Manager is as follows

$$M_{G-APPM} = \langle X, S, Y, \delta_\text{int}, \delta_\text{ext}, \lambda, t_a \rangle$$

$S = \{reqState, computeState\}$

$X = \{powerState_{\text{CLUSTER}}, (requestLoss, avWaitTime)\}$

$Y = reqState_{\text{serviceQueue}}, reqState_{\text{cluster}}, \{d \,|\, d \in d_{PM}\}, \{do\_A1, do\_A2, do\_A3,$
$do\_A4, do\_A5\}$, where $\{d \,|\, d \in d_{PM}\} = \{do\_A1, do\_A2, do\_A3, do\_A4,$
$do\_A5\}$ relative to the global cluster power states as shown in
Figure 21.6b.

$\delta_{ext} : S \times X \to S$

In response to input belonging to the input set X, predicts the *requestLoss*
and *avWaitTime* for the whole observation cycle, and then determines the
optimal power state for the cluster. The optimal state is determined by solving
an optimization equation, **PO**$_{\text{CLUSTER}}$ (shown below), heuristics or graph-
theoretic approaches such as shortest-path algorithms.

$$\delta_{\text{int}} : S \to S$$

Transitions from *'computeState'* to *'reqState'* after time t_a of
'timeToDecision'
Transitions from *'reqState'* to *'reqState'* after time t_a of 'observationCycle'
$t_a =$ 'timeToDecision' or 'observationCycle'
$\lambda : S \to Y$, is the output function

21.5.2.5.2 Cluster-Level Policy Optimization

Every t_{obs} observation cycle the Global Component Manager computes the
optimal power and performance state for the FE tier cluster. The optimal
decision is based on the current state of the FE tier cluster and the predicted
global request loss rate $B_M^G|_{\text{pred}}$ and predicted *global average wait time* $R_M^G|_{\text{pred}}$
constraints for the next interval. We depict the current decision interval as
the ith time interval. Assuming the cluster is in global state A_j, the Global
Component Manager determines the optimal state such that the sum of the
transition power cost c_{jk} from state A_j to A_k and the power consumption in
the new state p_k is minimum subject to two performance constraints — the
average wait time w_k and the *request loss rate* b_k in state A_k. The equation for
the transition time $t_{transjk}$ from state A_j to A_k follows from the fact that we
consider the cluster has reached a global state only after all the web server
atomic models have transitioned to the desired local states. We formulate
the optimization problem as an integer problem with decision variables x_{jk},
where j and k refer to states A_j and A_k, respectively.

Transition cost from global states A_j to A_k,

$$t_{trans_{jk}} = \underset{k:1 ton}{Max}[t_{trans_{jk}}]$$

PO$_{\text{CLUSTER}}$: Minimize energy consumption for i^{th} interval,

$$e_i = \sum_{k:0}^{n} (c_{jk} * t_{trans_{jk}} + p_k * t_{obs}) * x_{jk}$$

such that,

$$\sum_{k:0}^{n} w_k * x_{jk} <= R_M^G$$

$$\sum_{k:0}^{n} b_k * x_{jk} <= B_M^G$$

$$\sum_{k=0}^{n} x_{jk} = 1$$

$$\overset{n}{\underset{k:0}{\forall}} \, x_{jk} = 0|1$$

where,

R_M^G : global threshold for *average wait time*
B_M^G : global threshold value for *request loss rate*
w_k : average wait time in power state A_k
b_k : request loss rate A_k

$$w_k = \left[(r_{i-1} + \hat{\lambda}_i * t_{obs}) - \frac{t_{obs}}{(t_{trans_{jk}} + \hat{t_{proc_{ik}}}) * x_{jk}} \right] - qLen$$

$$b_k = \left(t_{trans_{jk}} + \frac{\sum_{p=1}^{N_i} [(p-1) * t_{proc_{ik}}]}{N_i} \right) * x_{jk}$$

$\hat{\lambda}_i$: predicted average request arrival rate for i^{th} interval
$\hat{t_{proc_{ik}}}$: predicted average job processing time for i^{th} interval in power state k

$\hat{t_{proc_{ik}}} = \frac{\hat{t_{proc_i}}}{n_s}$, where $\hat{t_{proc_i}}$ is the predicted average job processing time for i^{th} interval, n_s is the number of servers 'active' in state k

N_i : $\left[(r_{i-1} + \hat{\lambda}_i * t_{obs}) - \frac{t_{obs}}{(t_{trans_{jk}} + \hat{t_{proc_{ik}}}) * x_{jk}} \right]$ is the total number of jobs in queue at the end of the ith time interval

r_{i-1} : jobs in queue from $(i-1)^{th}$ time interval
n : total number of power states
p : the position of a job in the job queue
$qLen$: length of the queue

For the power managed FE tier,
Decisions d:$\{d|d \in d_{PM}\}$, $\{do\_A1, do\_A2, do\_A3, do\_A4, do\_A5\}$
States A_i: and A_j: $\in S_{FE-CLUSTER}$ and $S_{FE-CLUSTER} = \{A1, A2, A3, A4, A5\}$
In a similar manner, the power and performance optimizations are performed for each of the MT and BE server clusters.

21.5.2.6   *Server-Level Power and Performance Optimization*

The server-level power and performance optimization is performed by the Local Component Manager.

21.5.2.6.1 *Local Component Manager*

The Local Component Manager behaves similar to the Global Component Manager discussed in Section 4.2.5. While the Global Component Manager optimizes the cluster for power and performance, the Local Component Manager optimizes a single server within the cluster for power and performance. For example, if the Global Component Manager determines the optimal state for the cluster to be A1 where one server is in state 'Idle' (I) and the other servers are in state 'powerDown' (P) then the Local Component Manager for the 'Idle' (I) server S1will solve another optimization problem to determine the optimal state for the server based on its power states and transitions as depicted in Figure 21.6a. In solving the optimization equation it uses the performance values specific to the Local Queue. The DEVS atomic model formulation for the Local Power Manager is as follows

$$M_{PM} = \langle X, S, Y, \delta_{\text{int}}, \delta_{ext}, \lambda, t_a \rangle$$

$S = \{reqState, computeState\}$
$X = \{powerState_{SERVER}, (requestLoss, avWaitTime)\}$
$Y = reqState_{localQueue}, reqState_{SERVER}, \{d|d \in d_{LPM}\}$,

where $\{d|d \in d_{LPM}\}$= {doActive-A1, doActive-A2, doActive-A3, doActive-A4, doActive-A5, doIdle, doPwrdown} relative to power states of a server as shown in Figure 21.6a.

$$\delta_{ext} : S \times X \to S$$

In response to input belonging to the input set X, predict the *requestLoss* and *avWaitTime* for the whole observation cycle, and then determine the optimal power state for the server. The optimal state is determined by solving an optimization equation, heuristics or graph-theoretic approaches such as shortest-path algorithms.

$$\delta_{\text{int}} : S \to S$$

Transitions from '*computeState*' to '*reqState*' after time t_a of 'timeToDecision'
Transitions from '*reqState*' to '*reqState*' after time t_a of 'observationCycle'
t_a ='timeToDecision' or 'observationCycle'
$\lambda : S \to Y$, is the output function

21.5.2.6.2 *Server-Level Policy Optimization*

The optimization equation for the Local Component Manager looks similar to that of Global Component Manager. The only difference is that the performance values are obtained from the Local Queue and the power states and transitions correspond to that of a single server as depicted in Figure 21.6a.

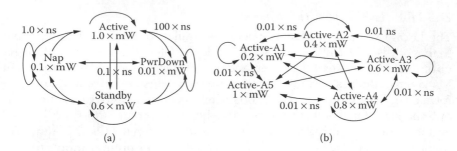

FIGURE 21.8

a) Figure shows the power states and transitions for an RDRAM memory module [ref]. b) Figure shows the power states and transitions for a hypothetical RDRAM module with five memory banks. Each state corresponds to number of active memory banks.

21.6   Evaluation of the Autonomic Power and Performance Management Technique for a Power-Aware Memory System

In this section we apply our autonomic management technique to collectively manage the power and performance of a power-aware memory system consisting of four *DRAM* chips. This case-study is motivated by the work performed by [30]. The power-aware techniques are based on the technology of *Direct Rambus DRAM (RDRAM)* [29].This technology delivers high bandwidth (1.6 GB per sec per device) using a narrow bus topology operating at a high clock-rate. As a result each *RDRAM* chip can be set to an appropriate power state independently. Each *RDRAM* chip supports several different power modes — *active, standby, nap, and power-down* in order of decreasing power consumption but increasing access time. The power management strategies adopted here make the assumption that data has been replicated across all the memory modules. This avoids the constraint on turning on/off RDRAM modules because each module can service all data requests. This helps us simplify the model to elucidate our approach better. As discussed in Section 4, we perform a global power optimization for the entire memory system and a local optimization for each RDRAM memory module. The modeling and simulation was performed using the DEVS modeling and simulation environment.

21.6.1   Static vs Dynamic Optimization Policies under Uniform Workload

Figure 21.9a compares the performance of the dynamic power optimization policy compared to the static policy where the RDRAM modules are always maintained in the '*active*' state. The Global Component Manager computes its decision for the optimal state given the predicted wait time and the predicted

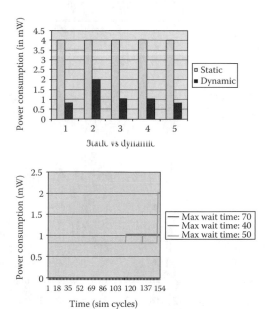

FIGURE 21.9

a) Figure shows a comparison of power consumed for static (no power management) vs dynamic optimization policies [29]. b) Figure shows the impact on power savings as we change the threshold values for the power optimization algorithm.

request loss values once every 20 simulation cycles. The values shown in Figure 21.9a are captured at the end of each decision period. On an average there is a savings of around 72%. However, these results depend on a number of parameters, such as the accuracy of the prediction mechanism, choice of cut-off values for the performance parameters R_M^G and B_M^G, and choice of the Global Component Manager's time to re-compute its decision. The impact of changing the value of R_M^G is shown in Figure 21.9b. As expected, with a maximum wait time of 70 simulation cycles, the RDRAM system transitions from state '$A5$' to state '$A1$' about 20 cycles later than with a maximum wait time of 50.

This means that the system can be in a low power state for a longer time while maintaining the performance values and thus leading to a greater savings in power. Applying the same concept to the server world means that we choose a realistic value of these performance parameters such that the servers operate at low power states as long as possible while meeting the performance constraints.

21.6.2 Static vs Dynamic Optimization Policies under Random Workload

The upper graph in Figure 21.10 shows the behavior of the workload generator and the lower graph shows the power consumption of the system as guided by the Global Component Manager. The dotted lines show that as the frequency

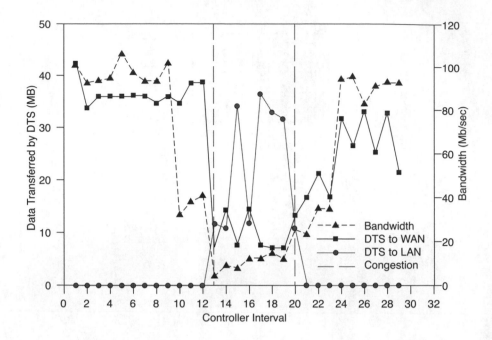

FIGURE 21.10

Figure shows impact on power consumption under a random workload. It can be seen from the figure that as the workload remains constant the system stays in the same power state (marked by circles 1 and 2). It also displays highly adaptive behavior towards changing workload.

of job generation increases the power consumption increases. However, in response to the second dotted section, the system is maintained at the same low power state.

21.6.3 Hierarchical Dynamic Optimization Policies

In Section 5.1, we make the Local Component Managers inactive. Thus, it is the Global Component Manager that performs optimization decisions. The individual RDRAMs may either remain in *'active'* or *'powerdown'* states. In this Section we demonstrate that additional power savings can be obtained by performing local optimizations together with global optimizations. We consider five hypothetical memory banks within one RDRAM module. Thus, we sub-divide the *'active'* state of an RDRAM module into five sub-states, corresponding to the number of banks ON at any given time. When the Global Component Manager determines the RDRAM to be *'active'*, the RDRAM may be in one of the five possible sub-states within the *'active'* state. The sub-states are shown in Figure 21.8b. The power consumption of each bank is computed equal to active state power consumption divided by the number of banks. In Figure 21.11, the Global Component Manager optimization determines state *'A1'* as optimal, where one RDRAM is in the *'active'* state. The power consumption in that state is uniformly 1.0 mW. The Local Component

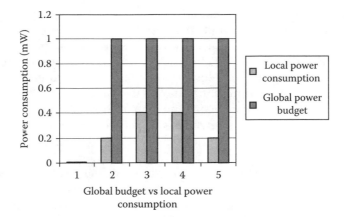

Global budget vs local power
consumption

FIGURE 21.11
Figure shows the reduction in power consumption due to local power optimizations. Local power optimizations lets an RDRAM operate at a lower power level then expected by the Global Power Manager.

Manager refines this decision and determines a local optimal sub-state with the *'active'* state among the five possible states *Active-A1, Active-A2, Active-A3, Active-A4, Active-A5*. This leads to an average additional savings of 69.8%
These results demonstrate a number of interesting concepts:

i) The utility of hardware features (such as supporting power states shown in Figure 21.8b) for exploiting greater power savings while maintaining performance.

ii) The hierarchical approach enables the Global Component Manager to optimize its solutions based on a smaller number of the local states (*'active'* and *'pwrdown'* in this case) among eight possible local states. The local optimization searches for an optimal state within the bounds of the global state. Thus the hierarchical approach reduces the search space for both global and local optimization.

iii) These results are interesting in terms of applying the framework for server cluster power management. The hierarchical approach lends itself well to handling the state explosion problem in the search spaces for global and local when performing the optimizations.

21.7 Conclusion

While most of the earlier research related to server and data center power management has dealt with the major objective of replacing static configurations with more dynamic and adaptive configurations of data center infrastructure, we attempt to standardize these efforts by putting into place a theoretical framework and general methodology for online power and performance management. We have already laid the foundation of the Autonomic Computing Paradigm modeled after the adaptive human autonomic nervous system [23]. We solve the data center power and performance management problem using the autonomic computing methodology.

We model power and energy consumption from a system level perspective that involves the complex interactions of different classes of device components such as processor, memory, network, and I/O. Current research focuses on a single component such as processor, or memory or disk for power savings. A system level power consumption perspective would present more opportunities for power savings since we can exploit the non-mutually exclusive behaviors of these components to set them at power states such that the global system power consumption is minimal. In addition this also enables the sharing of a fixed power budget between device components ensuring the same quality of service level while staying within the power budget.

Using the modeling and simulation methodology provides many advantages. We can experiment with power management techniques for a myriad of systems and devices without being hindered by limitations of what existing technology has to offer in terms of power savings. This can potentially lead to new directions for technology enhancement such as supporting new features in the hardware to enable additional power savings.

While most of the existing research has looked at servers and server cluster power management we look at a three tier data center servicing a myriad of requests. This difference in domain introduces the problem of scalability of solutions in data centers simply because of their sheer size and heterogeneity as opposed to a simple homogeneous server cluster. To address that problem we have adopted a hierarchical autonomic power and performance management approach. As discussed in Section 4, a hierarchical approach helps make the management more scalable.

Keeping in line with existing research, while we try to minimize data-center power consumption while maintaining performance constraints we adopt the mathematically rigorous optimization techniques for power management. While optimization always ensures the optimal solution we study its performance as an online power management technique. We try to establish the balance between computationally intensive optimization algorithms and the ensuing power consumed by these management algorithms to their effectiveness and speed in terms of real-time performance. This essentially means finding near optimal solutions that are as fast as heuristics and nearly as accurate as optimization algorithms. Unlike most other work that makes assumptions about data center traffic we plan to use accurate online prediction techniques such as Kalman filters [24] to make good guesses about the incoming workload which is used as the guiding parameter for the optimization algorithms. Another factor that impacts the runtime performance of power management algorithms is how often does one execute them? A very short interval would make the algorithm more adaptive to transient changes in the workload and more adaptive to sudden workload changes but on the other hand increases the overhead of frequent monitoring and management. This may potentially contribute to increased power consumption due to running the management algorithms very frequently. While a long decision interval can avoid this problem, it runs the risk of not being very adaptive to changing workload. While existing work has used different decision intervals there has not been much study to determine the relationship between an ideal observation time, the effectiveness of power management algorithms, and their runtime performance. We propose to achieve that objective with our modeling and simulation framework.

Drawing motivations from the domain of portable devices, we also try to bring in the idea of working with a fixed power budget in the server and data center world. As we have already outlined the issues related to data center growth and how the power requirements scale with that, working with a fixed budget might help check that problem. However, the challenge is to provide the same QoS while staying within the power budget even under peak load conditions. This research direction would require translating the power budget to compute and I/O power. This notion brings in a whole paradigm shift in terms of the way we traditionally look at application and workload requirements. For example, workload requirements characterization would require a power parameter in addition to compute, I/O, and performance parameters. Given that chipsets and processors are going to become very cheap, it only makes more sense to translate an application's compute and I/O requirements into power consumption in the platform instead of frequency for the processor and bandwidth for I/O.

The other unique feature provided specifically by data centers for dynamic power management is the Service Level Agreements (SLA) that establish its relationship to the customers. We seek ways to explore how much leeway these SLAs offer in terms of degrading performance in order to save power and yet keep the customers happy in terms of performance.

References

1. Q. Zhu, F. M. David, C. Devaraj, Z. Li, Y. Zhou, and P. Cao. "Reducing Energy Consumption of Disk Storage Using Power-Aware Cache Management." In HPCA, pp. 118–129, 2004.
2. R. Banginwar and E. Gorbatov, *Gibraltar: Application and Network Aware Adaptive Power Management for IEEE 802.11*, Proceedings of the Second Annual Conference on Wireless On-demand Network Systems and Services (WONS'05), 19–21 Jan. 2005, pp. 98–108
3. C-H. Hwang and A. Wu, "A predictive system shutdown method for energy saving of event-driven computation," Int. Conf. Computer-Aided Design, November 1997, pp. 28–32.

4. C. Hsu and U. Kremer, "The Design, Implementation, and Evaluation of a Compiler Algorithm for CPU Energy Reduction," PLDI'03, San Diego, CA, June 2003.

5. F. Douglis, P. Krishnan, and B. Marsh. "Thwarting the Power Hungry Disk," in Proceedings of the 1994 Winter USENIX Conference, San Francisco, January 1994.

6. G. Paleologo, L. Benini, A. Bogliolo, and G. De Micheli, "Policy optimization for dynamic power management," *IEEE Trans. Computer-Aided Design*, Vol. 18, June 1999, pp. 813–33.

7. Q. Qiu, Q Wu, and M. Pedram, "Stochastic modeling of a power-managed system-construction and optimization," *IEEE Trans. Computer-Aided Design*, Vol. 20, October 2001, pp. 1200–1217.

8. E. Chung, L. Benini, A. Bogliolo, and G. Micheli, "Dynamic Power Management for non-stationary service requests," *IEEE Trans. Computers*, Vol. 51, No. 11, November 2002, pp. 1345–1361.

9. T. Simunic, "Dynamic Management of Power Consumption," Power Aware Computing, edited by R. Graybill and R. Melhem, 2002.

10. J. D. Mitebell-Jacksun. Energy Needs in an Internet Economy: A Closer Look at Data Centers. Master's thesis, Energy and Resources Group, University of California at Berkeley, July 2001.

11. R. White and T. Abels. "Energy Resource Management in the Virtual Data Center," IEEE International Symposium on Electronics and the Environment. Conference Record., May 2004.

12. M. Elnozahy, M. Kistler, and R. Rajamony. "Energy Conservation Policies for Web Servers." In *Proceedings of the 4th USENIX Symposium on Internet Technologies and Systems*, March 2003.

13. J. Chase and R. Doyle, Balance of Power: Energy Management for Server Clusters, *Proceedings of the 8th Workshop on Hot Topics in Operating Systems (HotOS-VIII)*, May 2001, pp. 163–165.

14. APC White Paper #6, "Determining Total Cost of Ownership for Data Center and Network Room Infrastructure," 2003.

15. E. Pinheiro, R. Bianchini, E. V. Carrera, and T. Heath, "Load Balancing and Unbalancing for Power and Performance in Cluster-Based Systems," *Proceedings of the Workshop on Compilers and Operating Systems for Low Power*, September 2001; *Technical Report DCS-TR-440*, Department of Computer Science, Rutgers University, New Brunswick, NJ, May 2001.

16. The New York Times. There's money in housing internet servers. April 2001. http://www.internetweek.com/story/INW20010427S0010.

17. P. Bohrer, E. Elnozahy, T. Keller, M. Kistler, C. Lefurgy, C. McDowell, and R. Rajamony. *The case for power management in web servers. Power Aware Computing*, Kluwer Academic Publishers, 2002.

18. E.N. (Mootaz) Elnozahy, Michael Kistler, and Ramakrishnan Rajamony. Energy-efficient server clusters. In Workshop on Mobile Computing Systems and Applications, February 2002.

19. V. Sharma, A. Thomas, T. Abdelzaher, K. Skadron, and Z. Lu, "Power-aware QoS Management in Web Servers," Proceedings of the 24th IEEE International Real-Time Systems Symposium, p.63, December 03–05, 2003

20. T. Abdelzaher and V. Sharma. "A synthetic utilization bound for aperiodic tasks with resource requirements." In *Euromicro Conference on Real Time Systems*, Porto, Portugal, July 2003.

21. Y. Chen, A. Das, W. Qin, A. Sivasubramaniam, Q. Wang, and N. Gautam, Managing Server Energy and Operational Costs in Hosting Centers. In *Proceedings of the 2005 ACM SIGMETRICS International Conference on Measurement and Modeling of Computer Systems, SIGMETRICS 2005.*

22. L. Mastroleon, N. Bambos, C. Kozyrakis, and D. Economou, Autonomic Power Management Schemes for Internet Servers and Data Centers, *Proceedings of the IEEE Global Telecommunications Conference (GLOBECOM)*, November 2005.

23. S. Hariri, Bithia Khargaria, Manish Parashar, and Zhen Li, "The Foundations of Autonomic Computing," edited by Albert Zomaya, CHAPMN, 2005.

24. http://www.cs.unc.edu/~welch/kalman/

25. http://www.acpi.info/

26. http://www.intel.com/design/servers/ipmi/

27. M. Fleischmann. Dynamic Power Management for Crusoe Processors, January 2001. http://www.transmeta.com/.

28. Intel. Mobile Intel Pentium III Processor in BGA2 and MicroPGA2 Packages, 2001. Order Number 283653-002.

29. Rambus, RDRAM, 1999. http://www.rambus.com

30. A. R. Lebeck, X. Fan, H. Zeng, and C. Ellis. *Power Aware Page Allocation*. In ASPLOS, pages 105–116, 2000.

31. Cai, L. and Yung, L., *Joint Power Management of Memory and Disk*, IEEE, 2005.

32. Q. Zhu, F. M. David, C. Devaraj, Z. Li, Y. Zhou, and P. Cao. "Reducing Energy Consumption of Disk Storage Using Power-Aware Cache Management." In HPCA, pp. 118–129, 2004.

33. Banginwar, R. and Gorbatov, E., *Gibraltar: Application and Network Aware Adaptive Power Management for IEEE 802.11*, Proceedings of the Second Annual Conference on Wireless On-demand Network Systems and Services (WONS'05), 19–21 January 2005 pp. 98–108.

34. W. Felter, K. Rajamani, T. Keller (IBM ARL), and C. Rusu, A Performance-Conserving Approach for Reducing Peak Power Consumption in Server Systems, *ACM International Conference on Supercomputing (ICS), Cambridge, MA, June 2005.*

35. P. Rong and M. Pedram, "Hierarchical Power Management with Application to Scheduling," ISLPED (International Symposium on Low Power Electronics and Design), 2005.

36. S. Gurumurthi, A. Sivasubramaniam, M.J. Irwin, N. Vijaykrishnan, M. Kandemir, T. Li, and L.K. John, "Using Complete Machine Simulation for Software Power Estimation: The SoftWatt Approach." In *Proceedings of the International Symposium on High Performance Computer Architecture (HPCA-8)*, Cambridge, MA, February, 2002, pp. 141–150.

37. B. P. Zeigler, H. Praehofer, and T. G. Kim, *Theory of modeling and simulation*. 2nd ed. New York: Academic Press, 2000.

38. C. Amza, E. Cecchet, A. Chanda, A. L. Cox, S. Elnikety, R. Gil, J. Marguerite, K. Rajamani, and W. Zwaenepoel. Specification and implementation of dynamic Web site benchmarks. In *Proceedings of the 5th Workshop on Workload Characterization*, Austin, Texas, November 2002.

22

Trace Analysis for Fault Detection in Application Servers

Guofei Jiang, Haifeng Chen, Cristian Ungureanu, and Kenji Yoshihira

CONTENTS

Interest in monitoring and using traces of user requests for fault detection has been on the rise recently. In this chapter we propose novel fault detection methods based on trace analysis in application servers. One essential problem is how to represent the large amount of training trace data compactly as an oracle. Our key contribution is the novel use of varied-length n-grams and automata to characterize normal traces. A new trace is compared against the learned automata to determine whether it is abnormal. We develop algorithms to automatically extract n-grams and construct multiresolution automata from training data. Further, both deterministic and multihypothesis algorithms are proposed for detection. We inspect the trace constraints of real application software and verify the existence of long n-grams. Our approach is tested in a real system with injected faults and achieves good results in experiments.

22.1 Fault Detection and Diagnosis

The success of global networking has led to the wide use of various Internet services. Online searching, shopping, and transactions are increasingly becoming part of our daily life. While users see only a website of Internet services such as Google.com and eBay.com, the information system behind the scene is a large, dynamic, and distributed system and could consist of thousands of components, including servers, software, networking devices, storage equipments, etc. While each of these components is already complex enough by itself, the dynamic interaction between them introduces another magnitude of complexity. Further, Internet services are expected to run $24 \times 7 \times 365$ and maintain over 99.9% uptime. The complexity combined with the uptime requirement sets up a major system-management challenge.

Fault detection in such a system is a formidable task. Most current approaches for fault detection and diagnosis use event correlation [15]. This method collects and correlates events to locate faults based on known dependency knowledge between faults and symptoms. In practice, many runtime faults in an interconnected system are not very well understood. Since the runtime environments are very diverse, a fault may manifest itself in different ways. As a result, it is usually difficult to obtain such fault-symptom dependency knowledge precisely. As an example of a complex scenario, consider a class of problems in Internet services that are specific to individual user requests and not visible in collected events. Say for instance that the "checkout" button does not work after a customer has spent hours selecting products from a website. It is possible that this problem affects only that one customer and no one else. Such a problem may force us to trace how this individual request went through various components in the system and use such internal trace information to locate the fault.

Recently there has been much research activity in collecting and using traces for performance debugging and fault detection in distributed systems. Several research groups have developed tools to collect the traces of user requests. The Berkeley/Stanford Recovery-Oriented Computing (ROC) group modified the JBoss application server to monitor traces in the J2EE platform, and developed two methods to use collected traces for fault detection and diagnosis [6]. Aguilera et al. [2] developed application-independent passive tracing approaches in both RPC(remote procedure call)-style systems and message-based systems. Their method requires no modification to applications and middleware so that it can be widely used for performance debugging purposes. Magpie [3] uses low-overhead instrumentation to record fine-grained events generated by kernel, middleware, and applications. The Magpie request extraction tool then uses an application-specific event schema to correlate these events and precisely capture the control flow of every request. In addition, several commercial software packages, such as Hewlett-Packard's OpenView Transaction Analyzer [12], have been developed recently to monitor and trace transaction flows in distributed J2EE and .NET systems.

While these technologies enable us to monitor and collect traces in various distributed systems, in this chapter we focus on how to detect faults based on trace analysis. Internet services receive large numbers of user requests every day, which "probe and test" the system in a brute-force way. A fault or bug inside a system is likely to affect the traces of some user requests. In this chapter, we propose to use varied-length n-grams and automata to characterize and represent the normal traces compactly and then use the learned automata to determine whether a new trace is abnormal. We develop algorithms to automatically extract n-grams and build multiresolution automata from training trace data. Further, both deterministic and multihypothesis detection algorithms are proposed to detect abnormal traces. We inspect the trace constraints of real application software and verify the existence of long n-grams. Our approach is applied in a real system with injected faults and the experimental results demonstrate the effectiveness of our methods.

22.2 Abnormal Trace Detection

A trace records a sequence of components traversed in a system in response to a user request. The system architecture, functionality, and especially software control flows impose many constraints on the structure of traces. For example, components A-B-C-D always show up together and in that particular order because that is the only path starting from the component A. Such constraints are useful in fault detection to distinguish normal and abnormal traces. A fault inside the system is likely to affect some traces and cause these traces to violate some constraints. By detecting the abnormal traces, we expect to localize the cause of the abnormal traces — the faulty component itself.

If the dependency knowledge between specific faults and their symptoms is known, a fault can be detected and diagnosed directly based on its unique symptom. However, it is not realistic to characterize/model faults precisely in a large, complex, and dynamic system. Many runtime faults/bugs are not anticipated or well understood. In addition, we do not have sufficient data to model the faulty situation accurately because, in general, runtime faults are very rare. An alternative is to characterize/model *normal traces* rather than the faulty ones because large amounts of trace data from normal operation are available. This normal model can be used as an oracle to determine whether a new trace is acceptable or not.

A challenge here is that we do not know how much generalization capacity this model should have. We can only try to collect as many traces as possible and characterize/model these "known/seen" normal traces well. For a large system, however, it is difficult to include every normal trace in the training data. Further, both the distribution of "unknown/unseen" normal traces and the distribution of "faulty" traces affected by various faults are unknown. In addition, it is also hard to define the "similarity" metric in a conceptual trace space because it is not clear how traces are affected by various faults.

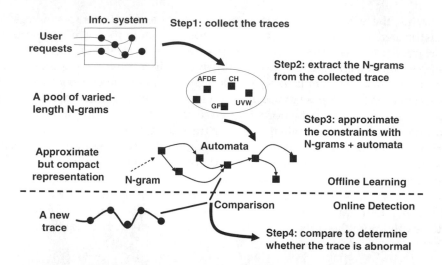

FIGURE 22.1
Steps of abnormal trace detection.

Two traces that look very similar in the structure could be thoroughly different with regard to whether they are normal or faulty. Conversely, two traces that look thoroughly different could both be normal traces. Thus, we have few clues about the direction of the trace space in which the normal model should generalize. Typically, high generalization capability would lead to high false negative rates (missed detections), while low generalization capability would lead to high false positive rates (false alarms). This motivates us to develop multiresolution algorithms for fault detection.

A trace includes a list of component names as well as the sequential order of these components. This component sequence order includes both the *local order* constraints between adjacent components and the *global order* constraints between nonconsecutive components. For example, in a trace ABCDEFG, the constraint that components A and B are consecutive is a local order constraint; the constraint that component A and E are three steps apart is a global order constraint. Meantime, depending on the monitoring mechanisms, a trace may or may not include call-return structure in its sequence. In this chapter, for the trace including call-return structure, calling a component and returning to a component are considered distinct.

N-gram is a commonly used natural language model [4]. It assumes that only the previous $n - 1$ words in a sentence have any effect on the probabilities for the next word. In this chapter we borrow this term to represent the contiguous component subsequences of a trace. For example, ABCD and EF are a 4-gram and a 2-gram, respectively. Automata are used to connect the n-grams to represent whole traces. Within n-grams, the components are bound together in that order and both the local and global order constraints remain. Between n-grams in the automata, only the local order constraints are captured. By controlling the length of n-grams, we can control how many local and global order constraints of a trace are represented in the model.

Figure 22.1 illustrates the basic steps of our approach. At first we collect traces, and our algorithm automatically extracts a series of varied-length n-grams from the training trace data based on frequency checking. These n-grams serve as the basic "genes" to construct any normal traces and are used as states in the construction of automata. Then the automata are built automatically by linking n-grams to represent whole traces. These steps may be computationally expensive but can be done offline. A new trace is compared against the learned automata to determine whether it is abnormal. This step is fast and able to support online detection.

22.3   N-gram Extraction

Theoretically, we can count the frequency of each trace and construct probabilistic automata to characterize the distribution of traces. In practice it is difficult to build robust probabilistic models because it is the highly dynamic user behavior that determines the distribution of traces. For example, after some items are on sale, user requests and their traces associated with these items could suddenly become much more frequent, even though there is no failure. Although it seems that gross user behavior in high-traffic Internet services is surprisingly predictable, showing clear diurnal and weekly patterns of behavior, we prefer to build models that do not fundamentally depend on the accuracy with which this behavior is captured. We rely on the fact that in spite of its dynamicity, the user behavior does not change the underlying reachability structure of the automata. In this chapter, we do not use probabilistic models to characterize how frequently a trace in the automata is visited. Instead we consider only whether the automata include the trace or not. The dynamics of user behavior will not affect the validity of our model, and as shown in our experiments, the reachability model is still good enough to detect many failures. Therefore, we use only the unique traces from the training data to extract n-grams, i.e., any trace only shows up once in our training data. Frequency checking and association rules are applied to determine which components are more tightly bound together than others. A threshold α is introduced to filter out those nonfrequent component combinations.

As shown in Figure 22.2, our n-gram extraction algorithm is similar to the classic sequential pattern mining algorithms [1]. Denote the number of unique traces in the training data as N. Algorithm 1 goes forward starting from $k = 1$. C_k is the set of n-grams with length $n = k$, i.e., k-grams. c_k^i is the ith k-gram in the set C_k. $f(s)$ is the number of times the sequence s appears in the set of the training trace data. Note that $f(s)$ has to be smaller than both $f(c_k^i)$ and $f(c_k^j)$ because the child sequence s subsumes both c_k^i and c_k^j, the parent sequences. Thus, we have $0 \le \alpha \le 1$, and the inequality $f(s) > \alpha \cdot min(f(c_k^i), f(c_k^j))$

Algorithm 1

Input: the set of unique traces
Output: the sets of varied-length n-grams.

C_1 = {the set of single components c^i_1 with $f(c^i_1) > 0$}.
$k = 1$
do
    for each two elements c^i_k, c^j_k, from the set C_k,
        if the last $k - 1$ component sequence of c^i_k equals
        the first $k - 1$ component sequence of c^j_k,
        then generate a new sequence
            $s = c^i_k$ plus the last component of c^j_k;
            count $f(s)$, the number of times that s appears in
            the trace data;
            if $f(s) > \alpha \cdot min(f(c^i_k), f(c^j_k))$,
                then put s into the set c_{k+1}.
    $k = k + 1$.
While C_k is not empty
return all C_j, for $1 \le j \le k - 1$

FIGURE 22.2
N-gram extraction algorithm.

implies that if the longer child sequence s can replace at least α percentage of one of its parent sequences, it is necessary to introduce the new sequence s as a $(k + 1)$-gram. There are several pruning techniques to reduce the set of n-grams. For example, if $f(s)$ is equal to $f(c^i_k)$, then we know that as long as the sequence c^i_k shows up in the traces, it has to exist as part of the longer sequence s. Since the longer sequence s has already represented the subsequence c^i_k here, c^i_k can be removed from the set C_k. The threshold α affects the length of extracted n-grams. As we will see in the experiments, the longest n-grams become much shorter as α increases. The traces in a large system are usually very diverse. As $\alpha \to 0$, the longest n-gram becomes the whole trace itself. Conversely, as $\alpha \to 1$, the longest n-grams become the single components, i.e., $n = 1$. Thus, by controlling α, we control the length of the extracted n-grams.

An example is given in Figure 22.3 to illustrate the extraction process. Assume that we have three traces: ABCDE, CDEA, and CDEBA, and the threshold is set to $\alpha = 0.6$. The number in the parentheses is the number of times that the associated sequence appears in the traces. At $k = 2$, the combined sequences marked with 'X' do not pass the threshold and they will not be put into the set C_2. The extraction process ends at $k = 3$ here and the length of the longest n-grams is three. If we apply pruning on these n-gram sets, C, D, and E at $k = 1$, and CD and DE at $k = 2$ will be pruned from the sets because the 3-gram CDE subsumes all of them with the same frequency number.

The implementation of Algorithm 1 is similar to the famous Apriori algorithm [1]. Assuming that the length of the longest n-grams is L, this algorithm

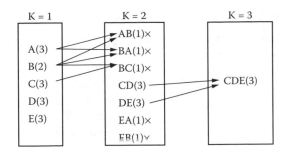

FIGURE 22.3
An example of extraction process.

needs $L+1$ data scans. The complexity of the algorithm is linearly dependent on the length of traces but is exponentially dependent on the number of components. However, in practice this algorithm could converge quickly, especially if the threshold α is not small.

22.4 Automata Construction

Algorithm 1 extracts a series of frequent n-gram sets C_k with varying length k. These n-grams are the basis for subsequent automata construction. Before constructing the automata, we sort the n-grams with same length (i.e., the n-grams in the same set) according to their frequency. The deterministic Algorithm 2 in Figure 22.4 is then used to disassemble whole traces into multiple n-grams for automata construction. The length of the longest n-grams is L, and then we have $1 \leq k \leq L$. E is the transition matrix of the constructed automaton. Algorithm 2 goes backward starting from $k = L$, following two deterministic rules to cut the whole trace:

- Rule 1: For different sets of n-grams, choose the set with longer n-grams first.
- Rule 2: For n-grams in the same set, choose the more frequent one first.

Denote the total number of n-grams extracted from Algorithm 1 as W, i.e., $W = \sum_{k=1}^{L} |C_k|$, where $|C_k|$ is the size of the set C_k. With N traces, it is straightforward to see that the complexity of Algorithm 2 is $O(WN)$. Before explaining the specific cutting process of Algorithm 2, it is useful to discuss the generalization ability that automata introduce to the model. Based on Algorithm 2, all training traces can be precisely represented by the constructed automata. The bottom line is that the set of single components C_1 is sufficient to construct any trace seen in the training data. However, the generalization ability of automata could regard other traces as normal traces. For example,

Algorithm 2

Input: the set of unique traces and the sets of n-grams
Output: the automaton E

set $E[m][n] = 0$ for any two n-grams m, n
for each trace T
    set $k = L$ and $l = T$'s length
    do
        for each k-gram c^i_k selected from C_k according to the sorted order (with the most frequent one first),

        search and replace all c^i_k in T with the assigned state number;

        if the length of the replaced part equals l,
            then break from the inner loop.
        $k = k - 1$.
    while the length of the replaced part $\neq l$ and $k \geq 1$.

    From left to right, set $E[m][n] = 1$ if an n-gram n follows another n-gram m contiguously in the trace T

remove the unused n-grams/states from E
return the matrix E

FIGURE 22.4
Automata construction algorithm.

for three traces ABCDE, CDEA, and CDEBA used in Figure 22.3, Algorithm 2 will build the automaton shown in Figure 22.5. Even if we restrict the sequence to lengths shorter than 6 to remove infinite loops, many traces such as ABABA and CDEAB are still regarded as normal ones because of generalization. States with multiple in and out edges in the automata usually introduce the opportunity of generalization.

Since not every normal trace is seen and collected in the training data, a certain capacity of generalization is desirable to reduce false positives in detection. For example, CDEA is a normal trace and we know that component A calls component B. Based on software's control flow, then it is possible that CDEAB is also a normal trace. The question is how much generalization the automata should have in order to achieve good performance in detection. As discussed in Section 22.2, in fact we hardly know the answer to this question. Moreover, it is also hard to know the direction of the trace space in which the automata should generalize. Therefore, our principle here is to keep the

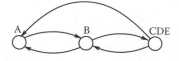

FIGURE 22.5
An example of automaton.

automata's representation as tight as possible to restrict the randomness in generalization. Later we will use the threshold α to control the generalization capacity of automata. To this end, heuristically, Rule 1 tries to link whole traces with the smallest number of n-grams and edges. In general more edges in the automata could lead to more generalization in the trace space. For a trace with fixed length, Rule 1 leads to a greedy optimization in cutting the trace. For the n-grams with the same length, Rule 2 implies that the more frequent one in the past (unique traces) is more likely to appear in this individual trace too.

22.5   Generalization of Automata

For each fixed threshold α, Algorithm 1 generates a series of varied-length n-grams. For different thresholds, it is clear that the automata built from these n-grams will be different. As $\alpha \rightarrow 0$, the longest n-gram becomes the whole trace itself and the automaton will include N states with no edges. These states are the original N traces in the training data. This automaton has zero generalization ability. If we use this automaton to distinguish normal and abnormal traces, the detection is an exact matching process. Conversely, as $\alpha \rightarrow 1$, the longest n-gram becomes the single component and the automaton is close to the control graph with single components as states. Therefore, this automaton introduces maximal generalization capacity because the single components cannot be further cut into smaller units. Given the same training data, as the threshold α increases, generally the length of n-grams becomes shorter and the generalization capacity of automata increases.

Given N traces and a threshold α, we extract n-grams and construct an automaton based on Algorithms 1 and 2, respectively. An interesting question is how many unique traces this automaton can generate. This number is a good metric to measure the generalization capacity of automata. Unfortunately, in case of automata with loops, the number or traces is infinite. An alternative is to count the number of unique traces with fixed length. Because we use varied-length n-grams in the automata, each state transition does not add the same length into the sequence. To this end, we have to introduce hidden states and augment the transition matrix E. Denote the length of each n-gram used in the automata as l_i, where $1 \leq l_i \leq L$. Denote the augmented transition matrix as $\overline{E}$ and the total length of n-grams used in E as $M = \sum_i l_i$. The augmentation algorithm is shown in Figure 22.6.

For each n-gram/state with length l_i, Algorithm 3 introduces l_i states to the new automaton. $f(i)$ is the state number for the first component of the n-gram i, and $e(i)$ is the state number for the last component of the n-gram i. Any two contiguous components within this n-gram contribute a related edge to the new automaton. If n-gram i has an edge to n-gram j in the original matrix E, the last component of the n-gram i has an edge to the first component of the n-gram j in the augmented automaton. Each state in the augmented

Algorithm 3

Input: the transition matrix E
Output: the augmented transition matrix $\bar{E}$

set $\bar{E}[k][l] = 0$, for $0 \le k, l \le M$
for each n-gram i in E
    $f(i) = e(i - 1 + 1$ // the first component
    $e(i) = f(i) + l_i - 1$ // the last component
    for each component j, where $0 \le j \le L_i$ in this n-gram i
        assign a new state number $f(i) + j$
        if $j \ge 1$ **then** set $\bar{E}[f(i) + j - 1][f(i) + j] = 1$

for each n-gram i in E
    for each n-gram j in E
        if $E[i][j] == 1$ **then** $\bar{E}[e(i)][f(j)] = 1$
return the matrix $\bar{E}$

FIGURE 22.6
Automata augmentation algorithm.

automaton represents only one component. Clearly, the size of $\bar{E}$ is much bigger than that of E.

Now given a fixed trace length, we can compute the number of traces that the automata can generate based on the dynamic programming algorithm. Let $I_t = (I_t^0, I_t^1, \ldots, I_t^M)^T$, where I_t^i is the number of traces whose length is t and the last state is i. Here T is matrix transposition. Let $I_0^i = 1$ if a trace could start from the state i and $I_t^i = 0$ otherwise. Based on dynamic programming, we can have the following recursive equations:

$$I_t^i = \sum_j \bar{E}[j][i] \cdot I_{t-1}^j \qquad (22.1)$$

$$I_t = (\bar{E}^T)^t I_0 \qquad (22.2)$$

Thus, the total number of the traces with length t is $\|I_t\|_1 = \sum_{i=1}^M I_t^i$.

Unfortunately, this number includes many duplicate traces that can be generated by different state sequences in the automata; it is not the total number of unique traces. As the length t of traces increases, this number could grow exponentially if there are loops in the automata. However, in some cases, a trace must start from certain components and end at certain components. It is clear that the augmented automaton is a nondeterministic finite automaton. Given a fixed length t, in general it is NP-hard to exactly count the number of unique traces that nondeterministic finite automata could generate. Though Gore et al. [8] proposed some algorithms to compute this number approximately, it is not clear whether there exist unbiased approximation algorithms with small standard deviations.

22.6   Detection Algorithms

Given the automaton constructed from the training trace data, the abnormal trace detection process is to determine whether a new trace is acceptable by this automaton. Both deterministic and multihypothesis detection algorithms are developed here but they have different conditions on "accepting" a trace as a normal one. Denote the set of n-grams included in the automata as C_a and the total number of n-grams in C_a as N_a; let $c_a^i \in C_a$, for $(0 \leq i \leq N_a)$ be an individual n-gram in the set C_a. In the deterministic algorithm, the same Rule 1 and Rule 2 in Section 22.4 are used to cut the new trace. The only difference is that the n-grams are chosen from C_a rather than from C_k, $(1 \leq k \leq L)$ in Algorithm 2. The deterministic algorithm is shown in Figure 22.7. Basically the new trace must satisfy the following two conditions to be classified as a normal trace:

- Condition 1: As per Rule 1 and Rule 2, the new trace could be completely cut into the n-grams that belong to C_a.
- Condition 2: The transitions of these n-grams follow a path in the automaton.

As shown in Figure 22.7, the algorithm basically tries to check whether a new trace can be interpreted as a specific state sequence in the automaton. The question is why we have to cut a new trace in this specific way. Underlying the structure of the automaton, essentially the abnormal trace detection is to run comparison of traces (i.e., compares the new trace with the pool of training

Algorithm 4

Input: the automaton and the new trace T
Output: true (normal) or false (abnormal)

set R = true, $l = T$'s length, and $m = 0$
for each n-gram $c_a^i \in C_a$ selected according to rules 1 and 2,

   search and replace all c_a^i in T with the state number;
   $m = m + 1$;
   if the length of the replaced part equal to l,
      then break the loop;

   else if $m == N_a$
         then R = false.
   if $R ==$ true, **then**
         from left to right, compare each state transition of T against the automaton;
         if any transition not found in the automaton,
            then R = false.
   return R

FIGURE 22.7
Deterministic detection algorithm.

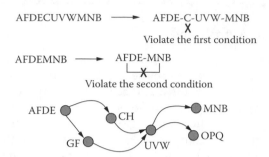

FIGURE 22.8
An example of condition violations.

traces). The deterministic cutting rules here are just like a hash function to convert traces to state sequences. Note that the same deterministic rules are used in cutting the training traces for automaton construction as well as the new traces for detection. As a result, all training traces will be detected as normal traces and the automaton also allows some strict generalization. An example of condition violations is shown in Figure 22.8.

The multihypothesis detection algorithm is developed to relax Condition 1. The algorithm searches all state sequences and keeps multiple hypotheses about how to cut the new trace. A new trace is acceptable as long as there is one state sequence that could generate this trace. Given a new trace and the automaton, the challenge is how to efficiently determine whether there exist such state sequences in the automaton. The multihypothesis detection algorithm is shown in Figure 22.9.

In Section 22.5, Algorithm 3 builds an augmented automaton $\bar{E}$ where each state represents only one component. Conversely, one component could be associated with multiple states. Denote the number of states in $\bar{E}$ as M and

Algorithm 5

Input: the matrix $\bar{E}$, O, and a new trace T
Output: true (normal) or false (abnormal)

at step $t = 0$ with the first component $T[0]$,
set $I^i_0 = O[i][T[0]]$ for $1 \leq i \leq m$.
For each step $t = k$, with the new component $T(k)$,
    $I^i_k = [\sum^m_{j=1} \bar{E}[j][i] \, I^j_{k-1}] . O[i][T[k]], 1 \leq i \leq M$;
    $k = k + 1$;
    if $k == l$, **then break** the loop.
$sum = \sum^m_{i=1} I_{i-1}$.
if $sum > 0$, **then** $R =$ true **else** $R =$ false.
return R

FIGURE 22.9
Multi-hypothesis detection algorithm.

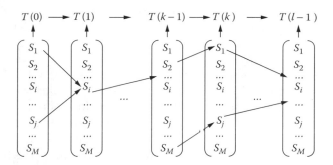

FIGURE 22.10
A Viterbi-like algorithm.

the number of components as P. When those hidden states are introduced in Algorithm 3, we use an $M \times P$ dimension matrix O to record the mapping relationship between the states and their associated components. Let $O[i][j] = 1$ if the state i represents the component j, otherwise let $O[i][j] = 0$. Therefore, we formulate our problem as a specific Hidden Markov Model (HMM), where $\overline{E}_{M \times M}$, $O_{M \times P}$ are the transition and emission matrix, respectively. The elements of the matrix $\overline{E}$ and O are either 1 or 0, which is different from the probability specification in classic HMMs. Denote the length of a new trace T as l. Following the sequential order (from left to right), denote each individual component that the trace T went through as $T[k]$, where $0 \leq k \leq l$, $0 \leq T[k] \leq P$.

Sum is the total number of the state sequences that match the new trace. If there is at least one state sequence (*Sum* > 0) that could match the trace, the new trace is accepted as a normal one. Otherwise the new trace is regarded as an abnormal one. As illustrated in Figure 22.10, the key idea behind Algorithm 5 is dynamic programming, and our algorithm is Viterbi-like. The s_i ($1 \leq i \leq M$) are the states of the automaton. Following the sequential order of the new trace, at each step $t = k$, at first we check which states are associated with the component $T(k)$ according to the emission matrix O, and then check which states at $t = k - 1$ could transfer to these current states according to the transition matrix $\overline{E}$. Finally, we count the total number of valid state sequences that could end at the current states.

22.7 Multiresolution Detection

As discussed in Section 22.5, given the same training trace data but different thresholds α, Algorithm 1 extracts different sets of n-grams and Algorithm 2 constructs different automata. These automata all accept the training traces precisely but they have different generalization ability. As $\alpha \to 0$, the detection algorithm becomes an exact-matching algorithm. Any trace unseen in the training data is regarded as an abnormal trace. Therefore, the learned automata are likely to introduce high false positive rates in detection. Conversely, as $\alpha \rightarrow 1$, the learned automata have the maximal generalization ability and the detection algorithm accepts many abnormal traces as normal ones. The automata are likely to cause high false negative rates. The threshold α determines a boundary in the trace space that the detection algorithm uses to separate normal traces from abnormal ones. In general, as α decreases, the detection boundary becomes tighter and the detection algorithm uses a finer resolution to determine whether a trace is abnormal. As mentioned in Section 22.4, for a fixed threshold, heuristically Algorithm 2 tries to draw this boundary as tight as possible to restrict the randomness in generalization.

In practice, it is usually difficult to choose an optimal threshold to balance false positives and false negatives. To this end, we propose to construct a series of automata with different thresholds and apply these automata concurrently or sequentially to support multiresolution detection. N-gram extraction and automata construction can be done offline so that the computational overhead of constructing multiple automata should not be a problem. New traces are compared against a set of automata instead of just one to generate alerts. The traces detected as abnormal by the automata with high thresholds are more likely to be true positives and should be analyzed first to locate faults. Conversely, the traces detected as abnormal only by the automata with low thresholds are more likely to be false positives. Meantime, the traces detected as abnormal by multiple automata are likely to be true positives. Therefore, we can rank these abnormal traces according to our confidence in their abnormality.

For a large distributed system, we do not have to choose one threshold for the whole system. Instead different thresholds can be chosen for different segments of traces. High threshold should be chosen for the less important, more reliable segments such as the segments that have been running for a long time and are stable. Low threshold should be chosen for the more important and less reliable segments. For example, the segments with new deployed software or equipments, the segments that have a lot of problems, and the segments that are critical for the service.

22.8   Trace Diversity: A Case Study

So far we have introduced our approach for abnormal trace detection. There is one important question that remains: Are there long n-grams existing in real software, and how diverse are those traces? If, by nature, software's traces are very diverse, we will not have many constraints in trace space, and any detection algorithms based on trace analysis cannot be very effective. This question motivated us to present a case study of real application software — Pet Store [13]. Pet Store is a blueprint program written by Sun Microsystems to demonstrate how to use the J2EE platform to develop flexible, scalable, cross-platform enterprise applications [13]. It has 27 Enterprise JavaBeans (EJBs), some Java Server Pages (JSPs), Java Servlets, etc. We use the same monitoring facility described in [6] to record traces of user requests, and the JBoss middleware is modified to support such functionality.

Given a threshold, Algorithm 1 extracts a series of frequent n-grams with varying length. Note that these n-grams are extracted from unique traces and the threshold implies how often they are shared by various traces. Thus, the distribution of these n-grams is a good metric to measure the diversity of traces. For a fixed threshold, in general if a system has higher percentage of long n-grams, the system is more deterministic. We collected most of the possible traces from Pet Store by emulating various user requests. Algorithm 1 is used to extract n-grams with various thresholds. The pruning technique mentioned in Section 22.3 is used to remove those n-grams which can be completely replaced by the longer n-grams that have the same frequency numbers.

Figure 22.11 shows how the total number of n-grams and the length of longest n-grams quickly decrease as the threshold increases. As the threshold approaches 1, the total number of n-grams is close to the number of components. Even for a relatively high threshold 0.5, there are n-grams as long as 50, which makes us believe that long n-grams do exist in real software. Because of the pruning, here $\alpha = 0.05$ has more n-grams than $\alpha = 0$. Figure 22.12 illustrates the cumulative distribution of varied-length n-grams. As shown in the figure, for various thresholds, the length of most n-grams (over 80%) is

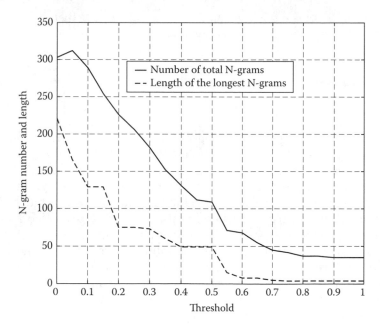

FIGURE 22.11
Number and length of n-grams.

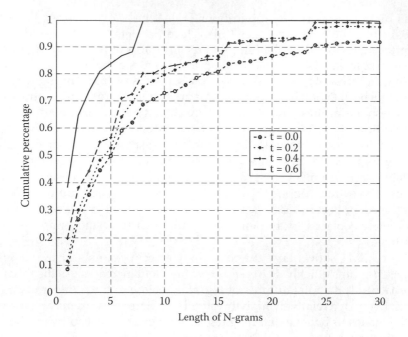

FIGURE 22.12
Cumulative percentage of n-grams.

less than 15. There is a big percentage of n-grams longer than 5. These long n-grams could impose many constraints in trace space.

22.9 Detection Experiments

Before our detection experiments, Algorithm 2 constructs a series of automata based on the extracted n-grams at different thresholds. Figure 22.13 illustrates the sizes of these automata. If we compare the number of states with the number of n-grams in Figure 22.11, it is clear that many extracted n-grams are not used in the automata, especially when the threshold is low. As the threshold approaches 0, the whole traces become the isolated states in the automata with no edges at all. It is interesting to see that the number of edges increases first and then keeps a relatively constant value. As we expected, the average edge per state (see the dotted line close to the x-axis in Figure 22.13) increases monotonely as the threshold increases. In fact this explains the growing generalization ability of the automata as the threshold increases.

In our detection experiments, two types of faults are injected into the components of the Pet Store software: "null call" and "expected exception" [10]. After a null call failure is injected into some component C, any invocation of a method in component C results in an immediate return of a null value

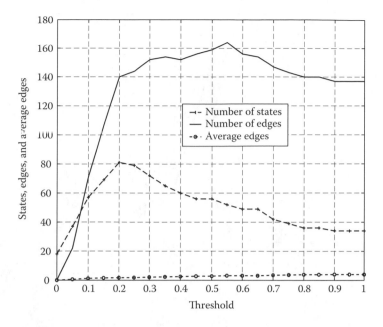

FIGURE 22.13
Size of automata.

(i.e., calls to other components are not made). The other failure type, expected exception, is injected into components that contain methods which declare exceptions. After the expected exception failure is injected into a component, any invocation of its methods declaring exceptions will raise the declared exception immediately (if the method declares many exceptions, an arbitrary one is chosen and thrown). Methods in that component, which do not declare exceptions, are unaffected by this injected failure. Other kinds of failures could also be injected (e.g., run-time exceptions, dead-lock, etc.), but typically these errors are easier to catch with many existing monitoring tools. In contrast, the two failures that we selected result in more subtle outcome, do not cause exceptions to be printed on the operator's console, and do not crash the application software. At the same time, these bugs can easily happen in practice due to incomplete, or incorrect, handling of rare conditions.

Both faults are injected to 15 EJB components of Pet Store resulting in a total of 30 cases. Note that the Pet Store package includes three applications and we only chose one in our experiment, which includes 15 EJB components. We applied both Algorithm 4 and Algorithm 5 in detection, and these two algorithms achieved the same detection results shown in Figure 22.14. Algorithm 5 has a much relaxed condition on accepting a trace and could reduce false positives in large systems where we may only be able to collect a small part of all traces for training. Since Pet Store is relatively simple software and the size of learned automata is small, the opportunity to have multiple trace

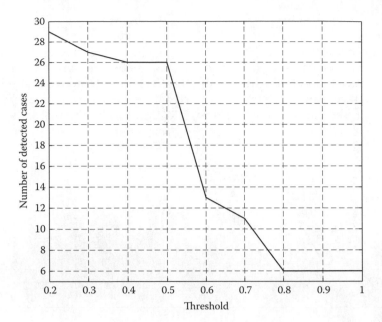

FIGURE 22.14
Detection accuracy of Pet Store.

hypotheses is low, and the two algorithms got the same detection results. As the threshold decreases, our approach can detect most cases. It is interesting to analyze the detection accuracy in Figure 22.14 with the length of the longest n-grams in Figure 22.11. From the threshold 0.5 to 0.6, the detection accuracy decreases steeply from 26/30 to 13/30 as the length of n-grams decreases steeply from 50 to 8. When the threshold is higher than 0.8, both detection accuracy line and n-gram length line are flat. This unveils an important fact that the constraints of long n-grams are critical in abnormal trace detection.

When the threshold is higher than 0.5, the abnormal traces detected by our algorithms are all true faulty traces (that go through the faulty components), though the detection accuracy is not very high. However, as the threshold is decreased below 0.5 the detection accuracy becomes much higher; in 15 cases, our algorithms detected some other traces as abnormal even though these traces did not go through the fault-injected components at all. At first we thought these traces were false positives. But after analysis, we found that these traces were affected by the fault even though they did not go through the fault-injected components. The traces of different user requests are not independent at some time. Because of the faulty component, the traces that went through the faulty component failed and changed some internal states of the system. Some other traces are affected by these states even if they do not go through the faulty component.

Therefore, after a fault happens in a system, usually we will detect a set of abnormal traces which include the traces that go through the faulty components as well as the traces that are affected by the fault but do not go through the faulty components. This phenomenon could make the detection process easier but the diagnosis process much harder. Because the fault is the root cause of these abnormal traces, we can use the timestamp of each trace to locate the true faulty trace. The timestamp is already logged for each trace. Among all traces detected as abnormal by our algorithms, the trace with the earliest timestamp is most likely to be the real faulty trace, because it is the failure of this trace that triggered others. After we applied this rule in detection, we have only one false positive in a total of 30 cases for $\alpha \leq 0.5$. For $\alpha > 0.5$, there is no false positive in our experiments. Since we collected most of the traces in the training data, the false positive rate is very low in our experiments. For large software, we believe that the false positive rate could also be low if training traces are collected for a sufficiently long time.

The situation can become very complicated if many requests go through the faulty components and interleave together with the affected traces. With the above rule, we can pick one faulty trace accurately with the earliest timestamp. But there are many following faulty traces that could mix up with the affected traces triggered earlier. Internet services receive many requests, and we can use normal traces to cross-validate whether an abnormal trace is a really faulty one. For example, if several normal traces cover each component of an abnormal trace, this abnormal trace is likely to be false positive because all of its components work well in other traces. Meantime, we can also artificially generate a "probing" request and force it to go through the suspicious area to verify whether an abnormal trace is a real faulty one.

22.10   Related Work

Our work is inspired by Berkeley/Stanford ROC group's success in using traces for fault detection [6, 10]. However, while they focus on developing the whole concept of recovery-oriented computing, in this chapter we are interested in developing specific machine learning technology to exploit the traces extensively. Compared with their tree structure and probabilistic model — PCFG — we proposed a thoroughly different structure, varied-length n-grams and automata, to characterize normal traces. As discussed in Section 22.3, we believe that it is difficult to build a robust probabilistic model to reflect dynamic user behaviors. Instead, our automata model is static as long as the application software is not modified. All other system changes including user behavior and load balancing will not invalidate our model, which enables us to collect a sufficiently large number of traces for training and further reduce false positives. More details about our approach are included in our longer paper [9]. In addition, Chen et al. [5] proposed an approach to dynamically track component interactions for fault detection using an online outlier detection engine — SmartSifter [14].

Anomaly intrusion detection is an active area in computer security research. Forrest et al. [7] proposed to use fixed-length n-grams to characterize the system call sequence of normal Unix processes. The short sequences of system calls are used as a stable signature in intrusion detection. Michael and Ghosh [11] proposed two state-based algorithms to characterize the system call sequences of programs and used a thoroughly different approach to build finite automata for intrusion detection. They proposed a probabilistic model to calculate the anomaly score. Both of these approaches use fixed-length n-grams and keep a moving window (move one system call each time) to cut the system call sequences. Our approach uses a thoroughly different method to extract varied-length n-grams and construct automata. Because our approach is used in fault detection rather than intrusion detection and the collected traces are not system call sequences, we are not able to compare our approaches with theirs.

References

1. R. Agrawal and R. Srikant. Mining sequential patterns. In *Proceedings of the International Conference on Data Engineering (ICDE)*, 3–14, Taipei, Taiwan, March 1995.
2. M. K. Aguilera, J. C. Mogul, J. L. Wiener, P. Reynolds, and A. Muthitacharoen. Performance debugging for distributed systems of black boxes. In *Proceedings of the 19th ACM Symposium on Operating Systems Principles*, 74–89, Bolton Landing, NY, October 2003.
3. P. Barham, A. Donnelly, R. Isaacs, and R. Mortier. Using Magpie for request extraction and workload modeling. In *6th Symposium on Operating Systems Design and Implementation*, 259–272, San Francisco, CA, December 2004.
4. W. B. Cavnar and J. M. Trenkle. N-gram-based text categorization. In *the 3rd Annual Symposium on Document Analysis and Information Retrieval*, 161–175, Las Vegas, NV, April 1994.
5. H. Chen, G. Jiang, C. Ungureanu, and K. Yoshihira. Failure detection and localization in component based systems by online tracking. In *The Eleventh ACM SIGKDD International Conference on Knowledge Discovery and Data Mining (KDD)*, Chicago, August 2005.
6. M. Chen, E. Kiciman, E. Fratkin, A. Fox, and E. Brewer. Pinpoint: Problem determination in large, dynamic internet services. In *International Conference on Dependable Systems and Networks 2002*, 595–604, Washington, DC, June 2002.
7. S. Forrest, S.A. Hofmeyr, A. Somayaji, and T.A. Longstaff. A sense of self for Unix processes. In *IEEE Symposium on Research in Security and Privacy*, 120–128, Los Alamitos, CA, 1996.
8. V. Gore, M. Jerrum, S. Kannan, Z. Sweedyk, and S. Mahaney. A quasi-polynomial-time algorithm for sampling words from a context-free language. *Information and Computation*, 134(1):59–74, April 1997.

9. G. Jiang, H. Chen, C. Ungureanu, and K. Yoshihira. Multi-resolution abnormal trace detection using varied-length n-grams and automata. In *IEEE Second International Conference on Autonomic Computing (ICAC)*, 111–122, Seattle,WA, June 2005.

10. E. Kiciman and A. Fox. *Detecting Application-Level Failures in Component-Based Internet Services.* Technical report, Computer science department, Stanford University, 2004.

11. C. C. Michael and A. Ghosh. Simple, state-based approaches to program-based anomaly detection. *ACM Transactions on Information and System Security,* 5(3):203–237, August 2002.

12. http://www.openview.hp.com/products/tran

13. http://java.sun.com/developer/releases/petstore

14. K. Yamanishi and J. Takeuchi. On-line unsupervised outlier detection using finite mixtures with discounting learning algorithms. In *The Sixth ACM SIGKDD International Conference on Knowledge Discovery and Data Mining (KDD)*, 320–324, Boston, MA, August 2000.

15. A. Yemini and S. Kliger. High speed and robust event correlation. *IEEE Communication Magazine*, 34(5):82–90, May 1996.

23

Anomaly-Based Self Protection against Network Attacks

Guangzhi Qu and Salim Hariri

CONTENTS

23.1 Introduction

The Internet has grown exponentially over the last few years and has expanded commensurately in both scope and variety. In addition to the increasing number and dependence upon Internet resources and services, there are also increasing interconnectedness and interdependence among large and complex systems; a failure in one sector can easily affect other sectors. Disruption of critical Internet services can be very expensive to businesses, life threatening for emergency services, and ultimately threaten the defense and economic security nationwide. In fact, the commission on critical infrastructure reports that the potential for disaster in the U.S. as a result of network attacks is catastrophic [53]. This raises an important question given the fragility of Internet infrastructures to small random failures (such as hardware failures, bad software design, innocent human errors, and environmental events). According to a recent study, the Internet is robust to random failures, however the Internet's reliance on a few key nodes makes it vulnerable to an organized attack and it has been shown that the average performance of the Internet would be reduced by a factor of two if only 1% of the most connected nodes are disabled. If 4% of them were shut down the network would become fragmented and unusable [41].

Network attacks have been launched at very high rates, almost daily, and the sophistication of these attacks is also increasing exponentially. Network attacks take many different forms such as denial of service, viruses, and worms. These malicious activities have become a significant threat to the security of our information infrastructure and can lead to catastrophic results. On the other hand, as emergency and essential services become more reliant on the Internet as part of their communication infrastructure, the consequences of denial-of-service attacks could even become life threatening. After the September 11 terrorist attack in the U.S., there is a growing concern that the Internet may also fall victim to terrorists. There are many indications that since September 11, the number of DoS attacks has greatly increased [11].

The existing techniques to respond to these attacks such as intrusion detection systems and firewall hardware/software systems are not capable of handling these complex interacting organized network attacks. The development of countermeasures (e.g., signatures for intrusion detection systems) is manually intensive activities and cannot detect future unknown attacks. Furthermore, the users are required to manually install/update the signature databases in order to protect themselves against the existing or new attacks.

23.2 Research Challenges

The main research goal is the development of advanced methodologies and effective technologies that can effectively detect network attacks in real time and configure the network and system resources to proactively recover from network attacks and prevent their propagations. However, automatic modeling and online analysis of the anomaly of the Internet infrastructure and services is a challenging research problem due to the *continuous change in network topology*, the *variety of services and software modules* being offered and deployed, and the *extreme complexity of the asynchronous behaviors of attacks*. In general, the Internet infrastructure can be considered as a structure varying system operating under a complex environment with a variety of unknown attacks. In order to develop an effective system for attack detection, identification, and protection, it becomes highly essential to develop systems that have the functionality of online monitoring, adaptively modeling and analysis tailored for real-time processing and proactive self-protection mechanisms.

Anomaly analysis involves detailed analysis of network attacks, identifying accurate network metrics and using these metrics for quantifying the impact of network attacks on various network traffic and network system components. The field of anomaly analysis is still in its infancy stage. A lot of work has focused on studying various network attacks on the network services and components, however very little has been done to quantify the impact of network attacks and to provide proactively protection against network attacks. As the uses of the Internet expand, the stakes and thus the visibility of vulnerability and fault tolerance concerns will rise [1,24]. This identifies the need for robust self-protection architectures that integrate vulnerability analysis and network survivability.

The capabilities of the Internet protocol such as Type of Service (ToS) and Quality of Service (QoS) are utilized for priority based and policy based routing. However, due to the increasing vulnerability, it is necessary to prioritize traffic based on security in addition to the traditional classification of applications based on their QoS. Furthermore, there are no standard protocols that can prohibit an attacker from consuming all available network bandwidth and resources. Due to lack of such protocols, packet-based attacks are not only possible, but they have become quite common. The vulnerability of the Internet could be primarily accounted to TCP/IP, because security was not a main design consideration when it was initially developed.

23.3 Classification of Network Attacks

Despite the best efforts of security experts, networks are still vulnerable to attacks. No amount of system hardening and traditional security measures can assure 100% protection against network attacks. William Stallings [60] describes a computer security attack as *"any action that compromises the security information owned by an organization."* Network attacks aim at utilizing network media and manipulating computing and/or network resources in order to severely degrade the performance of their services and eventually shut down the entire network. Some network attacks exploit the vulnerabilities in software and network protocols. For example, both the *Morris Internet worm* [43] and the *SQL Sapphire worm* [7] exploit buffer overflows, a common type of error occurring in many C programs in which the length of the input is not checked, allowing an attacker to overwrite the stack and consequently executing an arbitrary code on a remote server. Other attacks seek to survey a network by scanning and probing in order to find the vulnerabilities or holes in the systems. Based on Kendall's classification [29], we group network attacks into five major categories: Denial of Service (*DoS*), User to Root (*U2R*), Remote to Local (*R2L*), *Probe*, and *Worm/Virus*. In what follows, we describe briefly these types of network attacks.

23.3.1   *U2R* and *R2L* Attacks

U2R and *R2L* attacks exploit bugs or mis-configuration in the operating system to control the target system. For example, a buffer overflow or incorrectly setting file permissions in a SUID (Set User ID) script or program can often be deployed by this kind of attack. Password guessing is a typical *R2L* attack. Many computer users tend to choose weak or easily guessed passwords. An attacker could try common or default usernames and passwords. An attacker could use a brute force program to exhaustively test dictionary words. Any service requiring a password is vulnerable, for example, telnet, FTP, POP3, IMAP, or SSH.

23.3.2   Worm/Virus Attacks

Modern virus and worm attacks inherit *U2R* and *R2L* characteristics to exploit vulnerability in hosts. These vulnerabilities are exploited in large scale *DoS* attacks that are targeting any network resource found to be vulnerable. After being infected, the resources will degrade the overall network performance and their services as experienced in the *CodeRed*, *Nimda*, *SQL Slammer*, *RPC DCOM*, *W32/Blaster*, *SoBig* and other typical worm/virus attacks [19]. Their self-replicating and propagation characteristics complicate significantly the ability to protect and secure information infrastructure.

23.3.3   *DoS* Attacks

The different types of Denial of Service attacks can be broadly classified into *software exploits* and *flooding* attacks. In software exploits the attacker sends a few malformed packets to exercise specific known software bugs within the target's OS or applications. Examples of this type of attacks include *Ping of Death*, *Teardrop*, *Land*, and *ICMP Nukes*. These attacks are relatively easy to counter either through the installation of software patches that eliminate the vulnerabilities or by adding specialized firewall rules to filter out malformed packets before they reach the target system. In flooding attacks, one or attackers send continuous streams of packets aimed at overwhelming work bandwidth or computing resources. Although software exploit attacks are important, this focuses on flooding attacks, since they cannot be addressed by the operating system or application software patches. There are numerous DoS attacks in the flooding attacks category such as *TCP SYN* attack, *Smurf*, *UDP flood* attacks, and *ICMP flood* attacks.

23.3.4 Distributed Denial of Service Attacks

Based on the location of the observation point, flooding *DoS* attacks can be classified as a single source and multiple sources attack (also known as Distributed Denial of Service — *DDoS*). There are two common scenarios for *DDoS* attacks, the typical *DDoS* attack and the distributed reflector denial of service (*DRDoS*) attack [46].

In a typical *DDoS* attack, an attacker controls a number of handlers. A handler is a compromised host with a special program running on it. Each handler is capable of controlling multiple agents. An agent is a compromised host, which is responsible for generating a stream of packets that is directed toward the intended victim. *TFN* , *TFN2K*, *Trinoo* and *Stacheldraht* [17] are examples of distributed denial of service attack tools. These programs not only use TCP and UDP packets but also ICMP packets. Moreover, because the programs use ICMP_ECHOREPLY packets for communication, it will be very difficult to block packets without breaking most Internet programs that rely on ICMP. Since *TFN, TFN2K,* and *Stacheldraht* use ICMP packets, it is much more difficult to detect them in action, and packets will go right through most firewalls. The only way to destroy this channel is to deny all ICMP_ECHO traffic into the network. Furthermore, the tools mentioned above use any port number randomly; it is hard to prevent the port from malicious attack in advance using the fixed port scheme in current firewalls.

In the *DRDoS* attacks [20], routers or web servers are used to bounce attack traffic to the victim. Once the attacker controls multiple agents through the handlers, he/she does not command the agents to send traffic to the victim, rather the agents will send attack traffic to third parties (routers or web servers) with spoofed source IP address of the victim. The reply packets from the third parties to the victim constitute the *DDoS* attack [20].

23.4 Existing Network Security Techniques

As the number of network attacks increases significantly [39], many router-based defense mechanisms have been proposed, including ingress filtering [18,31], egress filtering and route filtering [31], router throttling [65], Pushback [26,34], Traceback [4,55,57,58,61], and various intrusion detection mechanisms

13,23,25,54,63]. Besides the reactive techniques discussed above, some tems take proactive measures to prevent DoS attacks. For example, distributed packet filtering [44] blocks spoofed packets using local routing information and SOS [30] uses overlay techniques with selective re-routing to prevent large flooding attacks.

The ingress/egress filtering mechanisms [18,31] take the advantage of the local topology information to block spoofed packets whose source/destination IP address do not belong to the stub networks known to the router. They have been proposed in the literature and implemented by router vendors but they were not able to mitigate the problem effectively. The route filtering [31] lowers the impact on other routers when the internal routing changes in an irrelevant way. A router throttle mechanism is proposed by Yau et al [65], which is installed at the routers such that they can proactively regulate the incoming packet rate to a moderate level and thus reduce the amount of flooding traffic that targets the victim resource. The pushback approach is to identify and control high bandwidth aggregates in the network [26,34]. The router could ask adjacent upstream routers to limit the amount of traffic from the identified aggregate. This upstream rate-limiting is called pushback and can be propagated recursively to routers further upstream.

Various traceback techniques [4,55,57,58,61] have been proposed to locate the attack source in order to protect from the DDoS attacks. Probabilistic packet marking [55] was proposed to reduce the tracing overhead at IP routers, which was refined by Song and Perrig in the reconstruction of paths and the authentication of encodings. Snoren et al. presented a hash-based IP traceback [57], which can track the origin of a single packet delivered by the network in an efficient and scalable way. Stone [61] built an IP overlay network for tracking DoS floods, which consists of IP tunnels connecting all edge routers. The topology of this overlay network is deliberately simple, and suspicious flows can be dynamically rerouted across the tracking overlay network for analysis. Then, the origin of the floods can be revealed. Our approach has different goals from the traceback by focusing more on protection of the network operations and services and thus reducing the impact of network attacks.

23.4.1 Modern Intrusion Detection Systems

Intrusion detection systems (IDS) have been an active area of research for quite some time. Kemmerer and Vigna [28] give a brief overview of IDS since the idea was originally proposed by Denning [15]. As network-based computer systems play an increasingly vital role in modern society and given the increased availability of tools for attacking networks, an IDS becomes a critical component of network security besides the existing user authentication, authorization, and encryption techniques. And it is an important component of the defense-in-depth or layered network security mechanisms. It aims at detecting early signs of attack attempts so that the proper response can be evoked to mitigate the impact on the overall network behavior. An IDS collects system and network activity data (e.g., tcpdump data and system log) and analyzes the information to determine whether there is an attack occurring with least false alarms and false negative alarms. Here intrusion refers to any set of actions that threatens the integrity, availability, or confidentiality of a network resource. Typically, based on the general detection strategy, intrusion detection systems can be classified into two categories – *signature-based* detection and *anomaly-based* detection.

A *signature-based* detection system finds intrusion by looking for activity corresponding to known intrusion signatures. It is the most widely used commercial IDS model. For example, IDIOT [13] and STAT [25] use patterns of well-known attacks or weak spots of the system to identify known intrusions. Most of the IDS systems are designed to detect and defend systems from these malicious and intrusive events depend upon "signatures" or "thumbprints" that are developed by human experts or by semi-automated means from prior known worms or viruses. For example, in the well known MS SQL Slammer attack, the worm sends 376 bytes to UDP port 1434 (*SQL Server Resolution Service Port*). *Signature-based* detection is essentially a rule-based approach. An IDS system based on this model needs continuous manual updates of the knowledge database. It is deployed in most of the current commercial security products/approaches. Sophos anti-virus application and Microsoft patch remote update system are typical examples of the misuse detection system. After a virus has been detected, a signature will be developed and distributed to client sites. A disadvantage of this approach is that signature based IDS can only detect intrusions that are following pre-defined patterns. In a novel attack, it becomes ineffective.

The second model is the *anomaly-based* intrusion detection systems, such as IDES [33], Stalker [54], Haystack [56], and INBOUNDS [63]. This approach uses statistical anomaly detection, which uses a reference model of normal behavior and flags any significant deviation from normal behaviors. The normal behavior with respect to one metric (response time, buffer utilization, etc.) is defined based on rigorous analysis of the selected metrics under normal and abnormal conditions. This approach can offer unparalleled effectiveness and protection against unknown or novel attacks since no a priori knowledge about specific intrusions is required. For example, the normal behavior of a web browsing could be profiled in terms of average frequency of requests, average connection durations, and so on. If significant increase in these measurement attributes is observed, an anomaly alarm will be raised. However, it may cause more false-positives than a signature based intrusion detection system because anomaly can be caused due to a new normal network behavior. The limitation of this approach is due to the difficulty in performing online monitoring and analysis of network traffic.

23.5   Related Works

In this section, we discuss briefly important related projects in the area of intrusion detection and protection.

EMERALD [48] is a distributed detection and response system that has been developed at the SRI research institute in 1997. EMERALD was conceived to apply mainly host-based detection techniques, but its architecture is also suitable for network-based detection. It applies both knowledge-based and anomaly-based detection methods. In terms of anomaly-based detection model, it uses historical records as its normal training data and it compares distributions of new data to the distributions obtained from those historical records and the differences between the distributions indicate an intrusion.

MULTOPS [21] is a MUlti-Level Tree for Online Packet Statistics defense mechanism. Routers can detect ongoing bandwidth attacks by searching for significant asymmetries between packet rates to and from different subnets. Statistics are kept in a tree that dynamically adapts its shape to reflect changes in packet rates, and avoid (maliciously intended) memory exhaustion.

Prelude IDS [5,6,69] is an open-source project that consists of three functional components: sensors, managers, and countermeasure agents. The sensors deploy knowledge-based techniques for both network and host-based detection. Anomaly detection is not applied. Several sensors are associated to a manager that receives alerts from sensors: it processes and correlates the alerts and decides appropriate countermeasures. More managers can be organized in a hierarchy manner. In Prelude system, both sensors and managers are involved in the detection process and the managers depend on the results delivered by the sensors.

D-WARD [37] is a network-based DDoS detection and defense system. The D-WARD architecture consists of fully autonomic subsystems that are deployed at the entry points of so-called source-end networks. The subsystems analyze and control the outgoing traffic with the goal to prevent hosts of the observed networks from participating in DDoS attacks. Therefore each subsystem autonomically imposes rate limits on suspicious flows. The detection method is based on predefined models of normal traffic.

COSSACK [42] is a distributed, network-based DDoS attack detection and response system. The COSSACK subsystems, called watchdogs, are located at edge networks. If a watchdog detects an attack against the associated edge network it multicasts an attack notification to the other watchdogs. Upon reception of such a notification, a watchdog checks if the attack flows or parts of it originate from its own edge network. If so, the watchdog attempts to block or rate-limit the corresponding flow by setting filter rules in routers or firewalls. COSSACK principally applies anomaly detection techniques.

In our *AUTONOMIA* framework [22], we developed online monitoring subsystem, which can collect system wide measurement attributes for analysis. We applied information theory to select the optimal feature set from multiple distinct features. Based on the chosen multiple features anomaly distance is calculated to detect the changes in the operational states for both network components and the system itself. We developed a proactive protection subsystem based on Quality of Protection (*QoP*) which takes actions on network traffic flows according to the anomaly analysis results.

23.6 Anomaly Based Control and Management Architecture

The overall goal of our approach is to formulate a theoretical basis for the construction of global metrics that can be used to analyze, manage, and control the operation of network systems under attacks. The model concept of our approach is illustrated in Figure 23.1. Any network or system's behavior can be characterized by using a set of focused analysis functions such as anomaly distance (*AD*). *AD* models quantify the vulnerability of the individual network components and network services. The dynamics of network services and complexity of network connectivity can be represented in three operational states (normal, uncertain, and abnormal) that can accurately characterize and quantify the operations of network services under various types of attacks. In fact, the concept of using index functions to quantify and characterize the behavior of networked systems and services is analogous to the biological concept that uses metrics such as temperature or blood pressure to figure out the activity of an organism [45]. We conclude that under average conditions, *AD* will indicate normal operational state whereas it indicates

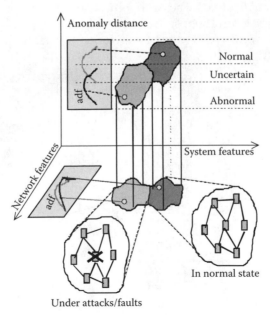

FIGURE 23.1
Principle of anomaly analysis.

uncertain and/or abnormal operational states when attacks occur. The *AD* value will be used as the mathematical basis to proactively respond to network attacks in a near real time mode. In our approach, any application and resource (network or pervasive system) is assumed to be in one of the following three states.

Normal State – when an application, computer system, or network node operates normally. We use the appropriate measurement attributes at each level to quantify whether or not the component operates normally. For example, if the number of unsuccessful TCP sessions in the network system is 0, we assume the network system is operating normally.

Uncertain State – In a similar way, we use the measurement attributes to describe the behavior of a component and/or a resource as being in uncertain state when the measurement attribute values associated with it are between the normal and abnormal thresholds.

Abnormal State – A component or a resource is considered to be operating in an abnormal state when its measurement attributes significantly deviate from the levels observed for normal operational states. For example, the number of incoming packets per second is an order of magnitude larger than the normal observed rate.

Our approach is based on developing real-time monitoring capabilities to achieve on-line anomaly analysis of network attacks and proactive self-protection mechanisms. The architecture shown in Figure 23.2 is our approach to implement the anomaly principle shown in Figure 23.1. In this activity, we leverage the services and the tools developed for the *AUTONOMIA* prototype [22] and *NVAT* [49–51]. The Autonomia control and management scheme is based on two control loops: *local control loop* and *global control loop*. The *local control loop* will manage the behavior of individual and local system

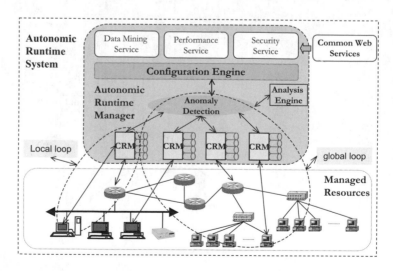

FIGURE 23.2
Anomaly based control and management architecture.

resources on which the components execute. This can be viewed as adding self-managing capabilities to conventional components/resources. This loop will control local algorithms, resource allocation strategies, distribution and load balancing strategies, etc. The local loop control is *blind* to the overall system behavior. Thus by itself it can lead to sub-optimal behavior. Unlike a local control loop that focuses on local optimal, the *global control loop* manages the behavior of the overall system or application and will define the knowledge that will drive the local adaptations. This control loop can handle *unknown* environment states and uses four system attributes for monitoring and analysis such as performance, fault-tolerance, configuration, and security.

The main software modules of the anomaly based control and management architecture include autonomic runtime system (*ARS*), common web services, autonomic runtime manager, and Managed resources.

23.6.1 Autonomic Runtime System (*ARS*)

The *ARS* exploits the temporal and heterogeneous characteristics of the applications, and the architectural characteristics of the computing and storage resources available at runtime to achieve high-performance, scalable, secure and fault tolerance operations. *ARS* will provide appropriate control and management services to deploy and configure the required software and hardware resources to run autonomously (e.g., self-optimize, self-heal) large-scale networked applications. The local control loop is responsible for control and management of one autonomic component, while the global control loop is responsible for the control and management of an entire autonomic application or system.

The *ARS* can be viewed as an application-based operating system that provides applications with all the services and tools required to achieve the desired autonomic behaviors (self-configuring, self-healing, self-optimizing, and self-protection). The primary modules of *ARS* are the following:

Common Web Services provides appropriate control and management Web services to deploy and configure the required software and hardware resources to run autonomously (e.g., self-optimize, self-heal etc.). These runtime services maintain the autonomic properties of applications and system resources at runtime.

Autonomic Runtime Manager (ARM) focuses on setting up the application execution environment and then maintaining its requirements at runtime. It also maintains the global control loop.

23.7 Anomaly-Based Analysis and Self Protection Engine

In this section, we discuss in further detail our approach to perform anomaly analysis in real-time and how to use the results of this analysis to achieve the desired self-protection services against any type of network attacks known or unknown.

23.7.1 Online Monitoring

The main goal of the online monitoring and analysis is to compute anomaly metrics that characterize and quantify the operational state of network systems at runtime. Hence, it is very important to identify effectively appropriate measurement attributes that can be used to reflect the operational state of the network components.

A Measurement Attribute (*MA*) denotes the value of some network attributes that can be measured online during the observation period. For example, *MAs* can be the rate of the outgoing *TCP SYN* packets ($TCP_{syn,out}$) of a network node, the total number of outgoing *UDP* packets (UDP_{out}) for the network system, and the *CPU* utilization (CPU_{util}) for a computer (either client or server).

Consequently, the *MAs'* values reflect the operational state of the network system and its components. We abstract the behavior of a network system in the model shown in Figure 23.3. During a given observation period *T*, the network component has an input *(I)* and an output *(O)* and a set of operational *MAs*. In order to characterize the operational state and the behavior of a large network, we have identified different *MAs* for all network and system components including of applications (FTP, Telnet, Web surfing, e-mails, etc.) and physical network devices (*routers, gateways, etc.*) as shown in Table 23.1. From the network security perspective, the input of the network node and its *MAs* should enable us to characterize the behavior of the system such that we can detect any abnormal behavior that might have been caused by network attacks.

The online monitoring engine is capable of monitoring multiple *MAs* for a single node or for the whole network. The online monitoring module includes two basic functional modules. *Host monitor*, which can run on any computing node (e.g., PCs with windows OS, Linux, etc.), records host resource usage statistics, for example, *CPU* utilization, memory usage. The *network monitor* reports network connection based flow information for the whole network. The flow information (e.g., packet rate, packet size) can be indexed by the source/destination IP address, source/destination port number, and

TABLE 23.1

Measurement Attributes

Protocols/Layers	Measurement Attributes
Application Layer HTTP, DNS, SMTP, POP3	NIP/NOP: number of incoming/outgoing PDUs IF: Invocation Frequency
Transport Layer TCP, UDP	NIP/NOP
Network Layer IP, ICMP, ARP	NIP/NOP AR: ARP Request Rate
System Resource CPU, Memory	CPU_{util}: CPU Utilization M_{use}: Memory Usage

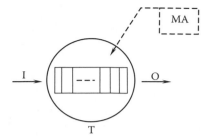

FIGURE 23.3
Minimal network system model.

protocol type. The network monitor is implemented based on the Cisco Net-flow specifications [8].

23.7.2 Anomaly Analysis Techniques

In this section we will focus on deriving the mathematical equations that characterize the operational state of network systems during a given time period. We use the anomaly distance function to quantify the deviation of the current network or system state from the perceived normal state.

23.7.2.1 Anomaly Distance Function

According to the anomaly analysis principle shown in Figure 23.1, an anomaly distance function can be used to quantify the operational states of various network services and resources. AD_{MA} can be calculated as in Equation 23.1. $D(\cdot, \cdot)$ is a function to compute the distance of two variables. This distance function could be a Euclidean [3], Manhanttan, Mahalanobis, Camberra, Chebychev, Quadratic, Correlation, and Chi-square distance metric [3,16,36,40]. $MA(t)$ is the current value of the MA at time t. MA_{normal} is the reference value when the monitored resource is in the normal state.

$$AD_{MA}(t) = D(MA(t), MA_{normal}) \qquad (23.1)$$

Figure 23.4 shows an example on how to use the AD metric to detect the occurrence of abnormal behavior. The main thrust of our research is in identifying the most appropriate MAs that can be used to accurately detect abnormal behaviors that might have been triggered by network attacks.

Anomaly Distance Function (ADF) is used to quantify the component/resource operational states (e.g. *normal, uncertain, and abnormal*) with respect to one or more measurement attributes (AD_{MA}). The ADF_{MA} can be calculated as the ratio of the current distance from normal level with respect to one MA

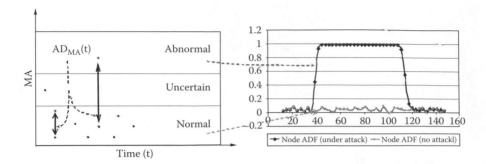

FIGURE 23.4
Anomaly distance w.r.t measurement attribute.

divided by the normal value for the *MA* as shown in Equation (23.2).

$$ADF(MA, t) = \begin{cases} 1 & if\ AD_{MA}(t) \geq \Delta_{MA} \\ \dfrac{AD_{MA}(t)}{\Delta_{MA}} & otherwise \end{cases} \tag{23.2}$$

$AD_{MA}(t)$ is the operational index computed online according to Equation 23.1 with respect to one measurement attribute (see Table 23.1). Δ_{MA} is the threshold that denotes the minimal distance from a normal operational state to the abnormal operational state with respect to one *MA*. As shown in Figure 23.4 when the network operates in normal state, the value of *ADF* is around 0.1. When the node endures an attack the *ADF* value will be close to 1. The *ADF* is in a sense analogous to the bio-logical metrics (body temperature, blood pressure, etc.) that have been used to characterize the state of biological systems.

To improve the detection accuracy, we use multivariate analysis because single network attribute may not accurately capture an abnormal behavior. For example, if we consider two attributes as shown in Figure 23.5, where UCL_i and LCL_i (i = 1 or 2) are the upper and lower thresholds for attributes x_1 and x_2, respectively. Assume that the state of a node or application is represented by point *A* of Figure 23.5, then for any single attribute x_1 or x_2 the node/application operates normally. However, when we combine the two metrics, the state *A* is outside the shaded oval area that designates the normal state with respect to the two metrics and thus it is abnormal. This example shows that it is possible that one *MA* cannot detect the occurrence of an abnormal behavior, but if we combine a few measurement attributes the detection of abnormal behavior can be significantly improved.

The multivariate analysis can be applied to study the behavior of network applications and the correlation between the measured attributes to determine whether or not they are operating normally. Hotelling's T^2 and Chi-Square multivariate analysis methods are deployed on the system logs in offline host intrusion detection [61,67]. Our approach for online multivariate

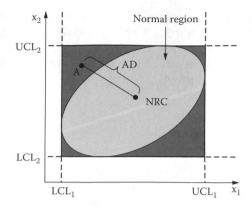

FIGURE 23.5
Multivariate analysis model.

analysis focuses on a network infrastructure and builds a new metric based on Hotelling's T^2 control chart to measure the abnormal behavior.

First, we need to determine the baseline profile. Suppose we have P measurement attributes. Then the normal behavior of the MA can be represented as:

$$\mathbf{MA} = \left\{ \begin{pmatrix} MA_1(t) \\ MA_2(t) \\ \vdots \\ MA_p(t) \end{pmatrix} \begin{pmatrix} MA_1(t+1) \\ MA_2(t+1) \\ \vdots \\ MA_p(t+1) \end{pmatrix} \cdots \begin{pmatrix} MA_1(t+M) \\ MA_2(t+M) \\ \vdots \\ MA_p(t+M) \end{pmatrix} \right\} \tag{23.3}$$

Based on these normal measurement attributes, two control limits — upper control limit (*UCL*) and lower control limit (*LCL*) can be determined using the **M** preliminary blocks to obtain in-control data to determine the mean $\bar{M}\bar{A}$ and covariance matrix S. From these values, we can then determine a normal region with respect to the P measurement attributes. The sample mean $\bar{M}\bar{A}$ determines the normal region center (*NRC*) and the sample covariance matrix S determines the shape of the normal region as shown in the shaded part of Figure 23.5. For a pre-defined Type-I error α, the upper and lower control limits are computed:

$$UCL = \frac{(M-1)^2}{M} B_{1-\alpha/2}[P, (M-P-1)/2] \tag{23.4}$$

$$LCL = \frac{(M-1)^2}{M} B_{\alpha/2}[P, (M-P-1)/2] \tag{23.5}$$

where $B_{\alpha/2}[P, (M-P-1)/2]$ is the $1-\alpha/2$ percentile of the β distribution with P and $(M-P-1)/2$ denotes the degrees of freedom.

The *AD* metric is used to quantify how far the current operational state of a component is from the normal state for a given attack scenario based on one or more **MAs**. Our initial research results show that the mean shifts of the selected network attributes can be effectively used for attack detection. The normalized anomaly distance AD_k with respect to an attribute MA_k is defined as

$$AD_k = [(MA_k(t) - \mu_{MA_k})/\sigma_{MA_k}]^2 \qquad (23.6)$$

Where, μ_{MA_k} and $\sigma^2_{MA_k}$ are the mean and variance under the normal operation condition corresponding to the measurement attribute k respectively. $MA_k(t)$ is the current value of a network attribute k.

A general definition of the *AD* metric with respect to a subgroup J of multiple correlated attributes can be defined using the statistic T^2 distance as:

$$AD_J = (MA_J(t) - \mu_{MA_J})^T \sum\nolimits^{-1}_{MA_J} (MA_J(t) - \mu_{MA_J}) \qquad (23.7)$$

Where μ_{MA_J} and $\sum_{MA_J}$ are the mean vector and variance matrix under the normal operation condition corresponding to the grouped attributes J, respectively $MA_J(t)$ is the current measurement of attribute group J.

23.7.2.2   Information Theory Based Anomaly Analysis

We have developed the *AD* metric based on Hotelling's T^2 method as described in Section 7.2.1. It is clear that when using multivariate analysis method to calculate the *AD*, the variance matrix has to be computed. This process requires computing the correlations between any pair of two features in the whole feature set. There are two concerns about the statistical multivariate analysis method. The first one is that all features are considered to have the same level of importance. In fact, they are not in most scenarios. Second, when the number of features is large, the calculation will significantly increase the computational overhead so it cannot be computed in real-time. Hence, we developed a new efficient approach that reduces the number of **MAs** to be analyzed in real-time based on information theory. This will help us identify the most important features in order to improve the prediction and accuracy of the desired data mining task i.e., detection of an abnormal behavior in a network service due to network attacks. After computing the optimal feature set, a linear classification function is trained by using a genetic algorithm (GA) on the weights, which helps differentiate among different types of attacks (*normal, dos, u2r, r2l, probe*). The linear function represents the weight summation of the discrete and/or discretized continuous feature values. We have validated our approach using the DARPA KDD99 benchmark dataset and competitive results have been achieved. For further information about the feature selection algorithm, classification algorithm, and the validation results, please refer to [52].

23.7.3 Quality of Protection (*QoP*) Framework

We focus on the development of anomaly metrics to quantify the behavior of network systems that can be used to detect abnormal operations caused by network attacks. Once this anomaly is detected, the corresponding traffic flows will be assigned lower scheduling priorities based on their anomaly distance. Hence, the legitimate traffic will experience the least impact from the network attacks.

Our approach complements the weakness in current QoS protocols that do not distinguish between normal and attacking traffic [47]. This weakness is the main reason for the denial of service to all other legitimate traffic. The *QoP* framework uses the *AD* metrics discussed previously to classify traffic into four classes as shown in Figure 23.6. Both probable normal and probable abnormal traffic are categorized in the suspicious attack traffic class.

i. Normal Traffic – The flow *ADs* for the packets belonging to this traffic class will be lower than a pre specified threshold ξ_1.

ii. Probable Normal Traffic – The flow *ADs* for the packets belong to in this class are larger than ξ_1 and less than ξ_2.

iii. Probable Abnormal Traffic – The flow *ADs* for the packets belong to in this class will be larger than ξ_2 and less than ξ_3.

iv. Attacking Traffic – The flow *AD* for the traffic flows in this class are larger than or equal to ξ_3.

For each traffic class described above, we have subclasses to allow fine granularities. For example, normal traffic flow can be divided into two subclasses (*A* and *B*) differing by 0.2 with respect to their Global Flow Anomaly Distance (*GFAD*) which is a network wide, comprehensive metric based on the information exchanged among the neighboring agents. In a similar way, we

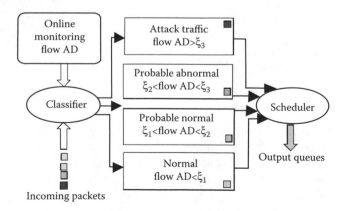

FIGURE 23.6
Queuing model based on flow *AD*.

classify attacking traffic and suspicious traffic into three subclasses with respect to *GFAD* that differs by 0.1 for each subclass. The monitoring and analysis agents that compute the anomaly metrics collaborate by exchanging their anomaly metrics for the suspicious traffic such that a global anomaly metrics such as *GFAD* can be determined. Hence, a malicious traffic flow is identified because upstream routers feel the pressure of the attack much earlier than downstream routers and consequently increase the *GFAD* value. By increasing the *GFAD*, the flow priority will be reduced and eventually will be dropped by the routers. The *GFAD* mitigates the impact of the attacking traffic at the earliest possible upstream nodes and thus improves the performance of legitimate traffic.

23.8   Experimental Results

We analyzed the benchmark KDD99 dataset [14] used in the Third International Knowledge Discovery and Data Mining Tools Competition to validate our approach. Lincoln Labs set up an environment to acquire nine weeks of raw TCP dump data for a local-area network (*LAN*) simulating a typical U.S. Air Force *LAN*. A connection is a sequence of TCP packets starting and ending at some well defined times, between which data flows to and from a source IP address to a target IP address. Each connection is labeled as either normal, or as an attack, with exactly one specific attack type. It is important to note that the testing data is not from the same probability distribution as the training data. There are 494021 records in the training dataset and the number of records in the testing dataset is about five million. The datasets contain a total of 22 different attack types. They fall into the following four main categories: *DoS, R2L, U2R*, and *Probe*. For each connection record, there are 41 features that are divided into discrete sets and continuous sets. In our approach, we use information theory to select the appropriate features for each type of attack. We have considered three different feature filtering methods: 1) Using discrete features; 2) Using continuous and discrete features; and 3) Using dependency among features.

23.8.1   Using Discrete Features

We analyzed all the nine discrete features from the original data set with respect to each type of network attack. We computed the mutual information of the discrete features with respect to the decision variable (to detect each type of attack). The four important features include *service, logged_in, protocol_type* and *flag* because they have the largest mutual information with respect to the decision variable. The other features are irrelevant in terms of their mutual information value. Hence, using the mutual information we can reduce the 41 features to only the four features mentioned above.

23.8.2 Using Both Discrete and Continuous Features

Usually, there are few distinct values for the discrete features, and therefore, it is straightforward to apply information theory. However, the value for continuous features has a wide range of distinct values. For example, the *src_bytes* feature has 3300 distinct values. It is obvious that different processing methods should be used to handle discrete and continuous features. For the discrete features, we assign a value to each nominal value. But for continuous features, it is not feasible to process each distinct value for two reasons. The first one is the high overhead in memory and computation if we consider all possible values. The second one is regarding the accuracy of the learning algorithm. If the testing dataset has different distributions in the training dataset there will be some distinct values that do not appear in the training dataset. Hence, this will reduce the accuracy of the analysis.

To deploy the same approach, we need to identify the features that could be used to detect each category of attacks. For example, the features *root_shell* and *num_file_creations* barely provide information on **DoS**, **Probe**, and **R2L** attack, but they are very important for the **U2R** attack.

We have implemented a genetic algorithm to discretize the continuous variable values into intervals to maximize the mutual information between the continuous features and the decision variable. Features are ranked in descending order according to their relevance to the final decision. When $\delta_{dos,1} = 0.4$, $\delta_{probe,1} = 0.31$, $\delta_{u2r,1} = 0.09$, $\delta_{r2l,1} = 0.08$, features are chosen as shown in Table 23.2 and Table 23.3 to reflect the removal of the irrelevant features. Without feature correlation analysis, we were able to get good detection rates only for **DoS** and **Probe** attacks as shown in Table 23.4, but poor detection rates for **U2R** and **R2L** attacks.

23.8.3 Using Dependency Analysis of Features

By applying feature selection algorithm [52] the features chosen for **U2R** and **R2L** attacks with respect to different threshold levels are shown in Tables 23.5 and 23.6, respectively. In Table 23.5, when $\delta_2 = 0.99$, the features chosen

TABLE 23.2

Ranked Features for *DoS* and *Probe* Attacks

DOS		Probe	
Feature	I(Y;X)/H(Y)	Feature	I(Y;X)/H(Y)
count	0.89973	src_bytes	0.617323
service	0.823221	service	0.508163
dst_bytes	0.711719	dst_host_diff_srv_rate	0.45012
logged_in	0.545972	dst_bytes	0.425978
dst_host_same_src_port_rate	0.531105	rerror_rate	0.343698
srv_count	0.475077	count	0.341383
protocol_type	0.43273	flag	0.329652
dst_host_count	0.428534	dst_host_srv_diff_host_rate	0.313887
src_bytes	0.403728	same_srv_rate	0.313486

TABLE 23.3

Ranked Features for *U2R* and *R2L* Attacks

U2R		R2L	
Feature	I(Y; X)/H(Y)	Feature	I(Y;X)/H(Y)
service	0.481635	service	0.559618
root_shell	0.37445	dst_host_srv_count	0.328744
dst_host_srv_count	0.281805	dst_host_same_src_port_rate	0.205641
duration	0.26578	dst_host_srv_diff_host_rate	0.183676
num_file_creations	0.255163	is_guest_login	0.159472
dst_host_count	0.177618	srv_count	0.149472
dst_host_same_src_port_rate	0.134272	dst_bytes	0.136806
srv_count	0.113392	dst_host_count	0.131907
dst_host_srv_diff_host_rate	0.091564	count	0.131043
src_bytes	0.086327	src_bytes	0.088246

TABLE 23.4

Results Comparison of Different Approaches

Class	Our Approach Using Continuous and Discrete Features	Our Approach Using Discrete Features Only	Winner Entry Using C5.0	CTree
Normal	98.45%	98.34%	99.5%	92.78%
Dos	99.93%	99.33%	97.1%	98.91%
U2R	75.34%	63.64%	13.2%	88.13%
R2L	41.34%	5.86%	8.4%	7.41%
PROBE	99.91%	93.95%	83.3%	50.35%

TABLE 23.5

Different Feature Subsets and Their Prediction for *U2R* Attacks

$\delta_{u2r,2}$	Subset (S)	e(S)	False Alarm	Detection Rate
0.99	{x1, x3, x5, x6, x7}	100.4%	0.007587	92.55%
0.9	{x1, x3, x5, x6}	94.05%	0.01431	96.23%
0.85	{x3, x4, x5, x6}	88.93%	0.01531	91.47%
0.8	{x1, x3, x6}	80.55%	0.019583	94.34%
0.7	{x1, x3, x5}	77.03%	0.067961	90.06%

TABLE 23.6

Different Feature Subsets and Their Prediction for *R2L* Attacks

$\delta_{r2l,2}$	Subset (S)	e(S)	False Alarm	Detection Rate
0.99	{x1, x3, x4, x8}	99.73%	0.092581	91.13%
0.9	{x1, x3, x4}	90.61%	0.09476	92.37%
0.85	{x1, x3, x4}	90.61%	0.09476	92.37%
0.7	{x1, x3}	76.53%	0.083524	92.46%

by the feature selction algorithm include *service* (x_1), *dst_host_srv_count* (x_3), *num_file_creations* (x_5), *dst_host_count* (x_6), and *dst_host_same_src_port_rate* (x_7). Training based on these features the learning algorithm determines the **U2R** attack classifier to be $f(x_1, x_3, x_5, x_6, x_7) = (-508)x_1 + (499)x_3 + (-908)x_5 + 480x_6 + (-90)x_7$. Applying this classifier on the testing dataset resulted in a detection rate of 92.5% with a 0.7587% false alarm. We also note that if we set $\delta_2 = 0.9$ the classifier based on feature set {*service, dst_host_srv_count, num_file_creations, dst_host_count*} can lead to a detection rate of 96.2% with a 1.43% false alarm. These results are significantly better than those obtained using the sequential feature selection approach [46].

For **R2L** attacks detection, the feature selection algorithm yields a feature subset that consists of *service* (x_1), *dst_host_same_src_port_rate* (x_3), *dst_host_srv_diff_host_rate* (x_4), and *dst_host_count* (x_8). Using these features in detecting **R2L** attacks, we get a 91.13% detection rate with a 9.258% false alarm. When training on features *service* (x_1) and *dst_host_same_src_port_rate* (x_3), we obtained a detection rate of 92.47% with 8.35% false alarm. The results are comparable to the optimal sequential selection of 8 features. However, the small number of features will result in a much faster learning process and it will reduce the overhead in collecting data when used in a real network environment.

23.8.4 *QoP* Evaluation

We validated our approach of using anomaly distance and **QoP** to proactively protect the network from DDoS attacks. We used the network topology shown in Figure 23.7 in our simulations as a proof of concept for our **QoP** methodology. The topology consists of five client networks and two server networks. There are 150 clients, 30 routers in all of the client networks and 12 routers and 30 servers in the server networks. The core backbone consists of 6 routers. The backbone links are of 100 Mbps, router-to-router links and router to client node links are 30 Mbps and 10 Mbps respectively. The server network links are 30 Mbps each. The routing of each network is configured with OSPF and the other protocols used are IP, TCP, UDP, and **QoP** (only during **QoP** evaluation). All the clients and servers in the topology are configured for simple file transfers using TCP/IP to generate constant traffic load at all times. Some of the client networks are configured with **DDoS** (TCP SYN) attacks that generate the attacking traffic at specified simulation time with set of zombies and adjusted attack intensity. The attacking traffic is scaled up to 100 times the regular traffic to be as close as possible to real-time attacking situations. The attacking traffic fills the bandwidth of various links in the core backbone network and every other link that comes in its path before they reach the victim. So in a normal scenario all the traffic competes for the same buffers and receives the same service rate.

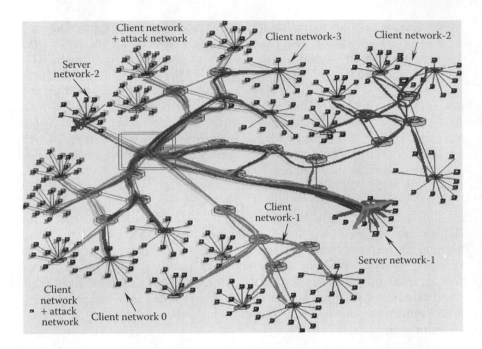

FIGURE 23.7
Network topology for simulation.

The effectiveness of the self-protecting mechanism is evaluated by quantifying the reduction in the end-to-end session time during an attack. The flow metrics used in our analysis include packet rate (PR), aggregate flow rate (AFR), and number of unsuccessful sessions (NUS).

The network topology shown in Figure 23.7 is used to evaluate the QoP framework. The topology consists of several networks and different sets of client-server clusters. There are three client networks numbered 0, 1, and 3 that are connected to servers in network 1. Multiple clients in networks 0 and 3 are configured to launch DDoS attack sessions by attacking two different servers in network 1 at time 60. The attack intensity of the clients in network 0 is higher than that of clients in network 3. Clients in network 2 are connected to network 2. The DDoS stops at time 150.

In analyzing the anomaly of the flows we use the input interface and the destination parameters of a flow because they cannot be modified by the attacker. To quantify the improvement that can be achieved by using QoP based self-protection, we evaluated the impact of the QoP framework on four types of traffic flows:

- Type 1: Normal traffic uses the same interface as the attacking traffic, but destines to other servers.

- Type 2: Normal traffic uses different interface from the attacking traffic and destines to the attacked server.
- Type 3: Normal traffic uses the same interface as the attacking traffic and destines to the attacked servers.
- Type 4: Normal traffic uses different interfaces from the attacking traffic and destines to other servers.

In what follows, we show how the *QoP* reduces the impact of the attacking flows on session times for the four flows mentioned previously and quantify the improvement level by comparing the results with the cases where nothing is done to encounter types of network attacks.

The *NUS* will be higher than 1 for attacking traffic and will be zero for normal flow traffic. The *AF* is defined as the aggregate flows for a given interface-destination pair. *AFR* will be high for the attacking flows. Normal traffic that uses the same interface-destination pair will have the same *AF* value as the attacking traffic. However, by using *NUS* value, which is equal to zero for the normal traffic, we can differentiate normal traffic from the attacking traffic. In our approach, we use multiple metrics (*PR, NUS,* and *AFR*) to improve the accuracy of our approach in isolating the attacking traffic from the normal traffic. For example, all the other types of normal flows that use different interfaces or different destinations have their *AF* values operating at normal rates.

In our approach, we evaluate anomaly distance for each flow with respect to the three flow metrics according to the network anomaly distance algorithm discussed in Section 7.3. Once the *GFAD*s are computed, we assign different priorities to each flow based on its *GFAD* value. For example, the attacking traffic from client network 0 whose *GFAD* is equal to 0.6 while the attacking traffic from client network 1 has *GFAD* to be equal to 0.8. Normal traffic that flows through the same interface and has the same destination as the attacking traffic has *QoP* priority of 2. All the normal traffic flows will be served at the default service priority 1 according to the operation system and network programming convention that lower number denotes higher priority.

The graphs in Figure 23.8 show the impact of attacks on normal traffic flows and also show that the impacts can be reduced when the *QoP* mechanism is deployed in each router. For example, the normal traffic for Type 3 and 4 flows are impacted by 90% while Type 4 flow is impacted by 60% when *QoP* is not deployed. For the same attack scenario, the impact on Type 3 flow is reduced from 90% to 40% while for Type 1 flow, it is reduced to only 20% when *QoP* mechanism is used. For Type 4 flow, the impact is reduced from 60% to only 5%. These results demonstrate that *QoP* is efficient in reducing the impact of network attacks on the legitimate traffic flows.

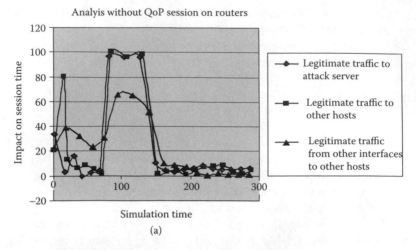

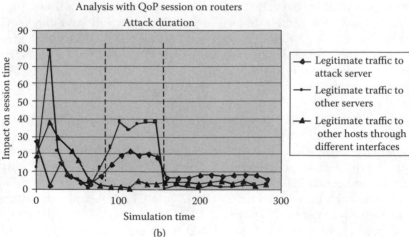

FIGURE 23.8

Performance comparison by applying *QoP* for self protection for DoS attack (a) on the impact of DDoS attacks on session time for different flows without *QoP*; (b) the impact on session time when *QoP* is used.

23.9   Summary

The research presented in this chapter is part of a large effort to develop an autonomic control and management environment (*AUTONOMIA*) that provides self-configuring, self-optimizing, self-healing and self-protecting services.

    In this chapter, we presented a theoretical framework and general analysis methodology to achieve (a) analyzing anomaly operations of networks and applications; (b) and self-protecting networks from a wide range of organized network and/or host attacks. We have showed how anomaly distance metric (*AD*) can be used to accurately detect a wide range of network attacks. To improve the detection rates and computational complexity of the monitored network features, we developed efficient filtering techniques based on information theory and dependency among features.

We evaluated our approach using the DARPA KDD99 benchmark dataset and the results showed that using the new decision dependent correlation metric we can detect efficiently *U2R* and *R2L* attacks. The best reported detection rates for *U2R* and *R2L* on the KDD99 data sets were 13.2% and 8.4% with 0.5% false alarm, respectively. For *U2R* attacks, our approach can achieve a 92.5% detection rate with false alarm of 0.7587%. For *R2L* attacks, our approach can achieve a 92.47% detection rate with false alarm of 8.35%. Furthermore, the feature importance analysis can be used to identify the optimal feature set that must be monitored and analyzed to determine whether the target system is under attack or not.

In addition, we have developed a *QoP* routing protocol that can automatically adjust the network traffic priorities according to anomaly metrics. This proactive network defense framework presented can be integrated with any existing Quality of Service (QoS) protocols by assigning high priority to normal traffic and low priority to abnormal traffic to minimize the impact of network attacks on various network services.

We are currently investigating efficient techniques to enable us to monitor in real-time large size networks, filter the huge amount of data generated by different network components, and detect and protect against network attacks. We are investigating techniques to continuously adapt the baseline modes that characterize normal operations as the network changes its topologies and normal operation characteristics.

References

1. R. Albert, H. Jeong., and A-L. Barabási, "The Internet's Achilles' heel: error and attack tolerance of complex networks," *Nature* 406 378, 2000.
2. D. Anderson, T. Frivold, and A. Valdes, "Next-generation intrusion detection expert system (NIDES): A summary," Technical Report SRI-CSL-95-07, Computer Science Laboratory, SRI International, Menlo Park, California, May 1995.
3. B.G. Batchelor, "*Pattern Recognition: Ideas in Practice*," New York: Plenum Press, pp. 71–72, 1978.
4. S.M.Bellovin, "ICMP Traceback Message," Internet Draft: draft-bellovin-itrace-00.txt, March 2000.
5. M. Blanc, L. Oudot, and V. Glaume, "Global Intrusion Detection: Prelude Hybrid IDS," Technical Report, 2003.
6. Botha, M. et al., "The utilization of artificial intelligence in a hybrid intrusion detection system," South African Institute for Computer Scientists and Information Technologists, pp. 149–155, 2002.

7. CERT, "CERT Advisory CA-2003-04 MS-SQL Serve Worm," http://www.cert.org/advisories/CA-2003-04.html, January 2003.

8. Cisco netflow, http://www.cisco.com/en/u/products/ps6601/products_ios_protocol_group_home.html.

9. CNN. "Cyber-attacks batter web heavyweights," http://archives.cnn.com/2000/TECH/computing/02/09/cyber.attacks.01/, February 2000.

10. CNN. "Immense network assault takes down yahoo," http://archives.cnn.com/2000/TECH/computing/02/08/yahoo.assault.idg/, February 2000.

11. CNN. "Denial-of-service attacks on the rise?" http://archives.cnn.com/2002/TECH/internet/04/09/dos.threat.idg/, April 2002.

12. S. Cowley and M. Williams, "Slammer slugs Internet, down but not out." Retrieved on April 20, 2003, from http://www.infoworld.com/article/03/01/25/030125hnsqlnetupd_1.html.

13. M. Crosbie, B. Dole, T. Ellis, I. Krsul, and E. Spafford, IDIOT — Users Guide, Technical Report TR-96-050, Purdue University, COAST Laboratory, September 1996.

14. DARPA KDD99, http://kdd.ics.uci.edu/databases/kddcup/task.html, 2005.

15. D.E. Denning, "An Intrusion Detection Model," *IEEE Transactions on Software Engineering*, SE-13: 222–232, 1987.

16. E. Diday, "Recent Progress in Distance and Similarity Measures in Pattern Recognition." *Second International Joint Conference on Pattern Recognition*, pp. 534–539, 1974.

17. D. Dittrich, "Distributed Denial of Service (DDoS) Attacks/Tools Page," from http://staff.washington.edu/dittrich/ddos/.

18. P. Ferguson and D. Senie, "Network Ingress Filtering: Defeating Denial of Service Attacks Which Employ IP Source Address Spoofing," RFC 2267, January 1998.

19. S. Gaudin, "2003 Worst Year Ever for Viruses, Worms," http://www.internetnews.com/infra/article.php/3292461, 2003.

20. S. Gibson. "Distributed Reflection Denial of Service," retrieved from http://grc.com/dos/drdos.htm, February 2002.

21. T.M. Gil and M. Poleto, "MULTOPS: a data-structure for denial-of-service attack detection," *In Proceedings of 10th USENIX Security Symposium*, August 2001.

22. S. Hariri, G. Qu, T. Dharmagadda, and R. Modukuri, "Impact Analysis of Faults and Attacks in Large-Scale Networks," *IEEE Security and Privacy*, September/October 2003 Vol. 1, No. 5.

23. J. Hochberg, et al., Nadir: An automated system for detecting network intrusion and misuse. *Computers & Security*, 12(23.3):235-248, 1993.

24. Y. T. Hou, Y. Dong and Z. Zhang, "Network Performance Measurement and Analysis Part 1: A Server-Based Measurement Infrastructure." Retrieved on April 20, 2003 from http://citeseer.nj.nec.com/249251.html.

25. K. Ilgun, A. R. Kemmerer, A. P. Porras, "State Transition Analysis: A Rule-Based Intrusion Detection Approach," *IEEE Transactions on Software Engineering*, 21(23.3), 1995.

26. J. Ioannidis and S.M. Bellovin, "Implementing Pushback: Router-Based Defense Against DDoS Attacks," Proceedings of NDSS'2002, San Diego, CA, February 2002.

27. J. Jin and J. Shi, "Automatic feature extraction of waveform signals for in-process diagnostic performance improvement," *Journal of Intelligent Manufacturing* 12, 257–268, 2001.

28. R. Kemmerer and G. Vigna. "Intrusion Detection: A Brief History and Overview," *IEEE Computer* 27–30, 2002.

29. K. Kendall, "A Database of Computer Attacks for the Evaluation of Intrusion Detection Systems," Master's Thesis, Massachusetts Institute of Technology, 1998.

30. A.D. Keromytis, V. Misra, and D. Rubenstein. "SOS: Secure Overlay Services." In *Proceedings of ACM SIGCOMM 2002*, August 2002.

31. T. Killalea, "Recommended Internet Service Provider Security Services and Procedures," *RFC 3013*, November 2000.

32. J. Lo, "Denial of Service or NUKE Attacks," http://www.irchelp.org/ irchelp/nuke/index.html, March 2005.

33. T. Lunt, A. Tamaru, F. Gilham, R. Jagannathan, C. Jalali, P. G. Neumann, H. S. Javitz, A. Valdes, T.D. Garvey, A real time Intrusion Detection Expert System (IDES) — Final Report, SRI International, Menlo Park, CA, February 1992.

34. R. Manajan et al., "Controlling High Bandwidth Aggregates in the Network," ICSI Technical Report, July 2001.

35. Matrix, "Slammer worm." Retrieved April 20, 2003, from http://www.matrix netsystems.com/ea/2003/20030130.jsp.

36. R.S. Michalski, R. E. Stepp, and E. Diday, "A Recent Advance in Data Analysis: Clustering Objects into Classes Characterized by Conjunctive Concepts," *Progress in Pattern Recognition*, Vol. 1, Laveen N. Kanal and Azriel Rosenfeld (Eds.). New York: North-Holland, pp. 33–56, 1981.

37. J. Mirkovic, G. Prier, and P. Reiher, "Attacking DDoS at the Source," *Proceedings of 10th IEEE International Conference on Network Protocol* (ICNP2002), 312–321, Paris, France, 2002.

38. D.C. Montgomery, *Design and Analysis of Experiments*, 5th Edition, New York: John Wiley & Sons, 2000.

39. D. Moore, G. Voelker and S. Savage, "Inferring Internet Denial of Service Activity," *Proceedings of USENIX Security Symposium, 2001*, August 2001.

40. M. Nadler and E. P. Smith, *"Pattern Recognition Engineering."* New York: John Wiley & Sons, pp. 293–294, 1993.

41. National Research Council, "Committee on the Internet under Crisis Conditions: Learning from the Impact of September 11," The National Academies Press, 2003. Retrieved from http://www.nap.edu/books/0309087023/html/.

42. C. Papadopoulos, R. Lindell, J. Mehringer, A. Hussain, and R. Govindan, "COSSACK: Coordinated Suppression of Simultaneous Attacks," *Proceedings of DARPA Information Survivability Conference and Exposition*, Washington, D.C., 2003.

43. B. Page, "A Report on Internet Worm," http://www.ee.ryerson.ca/~elf/hack/ iworm.html.

44. K. Park and H. Lee. "On the Effectiveness of Route-Based Packet Filtering for DDoS Attack Prevention in Power-Low Internets." In *Proceedings of the ACM SIGCOMM*, pages 15–26, San Diego, CA, August 2001. ACM.

45. M. Parashar, "Interpretive Performance Prediction for High Performance Computing." PhD thesis, Department of Computer Engineering, Syracuse University, 1994.

46. V. Paxson. "An analysis of using reflectors for distributed denial-of-service attacks," *Computer Communication Review 31(23.3)*, July 2001.

47. V. Paxson, "End-to-End Internet Packet Dynamics," Proc., SIGCOMM'97, September 1997.

48. P.A. Porras and P.G. Neumann, "EMERALD: Event Monitoring Enabling Responses to Anomalous Live Disturbances," *Proceedings of National Information Systems Security Conference*, 1997.

49. G. Qu, S. Hariri, et al., *"Using Abnormality Metrics to Detect and Protect Against Network Attacks,"* in the *Proceedings of IEEE/ACS International Conference on Pervasive Services* (ICPS 2004), Beirut, Lebanon.

50. G. Qu, S. Hariri, et al., "Online Monitoring and Analysis for Self Protection against Network attacks," International Conference on Autonomic Computing (ICAC 2004), New York, NY.

51. G. Qu, S. Hariri, X. Zhu, J. Jin, and Y. Mazin, "Multivariate Statistical Online Analysis for Self Protection against Network Attacks," AICSSA'05.

52. G. Qu, S. Hariri, and Y. Mazin, "A New Dependency and Correlation Analysis for Features," *IEEE Transactions on Knowledge and Data Engineering, Special Issue on Intelligent Data Preparation*, September, 2005.

53. Threat and Vulnerability Model for Information Security. Report to the President's Commission on Critical Infrastructure Protection 1997.

54. S.E. Smaha and J. Winslow, "Misuse detection tools," *Computer Security Journal* 10, 1, Spring, 1994, pp. 39–49.

55. S. Savage, D. Wetherall, A. Karlin, T. Anderson, "Practical Network Support for IP Traceback," In *Proceedings of ACM SIGCOMM'2000*, August, 2000.

56. S. Smash, "Haystack: An Intrusion Detection System," proceedings of the IEEE 4th Aerospace computer security Application conference, Orlando, FL. pp. 37–44, December 1988.

57. A.C. Snoren, C. Partridge, L.A. Sanchez, C.E. Jones, F. Tchakountio, S.T. Kent and W.T. Strayer, "Hash-Based IP Traceback," *Proceedings of ACM SIGCOMM'2001*, March 2001.

58. D. Song and A. Perrig, "Advanced and Authenticated Marking Schemes for IP Traceback," In *Proceedings of ACM SIGCOMM'2001*, March 2001.

59. E. H. Spafford, "The Internet Worm Program: An Analysis," Purdue Technical Report CSDTR-823, 1998.

60. W. Stallings, "Network Security Essentials" 2nd edition, Upper Saddle River: Prentice-Hall, 2003.

61. R. Stone, "CenterTrack: An IP Overlay Network for Tracking DoS Floods," Proceedings of 9th USENIX Security Symposium, Denver, Colorado, August 2000.

62. L.P. Swiler, C. Phillips, and T. Gaylor, (1998), "A Graph-Based Network-Vulnerability Analysis System." Sandia National Laboratories. F. Tsung and J. Shi, (1999), "Integration of Run-to-Run PID Controller and SPC for Process Disturbance Rejection," *IIE Transactions*, Vol. 31, pp. 517–527.

63. B. Tjaden, et. al, (2004) INBOUNDS: The Integrated Network-Based Ohio University Network Detective Service, retrieved from http://www.mts.jhu.edu/~marchette/ID04/Papers/SCI2000.pdf

64. "DoS Attack," Retrieved on April, 2004 from http://www.webopedia.com/TERM/D/DoS_attack.html.

65. D. Yau, J. Liu, and F. Liang, "Defending Against Distributed Denial-of-Service Attacks with Max-min Fair Server-centric Router Throttles," Proceedings of IWQoS'2002, Miami Beach, FL, May 2002.

66. N. Ye et al., "Hotelling's T^2 Multivariate Profiling for Anomaly Detection," Proceedings of the 2000 IEEE Workshop on Information Assurance and Security, West Point, NY, June 2000.

67. N. Ye and Q. Chen, "An Anomaly Detection Technique Based on a Chi-Square Statistic for Detecting Intrusions into Information Systems," *Quality and Reliability Engineering Journal*, vol. 17, p. 105–112, 2001.
68. K. Zaraska, "IDS Active Response Mechanisms: Countermeasure Subsystem for Prelude IDS," Technical Report, 2002.
69. K. Zaraska, "Prelude IDS: Current State and Development Perspective," Technical Report, 2003. (http://www.seclib.com/seclib/index.jsp?page=ids.general).

Index

A

Abstract mapping, 192, 193–195
Accord, 51, 57, 58, 62, 298, 417; *see also*
    Accord-CCA; Accord-WS
  adaptation support, 213–214
  autonomic elements, 58, 216–217
  dynamic composition, 59, 217–218
  prototype implementations, 218–219,
    233
  provisions, 212, 213, 233
  rules, 59, 60, 217, 218
  separations enabled by, 215–216
Accord-CCA, 62, 218–219
  application scenarios, 224–226
  enabling performance-driven
    self-management, 222–223
  experimental evaluation, 223
  frameworks used, 219
  managed components, 219–220
  rule execution, 221–222
  runtime infrastructure, 220–221
Accord-WS, 62, 219, 226
  an adaptive data-streaming application,
    229–232
  runtime infrastructure, 227–229
ACE (Accord Composition Engine), 197,
  217
ACT (Adaptive CORBA Template),
  177–178, 370
Actuators, *see* effectors
Adaptation, 57, 108, 193, 201–204, 319, 321;
    see also reconfiguration; service
    recipes
  conflict resolution, 201–204
  consistency-preserving, 371–372, 385
  control-based, 430
  coordination mechanisms, 111–114,
    197–198
  engine, 370, 371
  external *versus* internal, 190, 370–371
  ongoing properties, 107, 112
  rule-based, 430
  strategies for service optimization, 190,
    191–192, 196–197

  technologies, 215
  utility-driven, 299
Adaptive COBRA Template (ACT),
  177–178
Adapt-ready programs, 175, 176–177, 178
Addressing, 41, 55, 174, 181, 375
  using for network attacks, 497, 498, 504,
    510
  Web Service standards, 259, 260, 269
Ad-hoc networks, 101, 103–104, 105, 334
ADSS (Autonomic Data Streaming
    Service)
  adaptations, 426–427
  controller behavior for long-running
    simulations, 425–426
  controller design, 422–423
  data prediction accuracy, 425
  heuristic *vs.* control-based adaptation,
    428
  implementation, 423–425
  objectives, 421–422
  overhead, 428–429
  self-healing behavior, 231–232
  self-optimization behaviors, 229–231
Aggregation queries, 45
Aglets framework, 395
Alchemi Enterprise Grid Computing
    System, 370, 371, 382, 385
  applications, 377
  architecture, 377–378
  reconfiguring, 379–381
  scheduler, 378–379
Algorithms, 63, 115–116, 150, 155, 394
  addressing interoperability, 127
  component selection, 193
  detection, 430, 473, 474, 481–484
  deterministic, 137
  election, 135
  mapping, 285
  optimization, 199–200, 201
  scheduling, 379, 390
Alua, 54
American Power Conversion (APC), 437,
  438
AMRMesh, 224, 225